MOLECULAR MACHINES

MOLECULAR MACHINES

Edited by:

Maria Emilova Velinova

ARCLER
PRESS

www.arclerpress.com

Molecular Machines

Maria Emilova Velinova

Arcler Press
2010 Winston Park Drive,
2nd Floor
Oakville, ON L6H 5R7
Canada
www.arclerpress.com
Tel: 001-289-291-7705
001-905-616-2116
Fax: 001-289-291-7601
Email: orders@arclereducation.com

ISBN: 978-1-77361-086-3 (Hardcover)

DECLARATION

ABOUT THE EDITOR

Maria Velinova is Ph.D. holder in Quantum chemistry at the University of Sofia since April 2012. Her major research experience is in the field of Computational Chemistry, especially in statistical mechanics methods applied to different sorts of biomolecules. Member of the Laboratory of Quantum and Computational Chemistry at the University of Sofia.

TABLE OF CONTENTS

PART II MOLECULAR MOTORS

LIST OF CONTRIBUTORS

Maria Velinova

R. D. Astumian
Department of Physics, The University of Maine, Orono, ME 04469, USA.

Kristina Konstas
School of Chemistry, Monash University, Clayton, Victoria, 3800, Australia

Steven J. Langford
School of Chemistry, Monash University, Clayton, Victoria, 3800, Australia

Melissa J. Latter
Centre for Strategic Nano-fabrication, The University of Western Australia, Crawley, West Australia 6009 Australia

Kai Y Soo
Department of Biochemistry, La Trobe Institute of Molecular Science, La Trobe University, VIC 3086, Australia

Manal Farg
Department of Biochemistry, La Trobe Institute of Molecular Science, La Trobe University, VIC 3086, Australia

Julie D. Atkin
Department of Biochemistry, La Trobe Institute of Molecular Science, La Trobe University, VIC 3086, Australia
Centre for Neuroscience, University of Melbourne, Parkville, VIC 3010, Australia
Florey Neuroscience Institute, University of Melbourne, Parkville, VIC 3010, Australia

Robert P. Erickson
Department of Physics and Astronomy, University of California, Irvine, Irvine, California, United States of America,

Zhiyuan Jia
Department of Physics and Astronomy, University of California, Irvine, Irvine, California, United States of America,

Steven P. Gross
Department of Physics and Astronomy, University of California, Irvine, Irvine, California, United States of America,
Department of Developmental and Cell Biology, University of California, Irvine, Irvine, California, United States of America

Clare C. Yu
Department of Physics and Astronomy, University of California, Irvine, Irvine, California, United States of America,

Mattia Gazzola
Chair of Computational Science, ETH Zurich, Zurich, Switzerland,

Christoph J. Burckhardt
Institute of Zoology, University of Zurich, Zurich, Switzerland

Basil Bayati
Chair of Computational Science, ETH Zurich, Zurich, Switzerland,

Martin Engelke
Institute of Zoology, University of Zurich, Zurich, Switzerland

Urs F. Greber
Institute of Zoology, University of Zurich, Zurich, Switzerland

Petros Koumoutsakos
Chair of Computational Science, ETH Zurich, Zurich, Switzerland,

Kristen M. Bartoli
Department of Biological Sciences, University of Pittsburgh, 4249 5th Avenue, Crawford Hall 274, Pittsburgh, PA 15213, USA

Darryl L. Bishop
Department of Biological Sciences, University of Pittsburgh, 4249 5th Avenue, Crawford Hall 274, Pittsburgh, PA 15213, USA

William S. Saunders
Department of Biological Sciences, University of Pittsburgh, 4249 5th Avenue, Crawford Hall 274, Pittsburgh, PA 15213, USA

Yusuke V. Morimoto
Quantitative Biology Center, RIKEN, 6-2-3 Furuedai, Suita, Osaka 565-0874, Japan
Graduate School of Frontier Biosciences, Osaka University 1-3 Yamadaoka, Suita, Osaka 565-0871, Japan

Tohru Minamino
Graduate School of Frontier Biosciences, Osaka University 1-3 Yamadaoka, Suita, Osaka 565-0871, Japan

Johanna Haiko
Division of General Microbiology, Department of Biosciences, University of Helsinki, Finland

Benita Westerlund-Wikström
Division of General Microbiology, Department of Biosciences, University of Helsinki, Finland

Matthew A. Watson
EaStCHEM School of Chemistry, University of Edinburgh, Joseph Black Building, David Brewster Road, Edinburgh, EH9 3FJ, UK.

Scott L. Cockroft
EaStCHEM School of Chemistry, University of Edinburgh, Joseph Black Building, David Brewster Road, Edinburgh, EH9 3FJ, UK.

L. Niels Cornelisse
Functional Genomics, Center for Neurogenomics and Cognitive Research (CNCR), Department of Clinical Genetics, VUmc, Neuroscience Campus Amsterdam (NCA), VU University (VUA) and VU University Medical Center (VUmc), Amsterdam, The Netherlands

Evgeni Tsivtsivadze
Machine Learning Group, Intelligent Systems, Institute for Computing and Information Sciences, Radboud University, Nijmegen, The Netherlands,
Microbiology and Systems Biology Group, TNO, Zeist, The Netherlands

Marieke Meijer
Functional Genomics, Center for Neurogenomics and Cognitive Research (CNCR), Department of Clinical Genetics, VUmc, Neuroscience Campus Amsterdam (NCA), VU University (VUA) and VU University Medical Center (VUmc), Amsterdam, The Netherlands

Tjeerd M. H. Dijkstra
Machine Learning Group, Intelligent Systems, Institute for Computing and Information

Sciences, Radboud University, Nijmegen, The Netherlands,
Signal Processing Systems, Department of Electrical Engineering, Technical University Eindhoven, Eindhoven, The Netherlands

Tom Heskes
Machine Learning Group, Intelligent Systems, Institute for Computing and Information Sciences, Radboud University, Nijmegen, The Netherlands,

Matthijs Verhage
Functional Genomics, Center for Neurogenomics and Cognitive Research (CNCR), Department of Clinical Genetics, VUmc, Neuroscience Campus Amsterdam (NCA), VU University (VUA) and VU University Medical Center (VUmc), Amsterdam, The Netherlands

Taizo Mori
World Premier International (WPI) Research Center for Materials Nanoarchitectonics (MANA), National Institute for Materials Science (NIMS)

Ken Okamoto
World Premier International (WPI) Research Center for Materials Nanoarchitectonics (MANA), National Institute for Materials Science (NIMS)

Hiroshi Endo
World Premier International (WPI) Research Center for Materials Nanoarchitectonics (MANA), National Institute for Materials Science (NIMS)

Keita Sakakibara
World Premier International (WPI) Research Center for Materials Nanoarchitectonics (MANA), National Institute for Materials Science (NIMS)
JST, CREST

Jonathan P Hill
World Premier International (WPI) Research Center for Materials Nanoarchitectonics (MANA), National Institute for Materials Science (NIMS)
JST, CREST

Satoshi Shinoda
JST, CREST
Osaka City University

Miki Matsukura
Osaka City University

Hiroshi Tsukube
JST, CREST
Osaka City University

Yasumasa Suzuki
Kitami Institute of Technology

Yasumasa Kanekiyo
Kitami Institute of Technology

Katsuhiko Ariga
World Premier International (WPI) Research Center for Materials Nanoarchitectonics (MANA), National Institute for Materials Science (NIMS)
JST, CREST

Argyris Papantonis
Sir William Dunn School of Pathology, Medical Sciences Division, University of Oxford, Oxford, United Kingdom,

Joshua D. Larkin
Sir William Dunn School of Pathology, Medical Sciences Division, University of Oxford, Oxford, United Kingdom,

Youichiro Wada
Laboratory for Systems Biology and Medicine, Research Center for Advanced Science and Technology, The University of Tokyo, Tokyo, Japan

Yoshihiro Ohta
Laboratory for Systems Biology and Medicine, Research Center for Advanced Science and Technology, The University of Tokyo, Tokyo, Japan

Sigeo Ihara
Laboratory for Systems Biology and Medicine, Research Center for Advanced Science and Technology, The University of Tokyo, Tokyo, Japan

Tatsuhiko Kodama
Laboratory for Systems Biology and Medicine, Research Center for Advanced Science and Technology, The University of Tokyo, Tokyo, Japan

Peter R. Cook
Sir William Dunn School of Pathology, Medical Sciences Division, University of Oxford, Oxford, United Kingdom,

PREFACE

The studies of molecular machines and their application are very active at this moment because of their long-standing prospective. All that attention of the whole scientific community started with the award of the Nobel Prize for Chemistry to John Walker and Paul Boyer in 1997 for their work in elucidating the mechanism of the rotary generator of ATP in the mitochondrion. The expected benefits of their uses are now well known: ultimate size, possibility to design finely-tuned molecular devices and their low energy consumption. All above has inspired the design and synthesis of a variety of compounds that resemble macroscopic machinery. This book brought together different research works which discuss different strategies to synthesize and study molecules demonstrating original mechanical properties at the nanometre and at the atomic scale.

The book is divided on three parts, as the first one provides detailed introduction to the field of molecular machines and motors, and their advances and perspectives. The next part of the book commences with a description of the functions of the molecular motors, including kinesins and dyneins, as well as their putative roles in Amyotrophic Lateral Sclerosis, after which R. Erickson et al. presents the significant evidence that cargos in vivo are carried by multiple motors. Three studies on molecular microtubule motors then follow — by M. Gazzola et al., by K. Bartoli et al. and by Z. Katsimitsoulia et al. This part ends with two articles which discuss the structure and role of the bacterial flagellar motor.

The last third part gives a detailed review of real examples for natural and artificial molecular machines used in the practise. The area of knowledge for the molecular machine is expanding continuously. Therefore, molecular machines for sensing, for nucleotide recognition, and molecular machines involved in distinct steps in neurotransmitter release are also discussed here.

Editor

MOLECULAR MACHINES AND MOTORS

Maria Velinova

Despite the advances in molecular biology in the last century, the molecular basis of life – ranging from single cells to entire organisms – is still a mystery for scientists. The tiny molecular machines of nature are fundamentally different to the man-made machines of the everyday macroscopic world. These disparities are not just limited to the sizes – where the difference is a billion fold - but also the milieu in which these machine carry out their functions. However, both the molecular machines and the machines of our macroscopic world are similar in that they perform specific tasks efficiently and precisely. Nevertheless, our attempts to understand the mechanisms of molecular motion should be tempered by the fact that molecular machines function in a microcosm is completely different from the world that surrounds us. Of particular interest to us are the biomolecular motors equipped with catalytic systems that can convert chemical energy to produce stepwise linear or rotary motion [1].

Since the last few decades, scientists have been trying to replicate these complex biological machines ex-vivo. The major challenge in achieving this is the fabrication of artificial nanoscale motors and propulsion systems which are not just motile, but can also be targeted to precise locations through inbuilt navigation systems. Such high precision transient and spatial control of motion at the nanoscale in turn requires finely tuned and regulated motor activity and/or fuel supply [2].

Molecular machines in living cells

Living organisms employ a host of complex molecular machinery capable of functioning within the thermal noise of the microscopic world [3,4]. These biomolecular devices are vital to cellular functions like motility, mitosis and transportation of nutrients and toxins; in addition, molecular machines also enable cells to maintain an impressive level of coordination between these functions. Most of the natural molecular machines are made of protein, although scientists have synthesized some artificial DNA-based molecular machines as well. Instead of using electrical energy like the man-made machines, molecular machines convert the chemical energy stored in chemical bonds of macromolecules or as gradients across cell membranes into mechanical energy.

Nature has designed many elegant and self-assembled de-novo molecular machines – each with many types of tracks, different building blocks of organic molecules as well as different kinds of motors that move along them. Based on the direction of motion, biomolecular motors are broadly classified as either linear or rotary. The linear motors, also called the cytoskeletal motors, are grouped into three main families depending on the cytoskeletal proteins – kinesins, myosins, and dyneins [5]. The function of the linear biomolecular machine is illustrated in Figure 1.

The kinesin protein motor is a **processive motor** which is persistently contacted as it moves along microtubule filaments. Kinesin motors enable membrane transport and the movement of chromosomes during cell division. There are over 250 kinesin-like proteins that differ in their intra-cellular localization, structure and movement, but they all share a common catalytic portion known as the motor domain [6]. Other examples of linear processive motors include RNA polymerase and DNA helicase. The myosin family of linear motors on the other hand, is a **nonprocessive motor** which detaches from the microfilament tract and subsequently re-attaches, thus giving the appearance of hopping instead of walking. The myosin motors are generally involved in muscle contraction which they bring about by binding to actin filaments [7]. The energy required for myosin-actin motility is derived from ATP hydrolysis [8]. Although the members of the myosin superfamily share the basic structure and function of actin binding and ATP hydrolysis, not all of them have been successfully characterized.

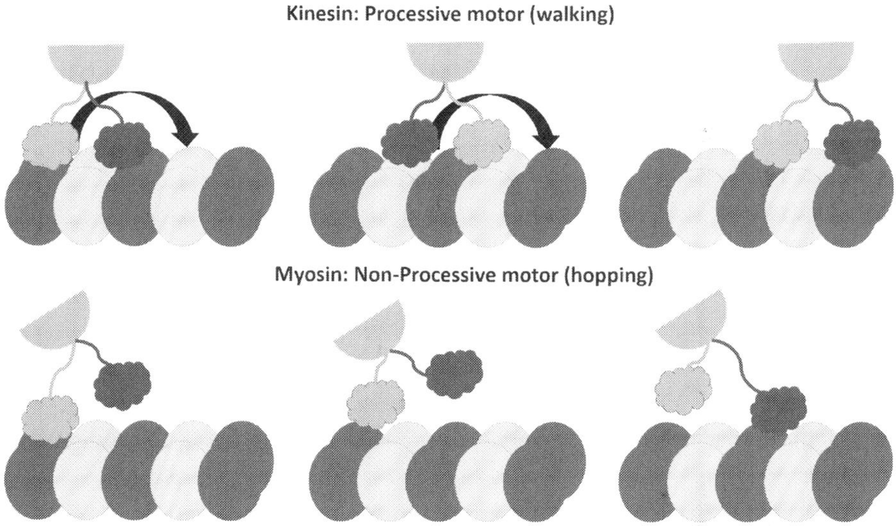

Figure 1. Biomolecular motors can move linearly in either a processive or a non-processive manner

The dynein superfamily of linear motor proteins move towards the minus-end of the microtubules - known as retrograde transport as it is in the direction opposite to kinesin [9]. The retrograde motion of the dynein motors is responsible for cellular transportation, mitosis, and ciliar and flagellar movements. Dyneins exist in two isoforms – cytoplasmic which is involved in cargo movement, and axonemal which produces bending motions of cilia and flagella [10].

The rotary biomolecular motor is best exemplified by the bacterial flagellum [11] as shown in Figure 2. Numerous bacterial species depend on flagella motors for motility: e.g., *E. coli, Salmonella enterica* serovar, Typhimurium (*Salmonella*), *Streptococcus*, *Caulobacter*, *Leptospira*, and *Bacillus*.

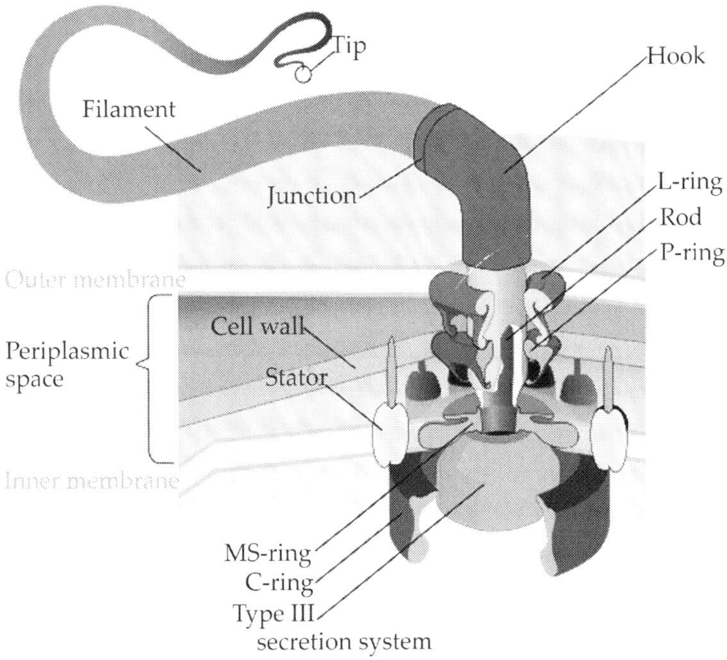

Figure 2. Propeller-like structure of a bacterial flagella motor [12]

The flagellum is the microbiological analogue of boat oars in terms of functionality. One flagellum unit consists of the flagella – the helical filaments that extend out of the cell into the medium – and the accompanying motor assembly. The latter in turn comprises of the rotary mechanism, the rate meters, particle counters, and gearboxes. The individual components of this machinery is precisely coordinated to help direct the cellular movement towards a gradient brought about changes in concentration of nutrients in the surroundings. Although the system is very powerful – generating a torque greater than 103 pN·nm – it is at the same time highly sensitive to environmental signals and rapidly disassembles into its individual protein units when removed from the cell membrane. This makes it particularly hard to study the exact operating mechanism of the flagellum in order to replicate its principle in an artificial nanodevice.

The second example of a rotary motor is the mitochondrial ATP synthase (ATPase) [13] which consists of two motor units – the membrane anchored F_0 unit and the free F_1 unit that projects into the mitochondrial matrix (Figure 3). ATPase is responsible for the synthesis of ATP and is powered by a proton gradient across the mitochondrial membrane created by proton pumps. The nano-propeller action of ATPase occurs on account of a 120° clockwise-rotation which is dependent on ATP binding – each time an ATP molecule is synthesized, the F_1 motor rotates. Likewise, an anti-clockwise rotation to the same degree can be driven by ATP hydrolysis, which helps pump protons against their electrochemical gradient. The detailed mechanism of the ATPase motor is still elusive as it is difficult to isolate form the membrane in a structurally and functionally intact form.

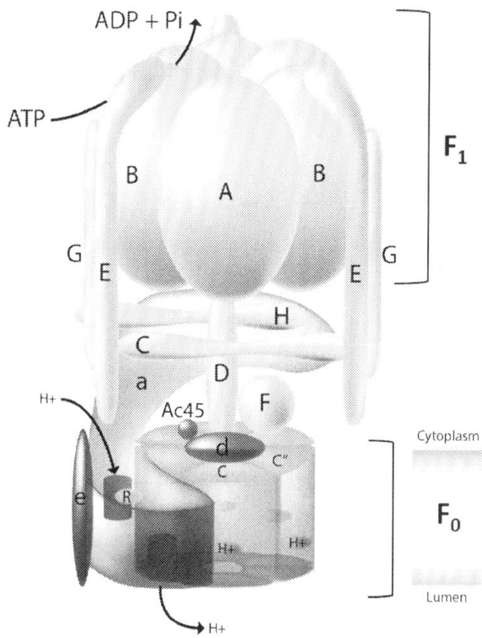

Figure 3. ATP synthase is composed of two rotary motor units (F_0 and F_1) connected by an asymmetric axle. Flow of hydrogen ions through the bottom motor, which is embedded in a membrane, turns the large cylindrical rotor (F_0). This is connected to the axle, forcing it to turn inside the second motor (F_1) at the top [14]

The molecular machines of the living cells are mainly responsible for cellular motility, transportation and structural stability. An unusual function of a bio-molecular machine is seen in the retinal photoreceptor cells. The human (and other vertebrate) eye is able to see an object when its image is focused on a single layer of photoreceptor cells on the back of the retina. Driven by photochemical energy through the absorption of a photon, the 11-cis-retinal moiety undergoes a cis-trans isomerization process [15] which involves a 180° rotation around a carbon-carbon double bond (shown in Figure 4).

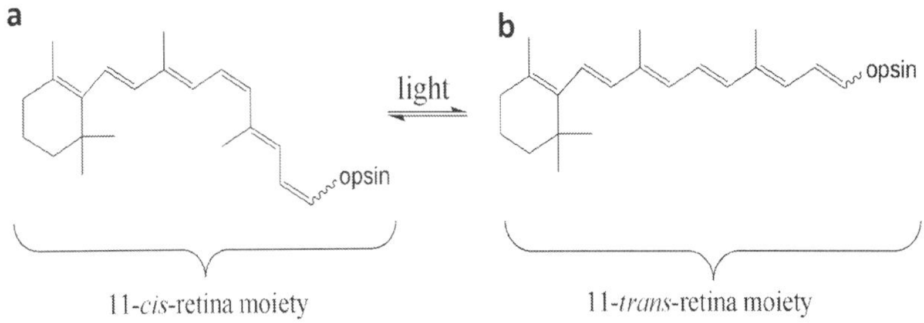

Figure 4. Cis-trans isomerization of molecule 11-cis-retinal moiety induced by light.

The 11-cis-retinal moiety of structure a () has an arch-like shape, which, upon conversion to the 11-trans isomer of structure b (), adopts a rod-like shape. The retinal moiety is covalently bound to a lysine molecule of the protein opsin via an imine bond – together forming the protein rhodopsin. Therefore, the conformational change in the retinal moiety brings about a change in the conformation of rhodopsin, which leads to the release of the 11-trans-retinal unit and a concomitant nerve pulse being sent to the brain to create the image. The free 11-trans-retinal is enzymatically reduced to trans-retinol (Vitamin A) and transformed to the cis moiety to start the vision cycle all over again. A high quantum yield () combined with high reversibility (> 106 cycles) truly makes the retinal cis-trans isomerization one of the most fascinating machines of the molecular living world.

DNA-based molecular motors

Unlike proteins which have a myriad of functions, the role of DNA is limited to that of an information carrier. It is probably for this reason that DNA has a significantly simpler structure compared to proteins – in contrast to the 20 amino acids and complex folding patterns that form the basis of protein structure, DNA comprises of just four nucleotide bases (adenosine, thiamine, guanine, and cytosine) attached in a linear fashion, which when paired with a complementary strand, takes the shape of a double helix. Ironically, while this structural simplicity limits the use of DNA in a living cell, this very feature makes it an attractive choice for the construction of artificial nano-machines.

The first such nano-machine to be developed was named DNA tweezers [16]. It consists of three strands of DNA – the central strand *A* to which the remaining strands *B* and *C* are partially hybridized (with overhangs on both ends), forming the so-called open conformation machine. A fourth auxiliary fuel strand (the *F* strand) is designed to hybridize with both overhang regions of strand *B* and *C* and when introduced to the open machine, the latter attains a closed conformation. The *F* strand can be removed from the system by adding its exact complement, thus switching the machine back to its original open conformation. In this way, the DNA tweezer machine can be easily switched on and off, and these states can be detected by attaching fluorescent tags to the two ends of strand *A*. In the prototype model, the 5' end was labeled with TET (tetrachloro-fluorescein phosphoramidite) and the 3' end with TAMRA (carboxy-tertamethylrhodamine). DNA tweezers are not just unique in their very conception but also in the use of DNA as the fuel – furthermore, since the 'fueling' depends on the principle of DNA strand complementarity, specific strand fuels will work with their machines and not trigger other non-specific machines surrounding them. This machine was later improved with the inclusion of a third intermediate state along-with the two conformational states [17]. In addition, a variation of the tweezers known as DNA-scissors, have also been developed [18].

Based on the same principles of branch migration and targeted fueling of DNA tweezers, a rotary nano-machine made of DX-DNA (double-crossover DNA) molecules was introduced [19]. Here the mechanics depended on the reversible transition between the paranemic crossover (PX) DNA and its topoisomer JX_2. The PX-DNA is formed by brown, green, and blue DNA strands of which the brown and green strands are connected to each other at the top and bottom double-helical regions by a single-stranded region. These single-stranded regions in turn are partially hybridized by blue strands with overhangs, which are complementary to incoming fuel strands. When the latter are introduced into the system, the blue strands are displaced, making room for the purple strands which in turn change the conformation of the PX molecule by a 180° rotation to the JX_2 state. The purple strands can be removed by adding complementary strands, followed by addition of fresh blue strands which results in another rotation to bring back the PX-conformation. In a very elegant complex molecular construction, the researchers attached half-hexagonal DNA structures formed by DX and ds-DNA onto one of the ds regions (brown or green) of PX motifs arranged in a linear array [20]. Due to the larger size of these structures, they could be visualized using an atomic force microscope to prove that the DNA rotary nano-machine indeed rotates.

Artificial molecular machines

Inspired by the de-novo construction of DNA machines as described above, synthetic chemists have developed a multitude of bio-machines from synthetic macromolecules which mimic the function of their biological counterparts [21]. The first of these machines were the artificial DNA-based, small-molecule 'walkers' that are able to walk uni-directionally over a track [22]. The ATPase system of eukaryotic cells has also inspired [23] a large number of synthetic uni-directional molecular motors that can be incorporated into larger constructs [24,25,26]. Although these synthetic nano-machines are more primitive in comparison to their natural analogous, they may be able to tolerate a diverse range of conditions and exhibit enhanced stability; synthetic systems therefore offer a net advantage in the development of complex nano-machinery [27]. Furthermore, artificial nano-machines can be controlled through various stimuli such as chemical, electrochemical and photochemical. This gives more tools to a scientist for designing and controlling nano-machines for a range of biological functions.

Molecular shuttle rotaxanes and catenanes

Rotaxanes and catenanes are interlocked organic molecular compounds with a distinctive shape and properties that guide their performance and control. They form the basis of many of the molecular machines described in this section.

Rotaxanes [28] are composed of a dumbbell-shaped molecule surrounded by a macrocyclic compound ("ring") and capped by bulky groups ("stoppers") that prevent disassembly (Figure 5a, b). They can be self-assembled using template-directed synthesis methods such as threading, clipping, and slippage.

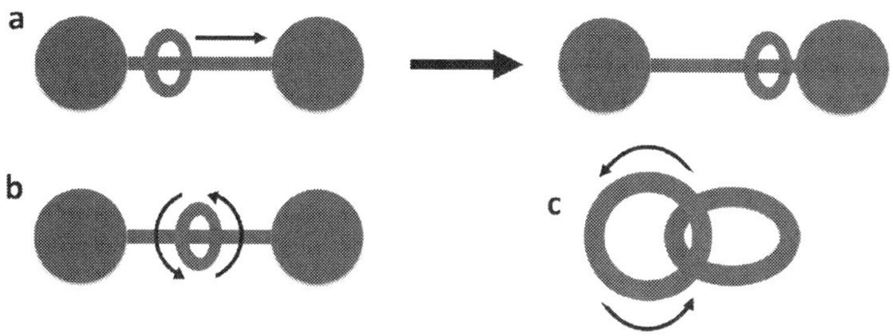

Figure 5. Schematic representation of the motions that can be obtained with simple interlocked molecular machines: ring shuttling in rotaxanes (a), and ring rotation in rotaxanes (b) and catenanes (c)

A rotaxane setup has been used to make a reversible molecular switch which can be operated with the help of two chemically active recognition sites – one low energy and the other high energy – in the neck region of the dumbbell. In the particular study cited, the thread was made of polyether marked with hydroquinol recognition sites and terminated at the ends by large triisoproplylsilyl groups. A tetracationic bead that interacts with the recognition sites was designed and self-assembled into the system. Upon application of a suitable stimulus – chemical, electrochemical, or photochemical – the recognition sites can be reversibly switched on and off. Furthermore, the stereo-electronic properties of the recognition sites can be altered by protonation/oxidation or deprotonation/reduction, thereby changing the affinity of the sites toward the macrocycle. In a recent study, light-induced acceleration of rotaxane motion was achieved by photo-isomerization [29]. Similarly, oscillating electric fields and various other controlling stimuli for switching rotaxane motion have been described.

The catenanes [25] on the other hand are made of (at least) two interlocked ring-like macrocycles (Figure 5c) that are non-covalently linked via a mechanical bond. Both macrocyclic components have recognition sites composed of atoms or groups of atoms that are either redox active or photo-chemically reactive. The rings can have different or similar recognition sites. In case of the latter, while one of the rings may rotate inside the other with conformations stabilized by noncovalent interactions, the two states of the inner ring, although differing by 180°, will be undistinguishable or degenerate [30]. Therefore for better control and distinguishable molecular conformations, it is desirable to have different recognition sites within the macrocycles, which would enable them to be controlled independently through their own specific stimuli. Like the rotaxanes, the catenanes can also be designed for chemical, photochemical, or electrochemical control. In addition, the stereo-electronic properties of a recognition site within a macrocycle can be modified such that at one point it has more affinity for the sites on the other ring. When that happens, the force balance will guide the rotating macrocycle towards a stable conformation requiring that particular site to be inside the other macrocycle. This affinity can be easily turned off or even reversed by changing the stimulus, along with an increase in the affinity of the second recognition site on the rotating macrocycle toward those on the static one. To achieve such a precise control over the molecular machinery, it is necessary to perform prior computational modeling and simulation.

Molecular switches

Molecular switches usually employ a cis-trans transformation along a double bond. In the unexcited or ground state, the molecular rotors are hindered due to the high energy barrier for cis-trans isomerization, but as soon as an excitation stimulus is given, they can overcome the barrier. Upon returning to the ground state a net rotation occurs bringing about the cis-trans isomerization. The common molecular switches employing this principle are e.g. stilbenes [31], rhodopsin [32] and azobenzenes [33]. Azobenzenes were among the first classes of molecular switches reported that undergo reversible photo-induced cis-trans isomerization [34]. Two stable states for azobenzene have been described (Figure 6) – the trans-isomer (a) and the cis-isomer (b).

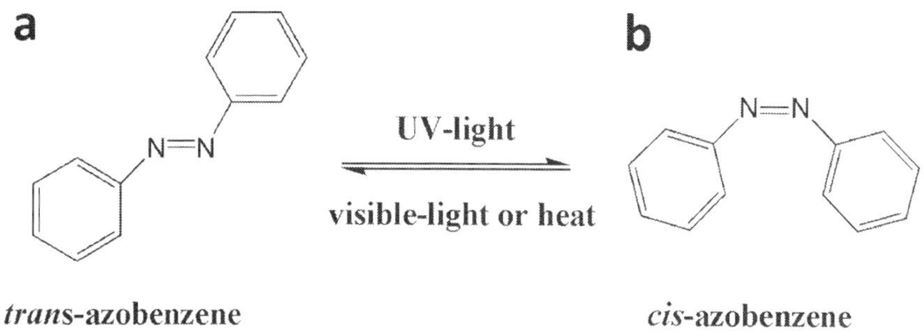

Figure 6. Cis-trans isomerization of azobenzene

The trans- to cis-isomerization is induced by UV irradiation while the reverse reaction can be completed by visible-light irradiation. A serious drawback of this system is the possible thermal isomerization from the cis- to trans-isomerization because of a low Gibbs free energy of activation.

Due to the interconversion of azobenzenes induced by irradiation at appropriate wavelengths, the conformation of molecules incorporating one or more azobenzene units can be altered reversibly by the controlled isomerization of these photoactive sites. Furthermore, these structural changes can be exploited to modulate the physical-chemical properties of either the molecule itself or their surrounding environment.

Molecular propellers and gears

The early examples of controlled molecular motions discussed in this section involved sterically hindered molecules, the rationale being that movement in one part of molecule is restricted by the presence of the other part of the molecule thus making their movements interdependent. Prominent examples are the molecular propeller [35] and molecular gear [36] as shown in Figure 7.

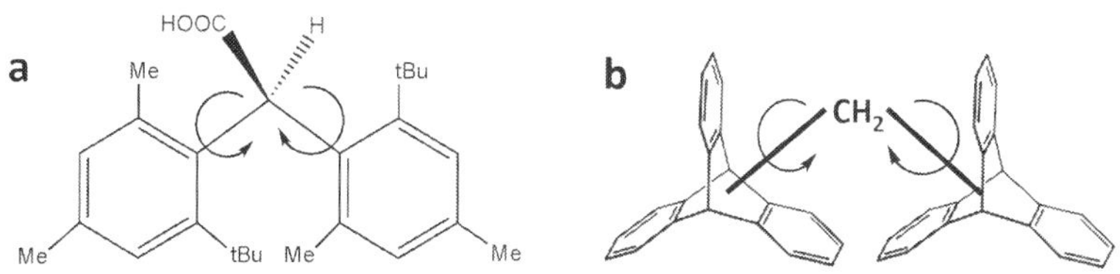

Figure 7. Example of molecular propeller (a) and molecular gear (b)

The molecules are composed of two bulky aryl rings that are linked to the same carbon atom and can therefore be regarded as the molecular equivalent of macroscopic propellers or gears. The diarylacetic acid derivatives (Figure 7a) has two identical aryl units (as blade) which cannot rotate independently of each other; when one ring rotates clockwise, it forces the other ring to rotate anti-clockwise (and vice-versa) through the single bonds connecting the aryl moieties to a carbon atom in the middle. Similarly, the molecular gear (Figure 7b) is based on the cooperative motion of two 9-triptycene moieties connected by a methylene group, with the triptycene rings acting as the notches of a bevel gear. As a result, when one ring rotates in one direction, the other rotates in the opposite direction. There is, however, no control over the direction and speed of rotation in these two examples.

Single-molecular-sized machines moving on surfaces

Most of the molecular machines have been designed and analyzed in solution phase or mounted on a surface; in both these scenarios, the molecules operate in a collective manner. It is highly desirable however to operate these motors in a unidirectional mode and with precise control at the single molecular level. The future development of artificial molecular transporters and nano robots will depend on the successful translation of a single molecule into a molecular machine with unidirectional motion along a surface.

Recent studies have achieved single molecule motion on surfaces through directed Brownian motion by free rotation of spherical molecular wheel [[37]], diffusion directed by surface geometry and specific binding [[38]], non-directional thermal spinning of molecules [[39]] and directional motion induced by pushing or pulling of molecules with an STM tip. The only drawback at this stage is that the movements cannot be controlled with precision as the molecules are still passive elements that diffuse along a preferential direction or dragged by the action of the macroscopic constituent (STM tip).

One example of a single molecular motor on surfaces has been reported by Sykes et al [[40]]. They were able to show that the molecule butyl methyl sulfide 49 adsorbed on a copper surface could be operated as a single-molecule electric motor. The molecular motor was powered by electricity from an STM tip using low-temperature scanning tunneling microscope (5K) which sent an electrical current through the molecule, thus directing the molecule to rotate in a preferred direction. After deposition of the molecule on the conductive copper surface, it was anchored to it via the sulfur atom. The initially achiral molecule becomes chiral upon anchoring to the copper surface and can then rotate along the central sulfur-copper bond. The rotating single molecule appears to be of hexagonal shape because of the three-fold symmetry of the molecule with timescales shorter than that of STM imaging. Different directions of the rotation have been observed depending on the intrinsic chirality of the STM tip.

How do we observe molecular machines?

The biological machines that we have discussed in the previous sections are macromolecular complexes of several globular proteins. The synthetic molecular machines on the other hand consist of one molecule and are therefore significantly smaller than their biological counterparts. Since their size falls below the diffraction limit of any optical microscope, one cannot rely on imaging to study most molecular machines. Other techniques such as spectroscopy have therefore been used to study molecular machines; these include atomic-force microscopy, X-ray reflectometry, circular dichroism, steady-state absorption spectroscopy, cyclic voltammetry, and NMR. Another limitation is that molecular motions typically occur on a picosecond time scale and the aforementioned spectroscopic methods, though generally sensitive to conformational structure, lack the temporal resolution to observe a fast conformational change. Determining the conformation of the device with a corresponding time resolution is vital for the understanding of nano-mechanics.

Conclusion

The recent explosion of research in nanotechnology, combined with important discoveries in molecular biology, has created a new interest in biomolecular machines and robots. The main goal of this field is to use various biological elements—whose function at the cellular level creates a motion, force, or a signal—as individual machine components that will perform the same function in response to the same biological stimuli, but in an artificial setting. With this technology, proteins and DNA both could act as motors, mechanical joints, transmission elements, or sensors. If all these components were assembled together they could form nanodevices which can move and manipulate objects in the nanoscale world, transfer information from the nano- to the macroscale world, and even travel in a nanoscale environment.

The future of molecular machinery is very promising. We are at the dawn of a new era in which many disciplines will merge, including robotics, mechanics, chemistry, biomedical engineering, biology, physics, and mathematics. Therefore, it is the need to design fully functional systems using principles of these different realms. However, bigger the promises of a new technology, greater the challenges. The more immediate ones include developing a complete database of different biomolecular machine components, the ability to interface or assemble different machine components and controlling and coordinating several biomolecular machines simultaneously.

REFERENCES

(Endnotes)

1. M. Schliwa (ed.), Molecular Motors, Wiley-VCH, Weinheim, Germany, 2003.
2. E. R. Kay, D. A. Leigh, F. Zerbetto, Angew. Chem., Int. Ed. 2007, 46, 72-191.
3. J. M. Berg, J. L. Tymoczko, L. Stryer, Biochemistry, W.H. Freeman and Company, New York, USA, 2002.
4. C. Mathews, K. van Holde, K. Ahern, Biochemistry, San Franciso, USA, 2000.
5. R. D. Vale, J. Cell Biol. 1996, 135, 291– 302
6. C. M. Farrell, A. T. Mackey, L. M. Klump, S. P. Gilbert J. Biol. Chem. 2002, 277, 17079–87
7. K. Kitamura, M. Tokunaga, A. H. Iwane, T. Yanagida, Nature 1999, 397, 129-134.
8. J. Howard, Nature 1994, 368, 98–99
9. H. C. Taylor, E. J. Holwill, Nanotechnology 1999, 10, 237-243.
10. T. A. Schroer, E. R. Steuer, M. P. Sheetz, Cell 1989, 56, 937–46.
11. D. J. DeRosier, Cell 1998, 93, 17-20.

12. By LadyofHats, PMID 17142059 (bend), Public Domain, https://commons.wikimedia.org/w/index.php?curid=6219592
13. K. Kinosita, R. Yasuda, H. Noji, S. Ishiwata, M. Yoshida, Cell 1998, 93, 21-24.
14. By NOchotny at the English language Wikipedia, CC BY-SA 3.0, https://commons.wikimedia.org/w/index.php?curid=45720713
15. R. R. Birge, Biochim. Biophys. Acta 1990, 1016, 293-327.
16. B. Yurke, A. J. Turberfield, A. P. Mills, F. C. Simmel, J. L. Neumann, Nature 2000, 415, 62–65.
17. F. C. Simmel, B. Yurke, Appl. Phys. Lett. 2002, 80, 883–85.
18. J. C. Mitchell, B. Yurke, DNA scissors. In DNA Computing. Proc. 7th Int. Meet. DNA-Based Computers, ed. SN Jonoska,NSeeman. Univ. So. FL, Tampa. Heidelberg: Springer Verlag, 2002.
19. N. C. Seeman, Nano Lett. 2001, 1, 22– 26.
20. X. Yang, L. A. Wenzler, J. Qi, X. Li, N. C. Seeman J. Am. Chem. Soc. 1998, 120, 9779–86.
21. E. R. Kay, D. A. Leigh, F. Zerbetto, Angew. Chem. Int. Ed. 2007, 46, 72–191.
22. M. J. Barrell, A. G. Campaña, M. von Delius, E. M. Geertsema, D. A. Leigh, Angew. Chem. Int. Ed. 2011, 50, 285–290.
23. R. A. Van Delden, M. K. J. ter Wiel, M. M. Pollard, J. Vicario, N. Koumura, B. L. Feringa, Nature 2005, 437, 1337–1340.
24. N. Koumura, R. W. J. Zijlstra, R. A. van Delden, N. Harada, B. L. Feringa, Nature 1999, 401, 152–155.
25. Vicario, J, Meetsma, A & Feringa, B. Controlling the speed of rotation in molecular motors. Dramatic acceleration of the rotary motion by structural modification. Chem. Commun. 5910–5912 (2005).
26. A. Cnossen, L. Hou, M. M. Pollard, P. V. Wesenhagen, W. R. Browne, B. L. Feringa, J. Am. Chem. Soc. 2012, 134, 17613–17619.
27. G. M. Whitesides, Sci. Amer. 2001, 285, 78–83.
28. J.-P. Sauvage, C. O. Dietrich-Buchecker (ed.), Molecular Catenanes, Rotaxanes and Knots, Wiley-VCH, Weinheim, Germany, 1999.
29. F. G. Gatti, S. Leon, J. K. Y. Wong, G. Bottari, A. Altieri, et al, Proc. Natl. Acad. Sci. USA 2003, 100, 10–14.
30. P. R. Ashton, T. T. Goodnow, A. W. Kaifer, M. V. Reddington, A. M. Z Slawin, et al. Angew. Chem. Int. Ed. Engl. 1989, 28, 1396– 99.

31. J. Saltiel, Y. P. Sun, Photochromism, Molecules and Systems in Studies in Organic Chemistry 40; H. Dürr, H. Bouas Laurent (ed.), Elsevier, Amsterdam, 1990, Chapter 3
32. R. R. Birge, Biochim. Biophys. Acta 1990, 1016, 293-327.
33. G. S. Hartley, Nature 1937, 140, 281.
34. H. Rau, Photoisomerization of Azobenzenes, in J. F. Rebek (ed.), Photochemistry and Photophysics, CRC Press, Boca Raton, 1990.
35. O. S. Akkerman, J. Coops, Recl. Trav. Chim. Pays-Bas 1967, 86, 755-761.
36. F. Cozzi, A. Guenzi, C. A. Johnson, K. Mislow, J. Am. Chem. Soc. 1981, 103, 957-958.
37. Y. Shirai, A. J. Osgood, Y. Zhao, K. F. Kelly, J. M. Tour, Nano Lett. 2005, 5, 2330-2334.
38. K. L. Wong, G. Pawin, K.-Y. Kwon, X. Lin, T. Jiao, U. Solanki, R. H. J. Fawcett, L. Bartels, S. Stolbov, T. S. Rahman, Science 2007, 315, 1391-1393.
39. O. P. H. Vaughan, F. J. Williams, N. Bampos, R. M. Lambert, Angew. Chem. Int. Ed. 2006, 45, 3779-3781.
40. H. L. Tierney, C. J. Murphy, A. D. Jewell, A. E. Baber, E. V. Iski, H. Y. Khodaverdian, A. F. McGuire, N. Klebanov, E. C. H. Sykes, Nat. Nanotechnol. 2011, 6, 625-629.

PART I: INTRODUCTION

CHAPTER

1

HOW MOLECULAR MOTORS WORK – INSIGHTS FROM THE MOLECULAR MACHINIST'S TOOLBOX: THE NOBEL PRIZE IN CHEMISTRY 2016

R. D. Astumian

Department of Physics, The University of Maine, Orono, ME 04469, USA.

ABSTRACT

The Nobel prize in Chemistry for 2016 was awarded to Jean Pierre Sauvage, Sir James Fraser Stoddart, and Bernard (Ben) Feringa for their contributions to the design and synthesis of molecular machines. While this field is still in its infancy, and at present there are no commercial applications, many observers have stressed the tremendous potential of molecular machines to revolutionize technology. However, perhaps the most important result so far accruing from the synthesis of molecular machines is the insight provided into the fundamental mechanisms by which molecular motors, including biological motors such as kinesin, myosin, F_0F_1 ATPase, and the flagellar motor, function. The ability to "tinker" with separate components of molecular motors allows asking, and answering, specific questions about mechanism, particularly with regard to light driven vs. chemistry driven molecular motors.

INTRODUCTION

Molecular motors drive almost all significant biological processes, yet very few if any man made technologies exploit controlled molecular motion. The work for which the 2016 Nobel prize in chemistry was awarded has begun to change this state of affairs by presenting inspired synthetic routes to facilely create a host of interlocking structures known as catenanes[1] and rotaxanes,[2]and directional rotors[3] that work with, rather than against, ubiquitous thermal noise.

At first, many systems operating as simple back and forth molecular switches were proudly proclaimed to be "molecular motors". Eventually, conceptual advances[4,5] that showed how omnipresent thermal noise can be exploited at the nano-meter scale filtered into the synthetic chemistry community. This insight led to synthesis of molecular machines with the ability to use external energy to power the cyclic motion necessary for creation of true molecular motors. Detailed study of different mechanisms – energy ratchets and information ratchets[6] – by which this occurs has led to important gains in understanding how biomolecular machines work. Specifically, the "power-stroke", presented as the mechanism by which most ATP and ion gradient driven molecular motors work in almost all leading textbooks of biochemistry and cell biology, is shown to be just plain wrong for these chemically driven motors. Instead, the key mechanism for molecular motors driven by a catalysed reaction such as ATP hydrolysis or ion transport is chemical gating in which the specificity for substrate vs. product depends on the state of the molecular motor, a mechanism known as an information ratchet.[6]

MOLECULAR ROTORS AND MOTORS

Consider two very different designs for a molecular rotor, one, a light driven rotor from Ben Feringa's group,[7] and the other an autonomous rotor driven by catalysis of a chemical reaction from David Leigh›s group.[8]

Light driven rotor

The light driven motor described by Feringa and colleagues[7] uses optical energy to populate an unstable state by a directional rotation followed by a thermal relaxation to a stable state in a cyclic fashion resulting in completion of a rotation about a carbon–carbon double bond. In Fig. 1 absorption of a photon by the stable (ground) states (M) cis isomer causes clockwise rotation as the molecule is promoted to an unstable (excited) state (M) trans isomer. This is followed by thermally activated helix inversion to the stable (P) trans isomer. Absorption of a photon by the (P) trans isomer leads to clockwise trans–cis isomerization to the unstable (P) cis isomer followed by thermally activated helix inversion to the starting (M) cis isomer. In the excited state the rotor freely rotates on one side of the plane of the stator but is hindered from helix inversion. In the ground state the barrier for helix inversion is lower than that for rotation about the carbon–carbon double bond, with the net effect being continuous rotation of the molecule when illuminated.

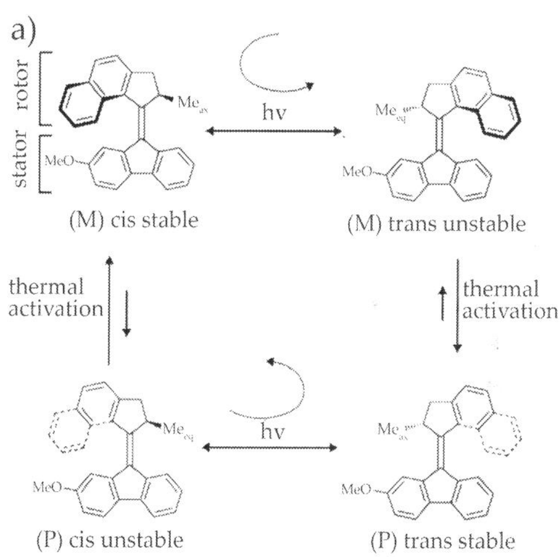

Figure. 1 (a) Light driven molecular motor cycle designed by Ben Feringa and colleagues. The design is based on over-crowded alkenes, where the notation (M) or (P) denotes left (Minus) or right (Positive) helicity. For convenience we designate the bottom part of the molecule as the "stator" and the top part as the "rotor". The subscript on the methyl group (Me) at the chiral carbon atom of the rotor denotes whether the orientation of the methyl is axial (ax) or equatorial (eq). (b) Ground (solid curve) and excited state (dashed curve) energy surfaces for motion of the rotor relative to the stator of the motor in (a).

The stability difference, ΔG, between unstable and stable states is determined by the intramolecular strain when the methyl group on the rotor is in the equatorial vs. axial positions. It is key that this ΔG be relatively large to ensure directional rotation. The relative likelihood of completing a clockwise vs. counter-clockwise rotation is

$$r = e^{2\Delta G} \qquad (1)$$

where ΔG is, as are all energies in this paper, expressed in units of the thermal energy.

While early light driven motors[3] were driven by cycles of illumination – heating – cooling – illumination, recent motors have been designed to allow for continuous operation at MHz rotation frequencies at a fixed temperature.[7] Similar ideas arise in stochastic conformational pumping mechanisms[9a,b] where time periodic external driving induces net cycling if some transitions involve a significant ΔG. The mechanical process in which the energy ΔG is dissipated in the environment is very reasonably termed a power stroke. Let us compare the light driven motor that requires a power stroke with a motor driven by catalysis of a chemical reaction.

Chemical catalysis driven rotor

Most biomolecular motors use energy from ATP hydrolysis or transport of protons from high to low electrochemical potential across a membrane rather than from light or from external modulations. A longstanding challenge has been to design a synthetic motor that similarly uses chemical energy from a catalyzed reaction to drive directed motion. This goal has recently been achieved[8] by David Leigh's group using catenanes, a type of the mechanically interlocked molecules held together by mechanical bonds for which Jean Pierre Sauvage[1] and Sir Fraser Stoddart[2] were awarded the Nobel prize. The rotor described by Wilson et al.[8] incorporates the main features of biological molecular motors – chemical catalysis, gating of the binding and release of substrate and product depending on the mechanical state of the motor, and conformational changes involving mechanical motion.

The molecular motor, a [2]catenane, comprises two interlocked rings of different sizes. The small blue ring is free to rotate relative to the larger ring, shuttling between the two recognition sites (shown in aqua) on the larger ring, one of which is labelled by deuterium. There are two catalytic sites (hydroxyl groups) represented by the small red cones, one near each recognition site, that catalyze the conversion of Fmoc-Cl to dibenzofulvene.

A kinetic cycle for the mechanism of the motor is shown in Fig. 2. The energy-releasing reaction is the conversion of the protecting group 9-fluorenylmethoxycarbonyl chloride (Fmoc-Cl) to dibenzofulvene. The free-energy released by this reaction is greater than that released by ATP hydrolysis. The key point is that the addition of the protecting group occurs as Fmoc-Cl and the removal of the protecting group is as dibenzofulvene, processes that are not the microscopic reverses of one another. The description of the catalytic conversion of Fmoc-Cl to bibenzofulvene is shown as a Michaelis–Menten catalytic mechanism at the bottom of the Fig. 2, thus emphasizing that the hydroxyl groups function as catalysts even though the focus is on the effects of the small-molecule reagents on the large moving ring.

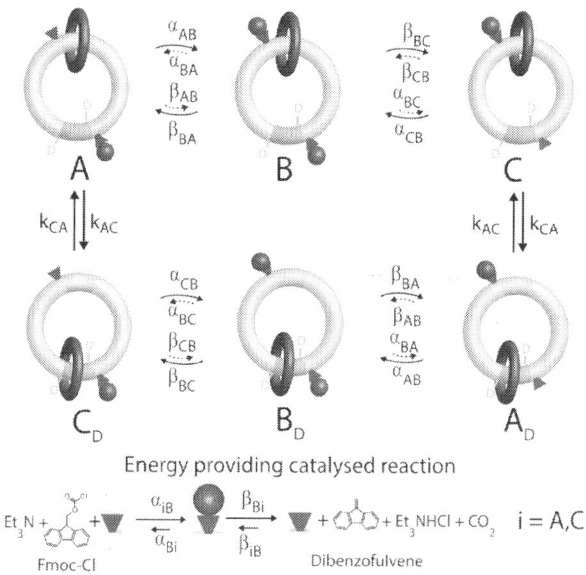

Figure 2. Kinetic cycle for the molecular motor described by Wilson et al.8 The deuterated recognition site serves as a fiduciary marker for determining directionality but does not otherwise influence the kinetics or thermodynamics of the motor.

It is very natural to imagine that, by analogy with Feringa's light driven motor, clockwise rotation of the blue ring from the perspective of the reader could be enforced by designing the system with a repulsive interaction between the bulky Fmoc protecting group (red sphere) and the dark blue ring, thereby engineering a "power-stroke" in the direct mechanical transitions such that $k_{CA} > k_{AC}$ and hence with ln $\left(\frac{k_{CA}}{k_{AC}}\right) = \Delta G > 1$ driving the transition C → A.

This intuitive idea however is wrong – the dark blue ring shown in Fig. 2 actually rotates (from the perspective of the reader) counter-clockwise irrespective of whether the interaction between the small blue ring and the Fmoc protective group is attractive ($\Delta G < 0$) or repulsive ($\Delta G > 0$). The directionality arises because the rate of attachment of the protecting group to the catalytic site is faster when the blue ring is at the distal rather than proximal recognition site ($\alpha_{CB} > \alpha_{AB}$). The rate of cleavage of the Fmoc group as dibenzofulvene, on the other hand, is nearly independent of the position of the ring ($\beta_{BA} \approx \beta_{BC}$). These two kinetic conditions on the rate constants reflect the fact that installing the protecting group as Fmoc depends on the location of the dark blue ring, but that removing the protecting group as dibenzofulvene does not. The net effect of these two kinetic conditions is to make the transition C → A less likely than the transition A → C, a seemingly slight bias which is nevertheless enough to provide directionality to the rotational motion although the experimentally observed velocity is very small.

It is essential at this point to make a very important distinction between the autonomous chemically fuelled motor of Wilson et al.[8] and previously described "chemically driven" molecular motors and pumps which were driven by external changes of the chemical environment – redox potential, pH, substrate concentration, etc. Such externally driven motors can, and for the most part do, work by a power stroke mechanism similar to the light driven motor described by Feringa and colleagues[7] and the biological pump driven by a fluctuating electric field described by Astumian et al.[9a]

The motor described by Wilson et al.[8] on the other hand works under a single reaction condition where substrate (Fmoc-Cl) associates and product (dibenzofulvene) dissociates because the chemical potential of substrate is large and that of product is small. This reproduces a major feature of ATP hydrolysis driven biomolecular motors. Autonomous motors driven by catalysis of a chemical reaction cannot operate by a power stroke mechanism, and instead function as information ratchets[6] where the key mechanistic design principle is the kinetic discrimination of the rate of the chemical reaction depending on the mechanical state of the motor by which $A_{CB} > \alpha_{AB}$.

In Fig. 2 rate constants for cleavage of the protecting group as Fmoc, and attachment as dibenzofulvene are shown, even though the rates for these processes are very small under the experimental conditions. This is done to allow for self-consistent examination of the kinetics and thermodynamics of this molecular motor.[8,10] The rate constants must obey the constraint

$$\frac{\alpha_{AB}\beta_{BA}}{\alpha_{BA}\beta_{AB}} = \frac{\alpha_{CB}\beta_{BC}}{\alpha_{BC}\beta_{CB}} = e^{\Delta\mu} \qquad (2)$$

where $\Delta\mu = \ln\left(\frac{[\text{Fmoc-Cl}][\text{Et}_3\text{N}]}{[\text{dibenzofulvene}][\text{Et}_3\text{NHCl}][\text{CO}_2]}K_{eq}\right)$ and K_{eq} is the equilibrium constant for the catalyzed reaction. An additional constraint

$$\frac{\alpha_{AB}\alpha_{BC}}{\alpha_{CB}\alpha_{BA}} = \frac{\beta_{AB}\beta_{BC}}{\beta_{CB}\beta_{BA}} = e^{\Delta G} \qquad (3)$$

must also hold. The coefficients α_{BA}, β_{BA}, α_{BC}, β_{BC} are first order "off" rate constants and the ratio between any two of them is independent of the free energies of the states A, B, and C and can be written as the exponential of the difference between two activation energies. The coefficients α_{AB}, β_{AB}, α_{CB}, β_{CB} on the other hand are pseudo-first order rate constants into which the bulk concentrations [Fmoc-Cl] and $[Et_3N]$, for the α›s, and [dibenzofulvene], $[Et_3NHCl]$, and $[CO_2]$, for the β›s, have been subsumed. The two relationships eqn (2) and (3) are thermodynamic identities that follow from the principle of microscopic reversibility.[11]

The directionality of the molecular motor in Fig. 1 is given by the ratio of the products of the net clockwise and counter-clockwise rate constants,

$$r = \left[\frac{(\alpha_{AB} + \beta_{AB})(\alpha_{BC} + \beta_{BC})k_{CA}}{(\alpha_{BA} + \beta_{BA})(\alpha_{CB} + \beta_{CB})k_{AC}}\right]^2 \qquad (4)$$

when $r > 1$ the blue ring rotates clockwise, and when $r < 1$ the blue ring rotates counter-clockwise, from the perspective of the reader.

A regrettable tendency in many recent papers is to point out that chemically driven motors "operate far from equilibrium" and then assign rate constants ad libitum for mathematical convenience rather than for thermodynamic consistency. This approach often leads to plausible but incorrect conclusions. As an example, consider the assignment $\alpha_{AB} = \alpha_{CB} = \beta_{BA} = \beta_{BC} = \gamma\, e^{\mu/4}$ and $\alpha_{BA} = \alpha_{BC} = \beta_{AB} = \beta_{CB} = \gamma\, e^{-\mu/4}$, where γ sets the time scale for the chemical transitions relative to the mechanical transition. With these expressions we immediately derive as the ratio of clockwise to counter-clockwise net rate constants the expression $r = e^{2\Delta G}$, seeming to suggest a critical role for the "power-stroke" just as for light driven motors. However, while this assignment for the rate constants is consistent with eqn (2), it is not consistent with eqn (3) and is thus fundamentally incorrect, being tantamount to ignoring both the steric interaction by which $\alpha_{CB} > \alpha_{AB}$, and any energetic interactions by which $k_{CA} \neq k_{AC}$ and which must be reflected in the α›s and β›s according to eqn (3). The negligibility of energetic interactions in the chemical rate constants is absolutely impossible even very far from equilibrium as pointed out first in the context of ion pumps.[9a]

When we carry out the slightly more mathematically complex imposition of the constraints of both eqn (2) and (3) we instead obtain the correct expression

$$r = \left[\frac{\left(1 + \frac{\beta_{BA}}{\alpha_{BA}} e^{-\Delta\mu}\right)\left(1 + \frac{\beta_{BC}}{\alpha_{BC}}\right)}{\left(1 + \frac{\beta_{BA}}{\alpha_{BA}}\right)\left(1 + \frac{\beta_{BC}}{\alpha_{BC}} e^{-\Delta\mu}\right)}\right]^2 = \left[\frac{(e^{-\Delta\mu} + q) + L_{uc}}{(1 + q\, e^{-\Delta\mu}) + L_{uc}}\right]^2 \tag{5}$$

where $q = \alpha_{BA}\beta_{BC}/\alpha_{BC}\beta_{BA}$ is a ratio of "off" rates and parametrizes the chemical gating. Thus we see that r is independent of ΔG. The term L_{uc} arises from un-coupled cycles[12] and setting $L_{uc} = 0$ represents the optimal directionality of the motor. Not surprisingly, if $\Delta\mu = 0$ then $r = 1$. More revelatory is the fact that if $q = 1$ then $r = 1$ irrespective of the value of $\Delta\mu$, i.e., if there is no chemical gating there is no coupled rotation despite continual chemical dissipation. Using the constraints eqn (2) and (3), with $\alpha_{CB} > \alpha_{AB}$ and $\beta_{BA} \approx \beta_{BC}$ we have for the motor in Fig. 2 that $q < 1$ and hence that $r < 1$, irrespective of ΔG.

There is a clear presumption in the literature that the ΔG for the mechanical stroke must be positive in the direction of motion. Howard comments that "If the structural states differ by a distance d, then the power stroke can do work against an external loading force F provided that $F \cdot d \leq \Delta G$, where ΔG is the decrease in free energy between the two chemical states".[13] and Eisenberg and Hill claim that the maximum efficiency can be no higher than $\Delta G/\Delta\mu$, where $\Delta\mu$ is the free energy released by the driving chemical reaction.[14] These assertions regarding the power stroke, based on the incorrect assumption that the "power stroke" must be energetically downhill (exergonic), are just wrong for chemically driven motors. In fact, the thermodynamic properties of a chemically driven motor – efficiency, stoichiometry, stopping force – do not depend on ΔG at all.[12,15] Perhaps having different stabilities for A and C affords a kinetic advantage? This idea, too, is incorrect. While the rate of the mechanical transition may indeed be enhanced when $\Delta G \neq 0$, the net rate of the chemically driven transitions between A and C through state B is inexorably decreased when the free energies of A and C are different from one another. In fact, the rate of rotation is optimized when the basic free energies of all states are identical, consistent with Jeremy Knowles theory for evolutionary optimization of enzyme catalytic function.[16]

Now consider the case that $q, e^{-\Delta\mu} \ll 1$ so that $q\, e^{-\Delta\mu}$ can be neglected in eqn (5). We recognize two distinct regimes of coupled transport.[12] When gating is very strong ($q \ll e^{-\Delta\mu}$) the motor operates in the thermodynamic control regime, also known as the tight-coupling regime, in which every conversion from substrate to product is accompanied by a counter-clockwise rotation, and application of torque sufficiently strong as to cause clockwise rotation induces conversion of product to substrate. An example is the F_1 ATP synthase where ATP hydrolysis drives rotation in one direction. When an external torque is applied to force rotation in the opposite direction ATP is synthesized.[17]

In contrast, when $e^{-\Delta\mu} \ll q$ the motor is regulated by the kinetic gating. In this regime the application of a sufficiently strong torque to cause a motor to move backward results in an increase in the rate of conversion of substrate to product, not in the conversion of product to substrate. This latter behaviour, surprising to many in the field of biological motors, was predicted by Astumian and Bier in 1996 (ref. 18) and demonstrated experimentally for kinesin by Nishiyama, Higuchi, and Yanagida[19] and by Carter and Cross[20] in 2002 and 2005, respectively.

PROSPECTIVE

Molecular motors can be designed to interact with a variety of inputs – light, chemical reactions, electric fields, external changes in pH and redox potential, etc. – and to provide a means for energy transduction between these inputs. Finding the right environment to harness these possibilities has, however, not been easy since many possible tasks for molecular machines require spatial and/or temporal coordination between numerous individual motors.

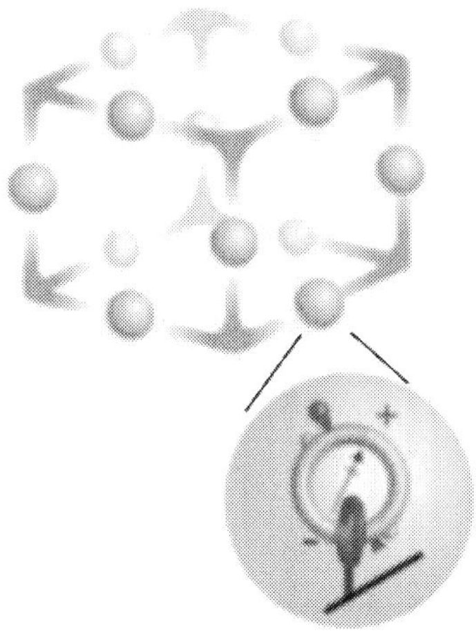

Figure 3. Schematic approach to "smart materials" that can interact with, and mediate energy exchange between, several modalities of energy input from the environment. The upper schema illustrates a single unit cell of a large periodic array, while the lower scheme shows an individual chemically driven molecular rotor similar to that describe in Fig. 2. Here, the catenanes localized on a metallic–organic framework (MOF) have dipole moments by which they can interact with an applied electric field, and chemical catalytic sites by which they interact with substrate and product in the environment. The yellow hemispheres are "molecular speed bumps" by which the height and position of the kinetic barrier to rotation can be engineered. In principle, energy can be transduced between an applied time-dependent electric field and the catalysed reaction in either direction. Cooperative interactions between the individual catenane rotors can give rise to very complex behaviour.

One approach to achieving this goal involves incorporating molecular motors in metallo–organic frameworks[21–23] (MOFs) as shown schematically in Fig. 3. This setup affords the possibility of using molecular motors to perform molecular tasks. In principle, the catalytic function performed by the two active sites shown could be designed at will, here described generically as S → P. In the example shown in Fig. 3 an electric dipole is incorporated in the design of the elementary molecular motors based on that of Wilson et al.[8] Then an applied oscillating electric field $\vec{E}(t)$ influences the motion[24,25] and can drive the catalysed reaction away from equilibrium, maintaining a steady-state chemical potential difference[24,26] $e^{\Delta\mu} = e^{\overline{\vec{E}(t)\cdot\Delta\vec{m}}}$ where $\Delta\vec{m}$ is the dipole moment difference between the states. Such coupled devices could in principle pave the way to new synthetic approaches for high energy (and high value) compounds as well as for new sensors.

There are of course approaches other than those based on a synthetic chemistry perspective, including those grounded in surface science[27–29] and self-organized assembly[30] in soft condensed matter.

CONCLUSIONS

The very different design principle necessary for a motor driven by chemical catalysis – gating – in comparison to that for a light driven or externally driven motor – power stroke – highlights the importance of microscopic reversibility for understanding the mechanism of chemically driven molecular motors.[31] 'In the IUPAC definition[32] of the principle of microscopic reversibility, "In a reversible reaction, the mechanism in one direction is exactly the reverse of the mechanism in the other direction. This does not apply to reactions that begin with a photochemical excitation", light driven processes are explicitly excluded from the constraint of microscopic reversibility'.

This is because in the absorption of a photon, and its reverse – stimulated emission – there is conversion between energy in one degree of freedom, the photon, and another single degree of freedom, the energy of the photo-chemically active molecule. In contrast, a thermal activation process, including binding of a high free-energy substrate, involves conversion between many degrees of freedom in the bath and only one (or perhaps a few) degrees of freedom in the molecule, irrespective of how far from equilibrium the reaction may be.

The importance of microscopic reversibility is also well illustrated in the context of a synthetic molecular pump.[33] When driven by external changes of the redox potential it is essential that the energy of an intermediate state depend on the oxidation state[34] – the pump acts as an energy ratchet. On the other hand, theoretical analysis of a model in which the same process is driven by a catalysed chemical reaction shows that the energy of the intermediate state is irrelevant and the sole determinant of the efficacy of pumping is chemical gating[34] – the motor functions as an information ratchet.

The concept of a power stroke – a viscoelastic mechanical relaxation from a high energy state to a low energy state – is given in almost all textbooks of biochemistry and cell biology to explain how molecular machines such as myosin, kinesin, and the flagellar motor convert chemical energy (e.g. ATP hydrolysis) or osmotic energy (e.g., a proton gradient) into mechanical motion and the performance of work. The recently synthesized molecular motor driven by chemical catalysis,[8] and the concomitant theoretical analysis, shows unequivocally that this mechanism is incorrect. Strain (or other energetic interactions between components) stored in a molecular motor, while of the essence for light driven motors and for motors driven by external modulation of the chemical environment, is irrelevant for molecular motors driven by chemical catalysis, including ion transport.

In the molecular world, equilibrium is a dynamic state in which all possible motions are explored with equal probability in the forward and reverse directions. The key principle for designing molecular motors that are driven by chemical catalysis is to use the chemical energy to selectively prevent unwanted motion, rather than to cause the desired motion. Any description of the effect of the catalysed reaction in terms of "deposition of chemical energy" or as causing violent kicks or judo throws is wrong and highly misleading.

The talk given by Richard Feynman in 1959, "Plenty of room at the bottom", represents a defining moment in the quest for making molecular motors, as does the 21st Solvay Conference on Chemistry held in Brussels in 2007 (see Fig. 4). At this meeting there was much emphasis on how chemists are inspired by the fabulous machinery of biology, and how, in turn, the synthetic molecular analogues are teaching much about the fundamental principles by which bio-molecular machines function. A very important philosophy, beautifully reflected in the work on molecular motors, is contained in a phrase found on Feynman›s blackboard at the time of his death – "What I cannot create I do not understand". The Nobel laureates in Chemistry for 2016, and others, have provided techniques by which to create and tinker with molecular motors. This breakthrough is not only the precursor to what promises to be an amazing technology, the applications of which have yet to be developed, but also the means by which to come to penetrating and detailed understanding of the fundamental principles by which the molecular motors of life carry out their essential functions.

Figure 4. Participants of the 21st Solvay Conference on Chemistry "From Non-covalent Assemblies to Molecular Machines", Brussels, Hotel Metropole, 30 November 2007.

ACKNOWLEDGEMENTS

It is a great pleasure to thank Chuyang Cheng, David Leigh, Shayantani Mukherjee, Cristian Pezzato, Fraser Stoddart, and Arieh Warshel for very useful discussions on molecular motors.

NOTES AND REFERENCES

1. C. O. Dietrich-Buchecker, J. P. Sauvage and J. P. Kintzinger, Tetrahedron Lett., 1983, 24, 5095–5098 .
2. P. L. Anelli, N. Spencer and J. F. Stoddart, J. Am. Chem. Soc., 1991, 113, 5131–5133 .
3. N. Koumura, R. W. J. Zijlstra, R. A. van Delden, N. Harada and B. L. Feringa, Nature, 1999, 401, 152–155 .
4. R. D. Astumian, Science, 1997, 276, 917–922 .
5. (a) J. V. Hernandez, E. R. Kay and D. A. Leigh, Science, 2004, 306, 1532–1537 ; (b) M. N. Chatterjee, E. R. Kay and D. A. Leigh, J. Am. Chem. Soc., 2006, 128, 4058–4073 .
6. (a) R. D. Astumian and I. Derényi, Eur. Biophys. J., 1998, 27, 474–489 ; (b) V. Serreli, C.-F. Lee, E. R. Kay and D. A. Leigh, Nature, 2007, 445, 523–527 .
7. J. Conyard, A. Cnossen, W. R. Browne, B. L. Feringa and S. R. Meech, J. Am. Chem. Soc., 2014, 136, 9692–9700 .
8. M. R. Wilson, et al. , Nature, 2016, 534, 235–240 .
9. (a) R. D. Astumian, P. B. Chock, T. Y. Tsong, Y. D. Chen and H. V. Westerhoff, Proc. Natl. Acad. Sci. U. S. A., 1987, 84, 434–438 ; (b) R. D. Astumian, Annu. Rev. Biophys., 2011, 40, 289–313 .
10. R. D. Astumian, Nat. Nanotechnol., 2016, 11, 582–583 .
11. R. D. Astumian, Nat. Nanotechnol., 2012, 7, 684–688 .
12. R. D. Astumian, S. Mukherjee and A. Warshel, ChemPhysChem, 2016, 17, 1719–1741 .
13. J. Howard, Curr. Biol., 2006, 16, R517–R519 .
14. T. L. Hill, E. Eisenberg, Y. D. Chen and R. J. Podolsky, Biophys. J., 1975, 15, 335–372 .
15. R. D. Astumian, Biophys. J., 2015, 108, 291–303 .
16. J. J. Burbaum, R. T. Raines, W. J. Albery and J. R. Knowles, Biochemistry, 2002, 28, 9293–9305 CrossRef .
17. H. Itoh, et al. , Nature, 2004, 427, 465–468 .
18. R. D. Astumian and M. Bier, Biophys. J., 1996, 70, 637–653 .
19. M. Nishiyama, H. Higuchi and T. Yanagida, Nat. Cell Biol., 2002, 4, 790–797 .

20. N. J. Carter and R. A. Cross, Nature, 2005, 435, 308–312 .

21. H. Deng, M. A. Olson, J. F. Stoddart and O. M. Yaghi, Nat. Chem., 2010, 2, 439–443 .

22. C. S. Vogelsberg and M. A. Garcia-Garibay, Chem. Soc. Rev., 2012, 41, 1892–1910 RSC .

23. A. Coskun, M. Banaszak, R. D. Astumian, J. F. Stoddart and B. A. Grzybowski, Chem. Soc. Rev., 2012, 41, 19–30 RSC .

24. R. D. Astumian and B. Robertson, J. Am. Chem. Soc., 1993, 115, 11063–11068 .

25. V. Bermudez, et al. , Nature, 2000, 406, 608–611 .

26. V. S. Markin, T. Y. Tsong, R. D. Astumian and B. Robertson, J. Chem. Phys., 1990, 93, 5062 .

27. J. Michl and E. C. H. Sykes, ACS Nano, 2009, 3, 1042–1048 .

28. H. L. Tierney, et al. , Nat. Nanotechnol., 2011, 6, 625–629 .

29. P. Mishra, et al. , Nano Lett., 2015, 15, 4793–4798 .

30. S. Sengupta, M. E. Ibele and A. Sen, Angew. Chem., Int. Ed., 2012, 51, 8434–8445 .

31. R. D. Astumian, Faraday Discuss., 2016 10.1039/c6fd00140h .

32. International Union of Pure and Applied Chemists (IUPAC), Compendium of Chemical Terminology, principle of microscopic reversibility, http://goldbook.iupac.org.

33. C. Cheng, et al. , Nat. Nanotechnol., 2015, 10, 547–553 .

34. C. Cheng, P. R. McGonigal, J. F. Stoddart and R. D. Astumian, ACS Nano, 2015, 9, 8672–8688

CHAPTER

2

ADVANCES TOWARDS SYNTHETIC MACHINES AT THE MOLECULAR AND NANOSCALE LEVEL

Kristina Konstas [1], Steven J. Langford [1] and Melissa J. Latter [2]

[1]School of Chemistry, Monash University, Clayton, Victoria, 3800, Australia

[2]Centre for Strategic Nano-fabrication, The University of Western Australia, Crawley, West Australia 6009 Australia

ABSTRACT

The fabrication of increasingly smaller machines to the nanometer scale can be achieved by either a "top-down" or "bottom-up" approach. While the former is reaching its limits of resolution, the latter is showing promise for the assembly of molecular components, in a comparable approach to natural systems, to produce functioning ensembles in a controlled and predetermined manner. In this review we focus on recent progress in molecular systems that act as molecular machine prototypes such as switches, motors, vehicles and logic operators.

Keywords: molecular machines; devices; switches; catenanes; rotaxanes; nanovehicles; rotary motors

INTRODUCTION

A machine is defined by the Oxford dictionary as 'an apparatus for applying mechanical power, having several parts each with a defined function'. Typically a machine requires at least one moveable component that leads to the generation of work. A molecular or nano-scaled machine can therefore be defined as an ensemble of molecular components or devices assembled to achieve a specific level of mechanical work (output) as a consequence of an external stimuli (input) [1–3]. Each molecular component performs a single function (structural or dynamic) whilst the entire assembly performs a more complex function leading to work. In a molecular system, the concept of movement (leading to work) is possible by molecular processes such as translational, rotational or geometric isomerism. Translational isomerisation involves movement of one component relative to another between two (or more) points differentiated by space. Rotational isomerism occurs in molecules (rotamers) about a specific axis if possible. This could be around a single carbon-carbon bond or in complex interlocked molecules. Geometric isomerisation particularly relates to the change in spatial orientation of two or more components constrained by an existing connectivity.

Scientists and engineers have learnt from the selectivity, specificity, precision and accuracy of biological processes and the ensembles so formed on a cellular and sub-cellular level and strive to apply these concepts in the laboratory to create molecular devices and machines. Natural machines such as ATP-ase, DNA-polymerase, opsin/rhodopsin, ribosomes etc. are all complex and fascinating examples of natures approach to nano-scaled machines. The synthesis and assembly of molecular building blocks capable of functioning in a controlled way and in a wholly synthetic sense is an achievable goal and, to this end, prototypical machines that demonstrate specific tasks or design features are being reported in ever increasing amounts. In fact, a simple web of science search identifies over 450 papers with the term 'molecular machine' in its title over the past decade.

To generalize the design principles of a molecular level machine, the following criteria have been identified as important features: 1) what energy input is required to make the machine perform work, 2) what is the type and level of movement performed by its components, 3) how will the machine's function be detected and monitored during operation, 4) does the system have a plausible repeat operation and can this establish a recurring process, 5) what is the timescale for a complete cycle of operation, and 6) what level and type of function is to be performed by the machine [1–4].

The earliest recognition and notion of building nano-scaled artificial molecular machines and devices was made by the Nobel Laureate Richard P. Feynman in 1959 [5,6]. His presentation entitled "There is Plenty of Room at the Bottom" to the American Physical Society sparked a great enthusiasm for scientists to pursue the construction of unimaginable miniaturized complex components in a complementary way to Nature. His argument is predicated on a vision that the 'Large-Downward (Top-Down)' approach to the nanometer-level for such construction by dismantling elemental arrays to this level will suffer through heat dissipation, resolution and the cost of fabrication. A better approach, termed the 'Small-Upward (Bottom-Up)' strategy, was argued based on natural examples - starting from small molecules and building up to a nanostructure through self-assembly and self-organisation processes [7]. This approach gives chemists the upper-hand in designing and developing strategies for the construction of molecular machines and devices. This review aims to highlight some of the advances made towards functional systems capable of molecular switching, motion and transport with a long-term goal of synthetic machinery examples at the molecular level.

INTERLOCKED MOLECULES – CATENANES

Mechanically-interlocked molecules have attracted considerable attention as the basis of molecular devices and nanoscale machines long after the first reported syntheses of these chemical curiosities in the mid 1960's [8,9]. Catenanes (from the Latin catena meaning chain) are the most common of this class and are defined by their interlocking molecular rings (Figure 1a). Rotaxanes (rota meaning wheel, axismeaning axle) are another common form (Figure 1b). With respect to nomenclature, the number (represented as n) is enclosed in square brackets, [n] and describes the number of molecular parts involved in mechanical bonding. For example, a [2]catenane denotes two interlocking rings, a [4]catenane denotes four interlocking rings.

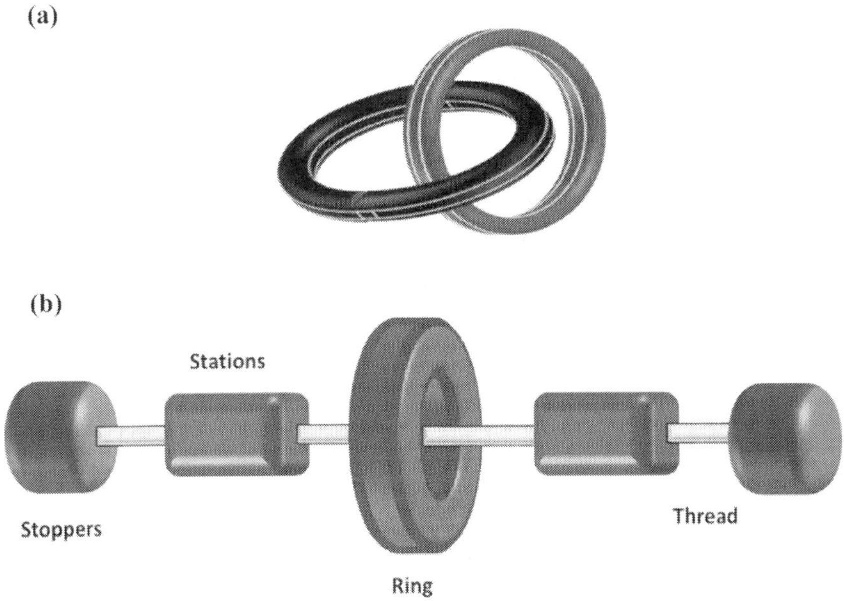

Figure 1. Schematic representation of two different types of interlocked molecules that have shown potential as molecular level machinery components. **(a)** Interlocked ring structures referred to as catenanes and **(b)** interlocked ring and thread terminated by bulky stoppers are termed rotaxanes.

The molecular dynamics of each component with respect to the other can be controlled and has been shown to be the basis of simple molecular devices. Recent developments in synthetic paradigms (template-direction etc) has allowed for the preparation of a variety of catenanes and more complex structures, such as trefoil knots in high yield [1,10–19]. The preparation of interlocked molecules involves two integral steps: (i) the molecular recognition and arrangement of the non-interlocked components in a threaded fashion – so-called self-assembly, and (ii) covalent bond forming reactions post molecular recognition, which mechanically-interlocks the two components. The syntheses of interlocking molecules were once deemed a great synthetic challenge, with often more than 20 multiple-step syntheses being employed using the covalent approach and ultimately leading to a statistical result with yields of less than one percent [9,20,21]. Synthetic strategies based on various non-covalent templates have been proposed by several research groups, leading to efficient preparative procedures with yields typically above 50%. Synthetic template-directed strategies are based on non-covalent interactions, such as hydrogen bonding, donor-acceptor interactions and metal complexation, which are used to assemble a pre-organized macrocyclic component and a linear component in a threading process, which after a bond-forming reaction, affords the interlocked molecules [22,23]. In addition, knotted interlocked molecules have also been prepared from single-stranded DNA amongst other molecules [24–26]. Of the non-covalent templates described in the literature, the most promising synthetic strategy are those utilizing transition metal templates or complementary π-electron-rich and π-electron-deficient recognition sites.

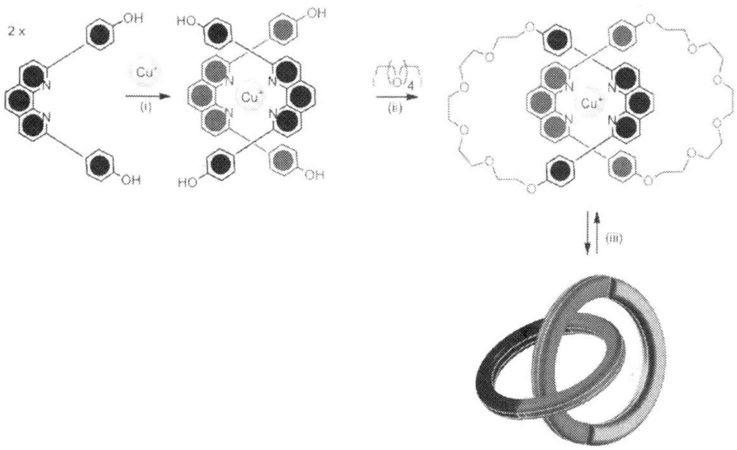

Figure 2. Sauvage's copper(I) template synthesis of an interlocked catenane in two sequential steps: **(i)** coordination of the disubstituted phenanthroline ligand to the metal; **(ii)** mechanical interlocking by a covalent bond forming reaction; **(iii)**Demetallation leads to a specific conformational change.

The metal ion template strategy was first developed by Sauvage and co-workers using copper(I) ions and phenanthroline ligands [27,28]. The shape and mutually orthogonal arrangement of the disubstituted phenanthroline ligands would encapsulate the tetrahedral copper(I) metal ion, generating the necessary cross-over point for catenane formation giving a pseudo-tetrahedral complex (Figure 2(i)). With the introduction of hydroxyl groups on the phenyl substituents of the phenanthroline ligand, the half-rings could undergo a Williamson ether macrocyclization reaction, as seen in Figure 2(ii). This synthesis provided a simplistic and efficient means for preparing catenanes but also rotaxanes and to a smaller extent knotted molecules [10]. By changing the oxidation state of the metal, or by competing ligation, the metal ion could be removed to give the free catenane. Sauvage's phenanthroline catenane undergoes a co-conformational change, which involves the circumrotation of both the macrocycles through the cavity of each other upon demetallation, leading to a conformation in which the phenanthroline ligands to be positioned away from each other (Figure 2(iii)). This is an example of a chemically controllable catenane.

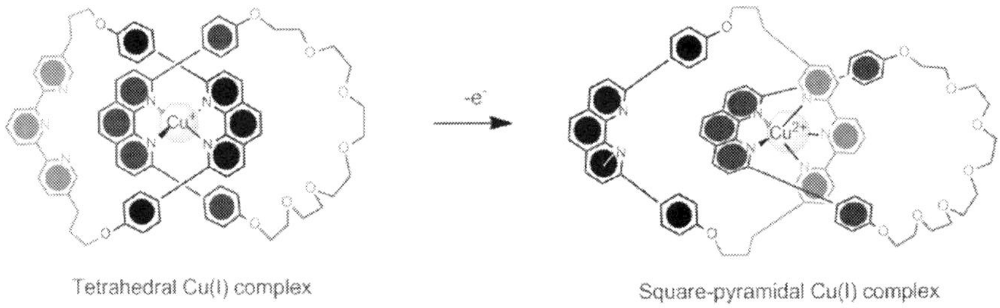

Figure 3. A model for a catenane-based molecular switch that is redox-switchable by control of the copper oxidation state [29].

In addition, the co-conformational motion can also be controlled electrochemically by reversible oxidation/reduction reactions of the metal centre and judicial choice of the ligands. For example, while copper(I) prefers a four-coordinate environment, copper (II) prefers a five-coordinate environment leading to ligands capable of stabilizing this ion upon oxidation. The example shown in Figure 3 is a model for a molecular switch based on a [2]catenane [29]. This model demonstrated in earlier work by Sauvage uses one of the macrocyclic rings to contain two discrete binding sites on opposing sides of the ring–terpyridyl (terpy) and diphenylphenanthroline (dpp). The second macrocyclic ring contains only a single dpp binding unit. Upon changing the oxidation state of the copper(I) to copper(II), the dpp/terpy conformation is adopted to favor the five-coordinate copper(II) preference. These manipulations have lead to the formation of molecular switches and muscles [30–33].

Sauvage's catenane is an early example of the processes that can be applied to interlocking molecules but a more amenable approach to switching behavior leading to rotational isomerism involves desymmetrization of the [2]catenane within the different ring components [34]. The synthesis of such compounds is challenging as they require the selection of pairs of complementary recognition units capable of distinctive binding strengths in both the neutral and reduced/oxidized states. The use of π-donors, namely tetrathiafulvalene (TTF), 1,5-dioxynaphthalene (DNP) and naphthalene diimide (NDI) moieties have been incorporated in a crown ether macrocyclic component that selectively holds the π-deficient tetracationic cyclophane around the donor TTF station. Rotation of the crown ether components in the TTF-containing [2]catenane-NDI interlocked molecule can be driven chemically, by oxidation and reduction cycles leading to a switching mechanism with an optical readout [34]. In this case, no translocation of the dicationic cyclophane occurs in response to the oxidative switching of the doubly charged species and the NDI unit remains outside of the electron rich crown ether as verified by ^{1}H NMR spectroscopy (Figure 4). These types of [2]catenane molecules are appealing and have the potential to act as machine-like molecules based on the movement of one component relative to the other, or as a molecular electronic device based on the electronic switching, a topic described in greater detail in Section 4.

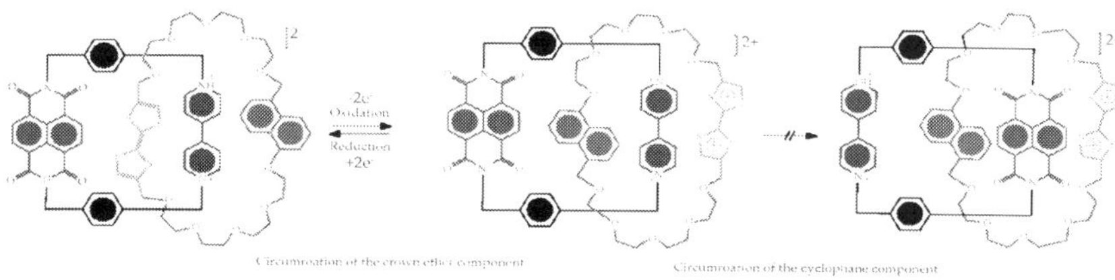

Figure 4. A representation of the desymmetrized donor-acceptor [2]catenane [34].

There are two significant challenges that need to be overcome in order to achieve a pragmatic molecular device or machine. While the catenanes discussed already display machine-like processes, it is all achieved in solution with properties of a bulk nature being measured.

As with natural systems, most machines are housed upon a platform or scaffold of some sort, be it a membrane, cell wall, or macroscopically, a factory floor. Furthermore, machines tend to be well-ordered so that their action and output can be used for work either as a sole entity, or in concert with other machine processes. There has been much achieved to date in each individual area, however few examples attempt to solve both necessities. One of the more recent attempts to overcome these challenges, involves embedding the catenanes into a metal-organic frameworks (MOF) in which the crystal lattice is utilized as the solid support [35,36]. The amalgamation of Stoddart and Yaghi's research has now led to new developments in metal-organic frameworks with a donor-acceptor catenane (Figure 5(a)) being crystallized with copper(I) to give a solid-state 2D network (Figure 5(b)) [37]. This structure demonstrates the first step towards the construction of new synthetic materials that contain dynamic behavior into robust crystalline scaffolds.

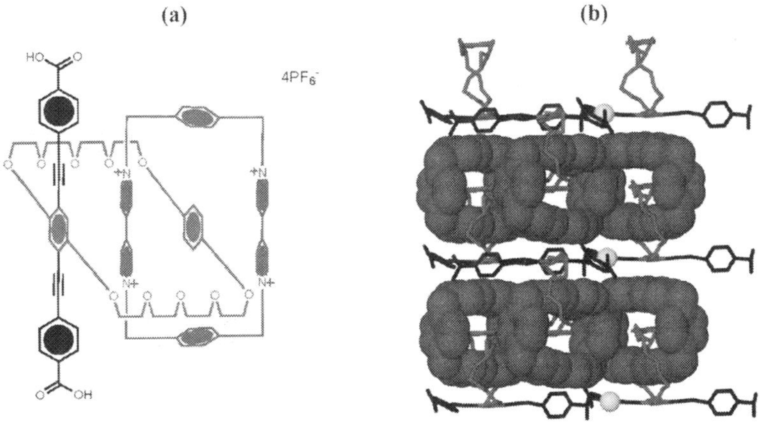

Figure 5. (a) Stoddart and Yaghi's catenane used in a metal-organic framework, **(b)** Side on view of a portion of the MOF along the a-axis [37].

INTERLOCKED MOLECULES – ROTAXANES

Rotaxanes differ from catenanes in that they are composed of both ring (cyclic) and axle (acyclic) components. A [2]rotaxane is formed when a thread, dumbbell or axle component is mechanically interlocked with a ring component. If the axle component is stoppered at both ends with bulky functional groups to prevent disassociation of the ring component at ambient temperature (Figure 1b), the system is called a rotaxane. If there are no stoppers or the stoppers are not of sufficient size to preclude dissociation at ambient temperature, the molecular assembly is called a [2]pseudorotaxane. A [3]rotaxane is typically formed when two rings are threaded onto an axle component, though two axles and one ring also satisfy the definition. There are three common methods to prepare rotaxanes, namely capping, clipping and slipping but a more recent improvement via a template effect has improved yields over what was previously achieved by statistical synthesis. The worth of rotaxanes as a molecular form of abacus or transport agent and their capacity to shuttle (molecular motion) and switch (molecular logic) has seen research in the area flourish over the past decade and much effort has been dedicated to advancing synthetic methodologies for enhanced functions [1,38–42].

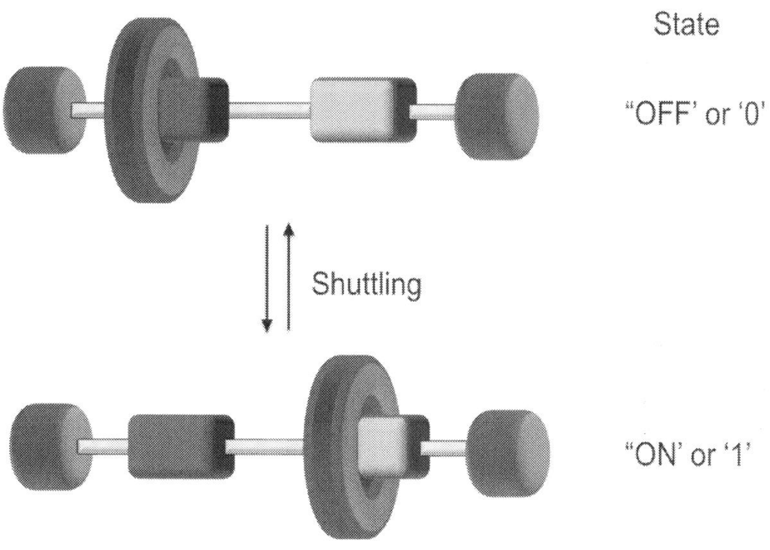

Figure 6. Rotaxanes are able to 'shuttle' between two different states indicating switching (ON/OFF) or binary processes (0/1). Shuttling is a form of translational isomerism. The challenge is controlling this motion.

The use of rotaxanes as a prototype for a class of molecular machines has arisen as a result of their ability to undergo translational isomerisation between two or more structures (Figure 6). This isomerisation is a result of the translational molecular motion of the macrocyclic ring along the acyclic component. The equilibration and dynamic behavior back and forth along the acyclic component is also known as shuttling in the same way a bus or tram may move between two stations at an airport. This shuttling can take place up to 40,000 times per second. Molecular recognition sites are built into both components to produce thermodynamic sinks resulting in points along the linear component where the two or more components predominantly reside through a bias. Again, an electronic complementarity between the components is vital for both synthesis and function. This complementarity often means that rotaxanes are often highly colored species.

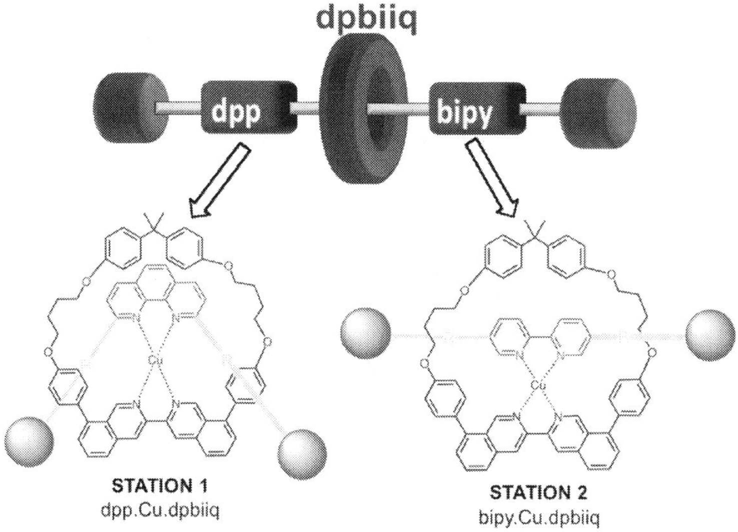

Figure 7. Copper-complexing [2]rotaxane comprising two different bidentate stations [43]. The diphenylbiisoquinoline ligand (dpbiiq) site in the macrocyclic ring structure is complementary to either the diphenylphenanthroline (dpp) or bipyridine (bipy) sites along the threading molecule and shuttling is possible upon electrochemically-driven complexation events with copper.

Metal complexing sites, quite often of different denticity (i.e., bidentate and tridentate) have been a popular choice for "stations" on axle components, with molecular shuttling driven by electrochemical stimulation. Sauvage and co-workers have since shown stations of the same denticity may also be incorporated into the axle structure and function as a [2]rotaxane via electrochemical means [43]. In this example, the axle binding sites include a highly shielding phenanthroline ligand and a non sterically hindering bipyridine chelate while the complementarity to afford copper complexation is provided by a bisquinoline unit in the macrocylic ring (Figure 7). Although the preparation involves a multi-step synthesis, the electrochemical-induced shuttling between the two stations is a fast, clean process and shows promise for multi-state machine developments in the future.

Extending the topology and complexity of linear motors, the stretching and contraction of skeletal muscle may also be replicated in a synthetic sense by one-dimensional molecular assemblies acting as molecular muscles upon the action of an external stimuli [44–48]. Using a transition metal complexation strategy involving a symmetrical, double-threaded topology Sauvage was able to demonstrate "artificial muscle" processes with changes in backbone length approximated to be between 65–85Å from molecular modelling. The key to their system was the use of two [2] rotaxanes to form a dimer, with this orientation allowing the individual thread of each rotaxane to move along the other, interconverting between a contracted and stretched confirmation akin to the working action of skeletal muscles. Each thread was synthesized to contain both bidentate and tridentate chelate "stations" with exchange between copper(I) and zinc(II) respectively facilitating motion (Figure 8) [49].

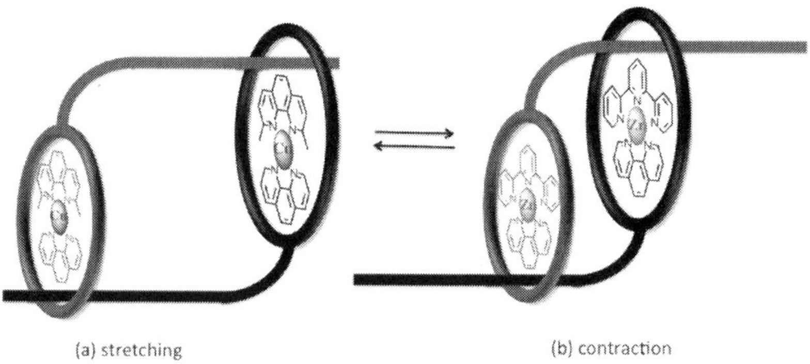

Figure 8. Schematic representation of Sauvage's "artificial muscle" using transition metal complexation to interchange between **(a)**stretched and **(b)** contracted geometries.

A more recent example of a molecular muscle reported from the Chiu research group utilizes a molecular cage that acts as the ring component of the [2]rotaxane with the additional muscle function possible by threading an axle component that is able to contract and stretch upon fluoride addition or removal respectively [50]. In their system, an axle bearing two ammonium stations, one pyridinium station and terminated by an alkyl bromide is threaded through the macrocyclic cage. Reaction of the threaded axle with a substituted pyridine yields a [2]rotaxane where the original thread reacted with the substituted pyridine to incorporate a fourth station (a second pyridinium). The new [2]rotaxane, comprised of the macrocyclic cage and thread with four stations (two ammonium stations and two pyridinium stations) was observed by NMR spectroscopy to stretch and contract through anion exchange. One of the biggest challenges in trying to mimic any natural processes in the laboratory is replicating the efficiency of the process so possibly of greater significance is the reported change of 36% between contracted and stretched conformations (based on molecular dynamic simulations), which is greater than the 27% change in human muscle.

Extending from the concept of linear motion demonstrated by rotaxane based machines that shuttle between multiple stations, Stoddart and co-workers have progressed this research area further to pioneer a two component machine that mimics the behavior of an elevator at the molecular level [51]. One component of the machine is comprised of a central aromatic platform appended with feet forming a tripod type arrangement. Incorporated within each of the three feet are two different binding sites: at the upper level, ammonium binding sites, and bipyridinium at the lower level both of which are able to interact with each of the three complementary crown ethers of the second component of the elevator, utilizing the knowledge and success of this binding motif in many of Stoddart's classical rotaxane examples [10]. Stoddart reports the molecular elevator has an approximate diameter of 3.5 nm and height of 2.5 nm with vertical movement (estimated distance of 0.7 nm) possible through acid-base manipulation [51]

What should be obvious to the reader by now is the essential requirement of multiple stations positioned along an axle for shuttling to occur in rotaxane machinery. As this research area has flourished with examples of such linear motion, some pivotal milestones have seen marked improvements with respect to molecular complexity (increased topology and threading). An even more recent development from the Leigh group has seen complexity rise further whereby more than one axle component is threaded through the same macrocycle yielding a [3] rotaxane in one pot as shown schematically in Figure 9 [53]. The success in achieving this more complicated interlocked system arises from the design considerations of the macrocycle. The positioning of a ligand in the marcocyclic structure functions in two ways: (1) act as a ligation site for a transition metal that will act as a catalyst in the bond formation of the axle components, and (2) provides a template for successive covalent bond formation of interlocked axle components. This high yielding methodology may be applicable to other higher order interlocked molecular assemblies which is important as molecular machinery design continues to develop in terms of complexity.

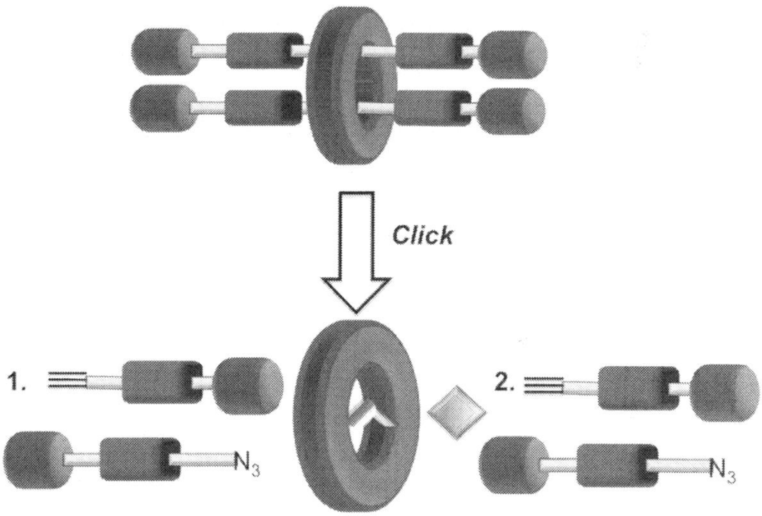

Figure 9. Template-assisted, one pot synthesis of a [3]rotaxane. Two, successive covalent bonds are formed (via the CuAAC click reaction) assisted by ligation to the ring yielding an interlocked structure consisting of two threaded components [53].

Interest has also extended to the potential biological applications mechanically interlocked rotaxanes may offer through the encapsulation or protection provided by the macrocyclic ring about the axle component. In particular, Leigh and co-workers reported a rotaxane comprised of a pentapeptide axle and macrocycle that protects against peptidase-catalysed hydrolysis over several days [54]. The use of oligopeptides axle components was also reported by Moretto et al. and the characteristic shuttling between stations was shown upon changing solvents (Figure 10) [55].

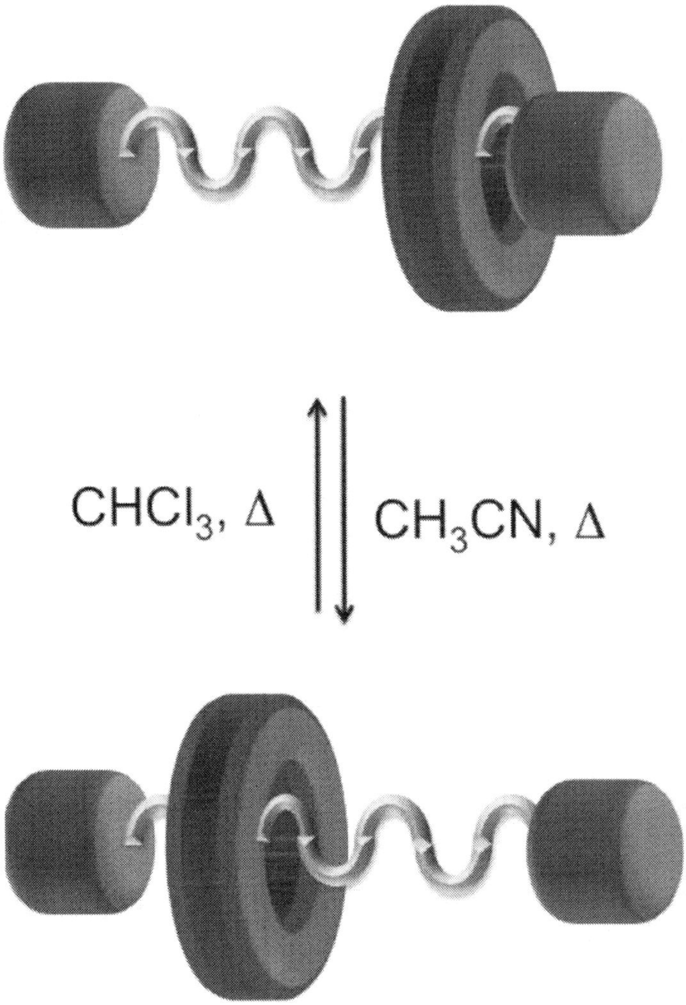

Figure 10. Schematic representation of a [2]rotaxane molecular machine that is comprised of an oligopeptide axle that results in a helical structure.

SWITCHES AND MOLECULAR LOGIC GATES

One variety of a molecular switch can be described as a nanoscale machine which undergoes reversible co-conformational shifts or molecular transformations (leading to an output) in response to protons, photons, electrons, temperature, or exogenous cations or anions (as inputs) between two or more states [56]. We have already seen simple prototypes of this design in the previous sections. As molecular switches convert input stimulations into output signals, these systems can be seen as performing simple logic operations, that is through ON/OFF or '0' and '1' type processes [57]. The use of two or more switches or gates can lead to Boolean operations including addition and subtraction. While simple molecular systems have led to displays of the simplest functions, such as YES or NOT, '1' is only with the advent of molecular recognition and self assembly, that mechanically interlocked molecules and supramolecular arrays have been efficiently used to prepare more complex logic gates including AND, XOR and INHIBIT. This section provides a small selection of supramolecular switches and logic gates.

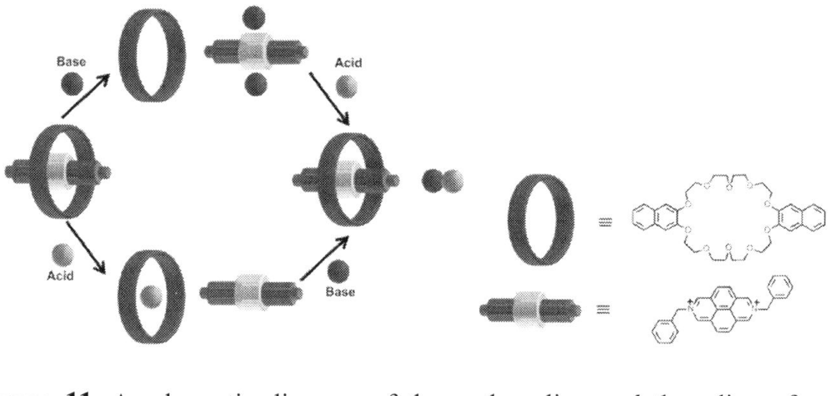

Figure 11. A schematic diagram of the unthreading and threading of a [2] pseudorotaxane, corresponding to an XOR logic function [58].

One of the earliest examples is that of a molecular XOR logic gate defined by the inclusion and exclusion of the two components of a [2]pseudorotaxane [58]. In this early example, Figure 11, the [2] pseudorotaxane contains a good electron acceptor (2,7-diazapyenium) which interacts with the electron donating group from an aromatic crown ether. The pseudorotaxane self-assembles due to the π-electron-acceptor-donor interactions of the two groups. In this case the unthreading and treading of the [2]pseudorotaxane occurs with the use of an acid and base stimuli. Addition of a suitable base (tributylamine, DABCO) unthreads the pseudorotaxane and forms the acceptor(base)$_2$ complex and with the addition of acid the pseudorotaxane is formed again along with a salt.

As a result of advancements made towards the synthesis of interlocking molecules, several rotaxanes and catenane–based molecular switches have been developed. Several interesting assemblies have been recently published, one of which demonstrates a [2]catenane molecule which operates as a single push-button molecular switch [59]. The resulting mechanically interlocked molecule exists in one translational form in its ground state that is mechanically switched to another translation form in the excited oxidized state (Figure 12).

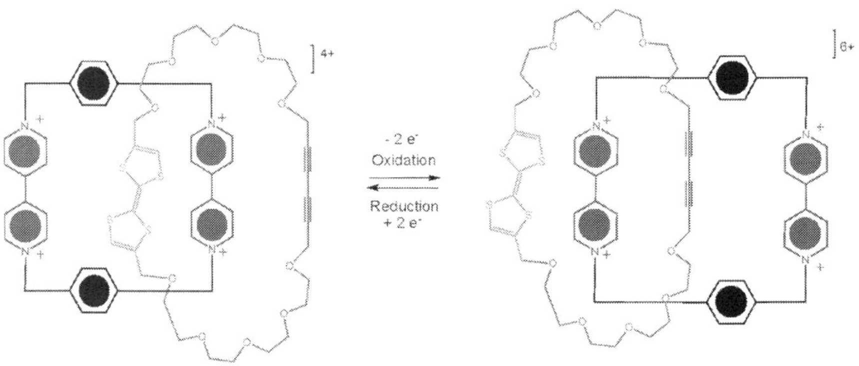

Figure 12. A representation of the push-button molecular switch [59].

Recently developed by Chiu is this visible-active squaraine-based [2] rotaxane molecular switch (Figure 13) [60]. This molecular switch operates through the selective encapsulation and exposure of the squaraine station of a two-station [2]rotaxane upon the addition and removal of sodium ions, respectively. Long-waved-fluorescence signals are produced and can be seen when the sample is irradiated with 365 nm UV light. The co-conformational state is reversibly upon switching with the addition of a sodium ion source and in this particular example sodium perchlorate was used for this purpose.

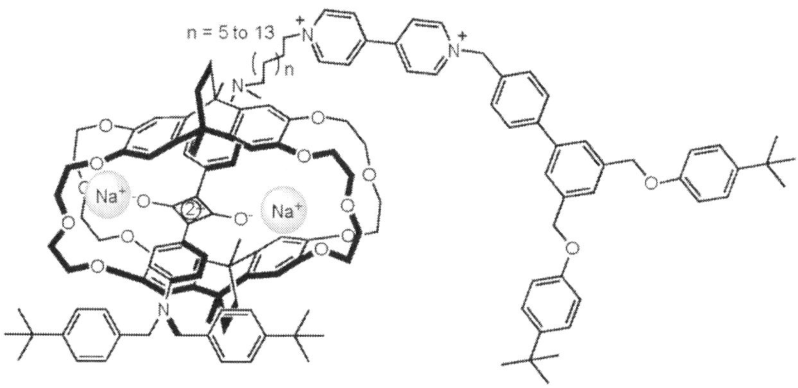

Figure 13. A representation of the squaraine-based [2]rotaxane molecular switch of Chiu [60].

MOLECULAR ROTARY MOTORS

Biological motors consume chemical energy that drive sequential changes with other molecules to perform highly specialized tasks and functions essential for life. The design principles that facilitate the complex and precise actions of such motors is a significant challenge for their successful application in synthetic environments [61,62]. The following two sections will describe efforts to create rotary motors followed by translation motion towards the transport of cargo at the molecular level. The Feringa laboratory have been successful at demonstrating repetitive, unidirectional rotation (of MHz frequencies) using organic frameworks which have a fixed stator component connected to a rotor through an axle which operate upon external stimulus [63–65]. Exploitation of the light induced cis-transisomerisation has proven to be a popular design and a variety of rotary motors are possible through combinations of different organic frameworks flanking a central alkene axle (Figure 14). The upper and lower organic units may be identical (1st generation rotary molecular motor) or differ in structure (2ndgeneration rotary molecular motor) and choice of substituents may further assist in controlling the rotational speed of these motors through steric interactions [66,67].

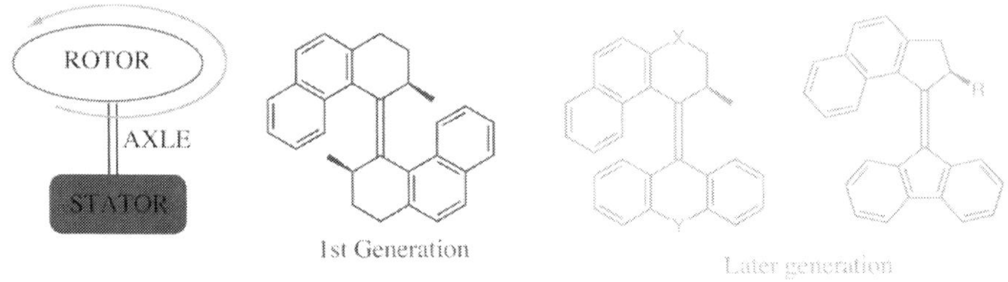

Figure 14. Schematic representation of the general design components of the Feringa rotary molecular motor with evolutionary design [61].

A recent development in controlling the rotary motion of the Feringa motors involved the introduction of a self-complexing lock which functions in a similar way to the rotaxane threading process (see Section 3) [68]. As shown schematically in Figure 15, the rotor component has been covalently modified with a functionalized thread while the stator was appended with a macrocyclic crown ether ring. The threading process is under acid-base control due to the non-covalent interaction between the crown ether and ammonium salt in its protonated state and results in a locked or unlockedstate for the rotary motion of the rotor component.

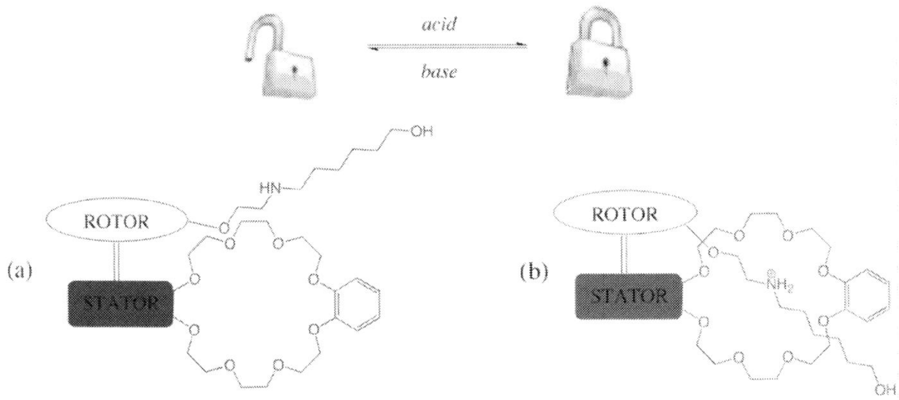

Figure 15. Schematic representation of a molecular rotary motor with self-complexing lock. The well known complementarity between crown ethers and ammonium salts provides a non-covalent interaction to produce a (a) locked and (b) unlocked state for the molecular rotary motor.

MOLECULAR TRANSPORT

The ability to control mechanical motion and potentially transport cargo are challenges being actively explored [69–71]. The original report on single-molecule nanocars detailed the synthesis of a semi-rigid organic chassis appended with four fullerene wheels via alkynyl axles which moved in a directional rolling motion across a gold surface (observed by scanning probe microscopy) [72]. While the nanocar was a significant achievement in establishing proof-of-concept for the ability to roll (rather than slide) molecular wheels across a surface (upon heating the gold surface) it also highlighted the essential design components to include: (1) a chassis, (2) multiple wheels and (3) axles which link the wheels to the chassis (Figure 16). Subsequent modifications and improvements with regards to these three components has since paved the way for later generation nanovehicles [71].

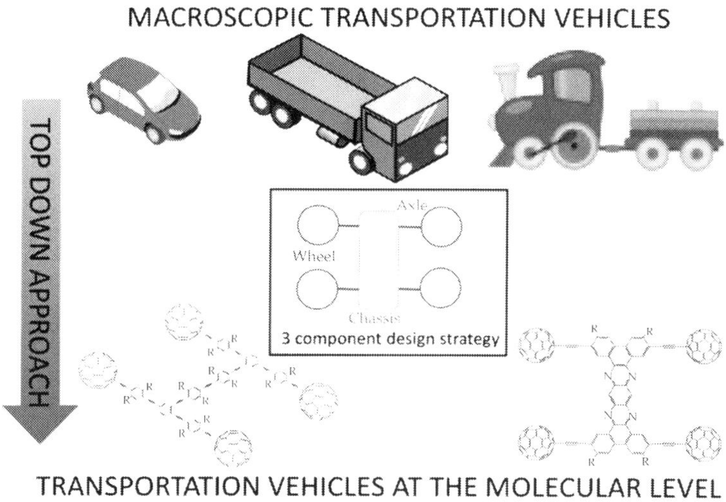

Figure 16. The concept of a "top-down" approach for the construction of vehicles at the molecular level as developed by the Tour group. Using a three component design strategy analogues of macroscopic transportation vehicles (e.g., cars, trucks, trains) have been created such as nanocars and nanotrucks. R = solubilizing groups.

Modification of the chassis fragment to incorporate a planar polyaromatic spacer yielded a construct referred to as a nanotruck which was designed to provide a suitable "loading bay" for either acid or base binding to the aza-aromatic spacer although STM imaging to demonstrate transport across a metal surface was hindered by surface impurities and possible stability issues relating to surface chemistry [69]. Another structural variation of the nanotruck has been prepared and is comprised of a porphyrin core with carborane wheels [73]. If future surface imaging confirms the expected rolling motion, this nanotruck also has the ability to transport metals via coordination within the porphyrin macrocycle and may realize the goal of cargo delivery between specific locations. Improvements have also included the replacement of fullerene wheels to include ruthenium-based wheels [74] which not only improve solubility compared to the fullerenes but are also less restrictive in synthetic manipulations.

The Feringa unidirectional motor (see Section 5) as powered through light stimulus was chosen to power the 'nanocar' but due to incompatibilities with the original fullerene wheel design was adapted to an improved nanocar model with carborane wheels [75]. Three-wheeled nanocars [69] and pinwheel [76] arrangements have also been synthesized and characterized fully. While such designs offer limited translational motion they show much greater potential for a pivoting motion about a central axis, although conclusive imaging data to support this has not been reported to date. To truly 'make the molecules do the work' at the nanoscale and produce nanovehicles in an assembly line type manner self-assembled structures have also been examined in this research program [77].

The biggest advantage to such approach lies in the reduced synthetic effort, which for the examples discussed above is multi-step, time consuming, low yielding and even with the modular, three component design protocol does not allow for structural modification as easily as a self-assembled approach. Both metal coordination and hydrogen bonding interactions were trialed to assemble fragments successfully into carborane-wheeled 'nanocars' although stability under surface imaging conditions has yet to be determined. The assembly line methodology has also been extended to create a linear nanotrain through the assembly of multiple fragments of a 2-pyridone which contains two hydrogen bonding site located on opposite ends of the monomer [78]. It seems a logical development that the assembly line could also be a useful approach for assembly 'carriages' of varying components which individually could serve different functions to the transportable vehicle.

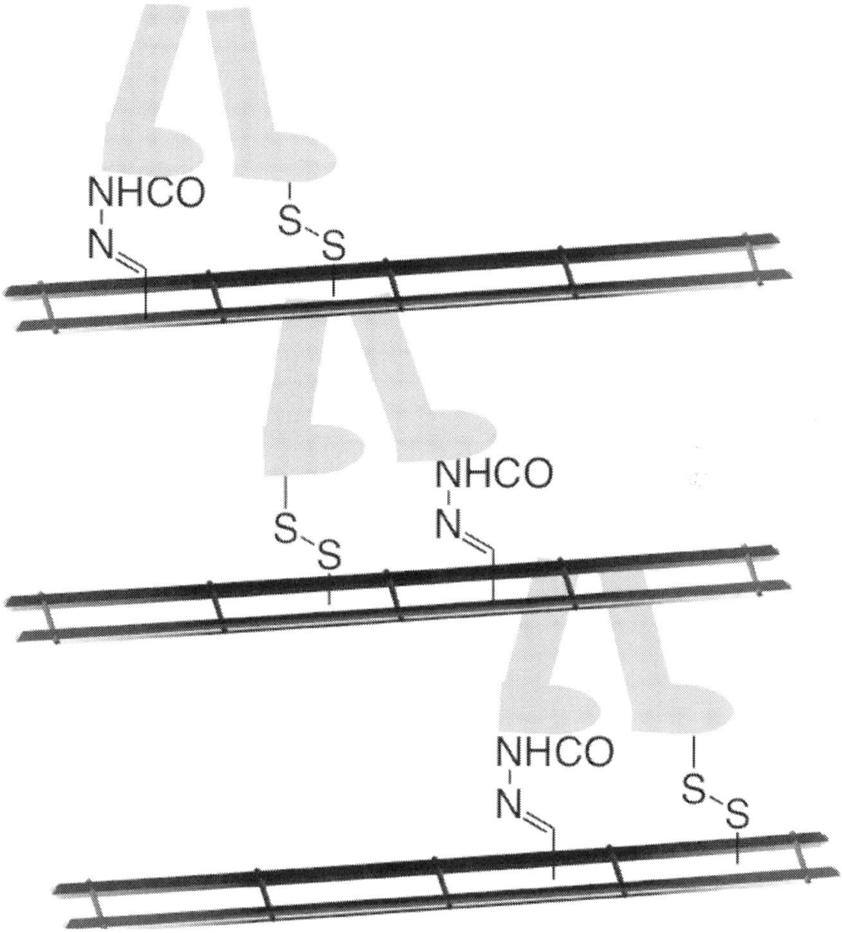

Figure 17. Teaching molecules to walk. The use of a scaffold containing functionality designed to make disulfide and hydrazone links has been used to allow a molecule to transverse the walkway in a controlled fashion.

An advancement on molecular motion in a different structural form to Tours nanovehicles arose from the Leigh laboratory recently where a small molecule (synthetic) was shown to walk along a track comprised of four foot holes in a repetitive manner through changes to the chemical environment (Figure 17) [79]. The walking motion is possible due to careful design of the interactions between the feet and track. Specifically, the use of disulfide and hydrazone units allows for kinetic control between the feet and track upon different chemical stimulus (acid or base). Taking inspiration from biological motor proteins further studies are being conducted to investigate whether the walking molecule is able to carry and deliver a cargo and over what distance this linear motion may be achieved by extending the track using polymeric units.

CONCLUSIONS

In reviewing the current literature relating to molecular machine research we aimed to illustrate fundamental developments and progress relating to typical functions carried out by macroscopic machinery including transport, rotation and switching. While each of those alone does not create an artificial machine the progress does make a significant contribution to unlocking the key processes relating to control and function at the molecular level. Biological molecular machines far outweigh any artificially constructed machine both in complexity and function which will continue to inspire and challenge scientist for many years to come, one that will no doubt benefit through multi-disciplinary cooperation.

REFERENCES

1. Balzani, V; Credi, A; Raymo, FM; Stoddart, JF. Machines at the Molecular Level. Angew. Chem. Int. Ed 2000, 39, 3348–3391.
2. Balzani, V; Credi, A; Stoddart, JF. The Bottom-Up Approach to Molecular-Level Devices and Machines. Chem. Eur. J 2002, 8, 5524–5532.
3. Ballardini, R; Balzani, V; Credi, A; Gandolfi, MT; Venturi, M. Artificial Molecular-Level Machines: With Energy to Make them Work? Acc. Chem. Res 2001, 34, 445–455.
4. Balzani, V; Venturi, M; Credi, A. Molecular Devices and Machines; Wiley-VCH: Weinheim, Germany, 2003; pp. 269–277.
5. Feynman, RP. There's Plenty of Room at the Bottom. Eng. Sci 1960, 23, 22–36.
6. Feynman, RP. The Wonders that Await a Micro-Microscope. Sat. Rev 1960, 43, 45–47.
7. Balzani, V; Credi, A; Venturi, M. The Bottom-up approach to Molecular-Level Devices and Machines. Chem. Eur. J 2002, 8,

5524–5532.

8. Vetter, W; Schill, G. Directed Synthesis of Catena Compounds. IX. Mass spectrum of a Catena Compound. Tetrahedron 1967, 23, 3079–3093.
9. Schill, G. Catenanes, Rotaxanes and Knots; Academic Press: New York, NY, USA, 1971.
10. Stoddart, J-F. The Chemistry of the Mechanical Bond. Chem. Soc. Rev 2009, 38, 1802–1820.
11. Crowley, JD; Goldup, SM; Lee, AL; Leigh, DA; McBurney, RT. Active Metal Template Synthesis of Rotaxanes, Catenanes and Molecular Shuttles. Chem. Soc. Rev 2009, 38, 1530–1541.
12. Yamashita, K; Kawano, M; Fujita, M. Photoswitchable Molecular Lock. One-Way Catenation of Pt-(II)-Linked Coordination Ring via the Photolabilization of a Pt(II)-Pydridine Bond. J. Am. Chem. Soc 2007, 129, 1850–1851.
13. Collin, J-P; Heitz, V; Sauvage, J-P. Metal-Containing Rotaxanes and Catenanes in Motion: Towards Molecular Machines. Top. Curr. Chem 2005, 262, 29–62.
14. Megiatto, JD, Jr; Schuster, DI. Introduction of Useful Peripheral Functional Groups on [2]Catenanes by combinging Cu(I) Template Synthesis with "Click" Chemistry. New. J. Chem 2010, 34, 276–286.
15. Sauvage, J-P. Transition Metal-Containing Rotaxanes and Catenanes in Motion: Towards Molecular Machines and Motors. Acc. Chem. Res 1998, 31, 611–619.
16. Peinador, C; Blanco, V; Quintela, JM. A new Doubly Interlocked [2]Catenane. J. Am. Chem. Soc2009, 131, 920–921.
17. Dietrich-Buchecker, CO; Sauvage, J-P. A Synthetic Molecular Trefoil Knot. Angew. Chem.: Int. Ed. Engl 1989, 28, 189–192.
18. Dietrich-Buchecker, CO; Guilham, J; Pascard, C; Sauvage, J-P. Structure of a Synthetic Trefoil Knot Coordinated to Two Copper(I) Centers. Angew. Chem.: Int. Ed. Engl 1990, 29, 1154–1156.
19. Carina, RF; Dietrich-Buchecker, CO; Sauvage, J-P. Molecular Composite Knots. J. Am. Chem. Soc1996, 118, 9110–9116.
20. Schill, G; Schweickert, N; Fritz, H; Vetter, W. Synthesis of

[2]-Catenanes from [2]-Rotaxanes. Chem. Ber 1988, 121, 961–970.

21. Schill, G; Doerjer, G; Logemann, E; Fritz, H. Studies on the Synthesis of Molecules with Knot Structure. Fourfold bridged 5,6-diamino-1,3-benzodioxole derivatives. Chem. Ber 1979, 112, 3603–3615.
22. Sauvage, J-P; Dietrich-Buchecker, CO. Molecular Catenanes, Rotaxanes and Knots; Wiley-VCH: Weinheim, Germany, 1999.
23. Philp, D; Stoddart, J-F. Self-Assembly in Natural and Unnatural Systems. Angew. Chem.: Int. Ed. Engl 1996, 35, 1155–1196.
24. Du, SM; Seeman, NC. Synthesisof DNA Knot Containing Both Positive and Negative Nodes. J. Am. Chem. Soc 1992, 114, 9652–9655.
25. Du, SM; Tse-Dinh, Y-C; Seeman, NC. Topological Transformations of Synthetic DNA Knots. Biochemistry 1995, 34, 673–682.
26. Du, SM; Stollar, BD; Seeman, NC. A Synthetic DNA molecule in Three Knoyted Topologies. J. Am. Chem. Soc 1995, 117, 1194–1200.
27. Dietrich-Buchecker, CO; Sauvage, J-P. New Family of Molecules: The Metallo-Catenanes. Tetrahedron Lett 1983, 24, 5095–5098.
28. Dietrich-Buchecker, CO; Sauvage, J-P. Interlocking of Molecular Threads: From the Statistical Approach to the Templated Synthesis of Catenands. Chem. Rev 1987, 87, 795–810.
29. Livoreil, A; Sauvage, J-P; Armaroli, N; Balzani, V; Flamigni, L; Ventura, B. Electrochemically and Photochemically Driven Ring Motions in a Dissymmetrical Copper[2]-Catenate. J. Am. Chem. Soc1997, 119, 12114–12124.
30. Dietrich-Buchecker, C; Jimenez-Molero, MC; Sartor, V; Sauvage, J-P. Rotaxanes and Catenanes as Prototypes of Molecular Machines and Motors. Pure Appl. Chem 2003, 75, 1383–1393.
31. Sauvage, J-P. Transition Metal-Complexed Catenanes and Rotaxanes as Molecular Machine Prototypes. Chem. Comm 2005, 12, 1507–1510.
32. Miljanić, OŠ; Dichtel, WR; Aprahamian, I; Rohde, RD; Agnew, HD; Heath, JR; Stoddart, JF. Rotaxanes and Catenanes by Click Chemistry. QSAR Comb. Sci 2007, 11–12, 1165–1174.
33. Goldup, SM; Leigh, DA; Long, T; McGonigal, PR; Symes, MD;

Wu, J. Active Metal Template Synthesis of [2]Catenanes. J. Am. Chem. Soc 2009, 131, 15924–15929.

34. Cao, D; Amelia, M; Klivansky, LM; Koshkakaryan, G; Khan, SI; Semeraro, M; Silvi, S; Venturi, M; Credi, A; Liu, Y. Probing Donor-Acceptor Interactions and Co-Conformational Changes in Redox active Desymmetrized [2]Catenanes. J. Am. Chem. Soc 2010, 132, 1110–1122.
35. Li, Q; Zhang, W; Miljanić, OŠ; Knobler, CB; Stoddart, JF; Yaghi, OM. A Metal-Organic Framework Replete with Ordered Donor-Acceptor Catenanes. Chem. Commun 2010, 46, 380–382.
36. Murray, LJ; Dincă, M; Long, JR. Hydrogen Storage in Metal-Organic Framework. Chem. Soc. Rev2009, 38, 1294–1314.
37. Li, J-R; Kuppler, RJ; Zhou, H-T. Selective Gas Adsorption and Separation in Metal-Organic Frameworks. Chem. Soc. Rev 2009, 38, 1477–1504.
38. Rescifina, A; Zagni, C; Iannazzo, D; Merino, P. Recent Developments on Rotaxane-Based Shuttles. Curr. Org. Chem 2009, 13, 448–481.
39. Ma, X; Tian, H. Bright Functional Rotaxanes. Chem. Soc. Rev 2010, 39, 70–80.
40. Silvi, S; Venturi, M; Credi, A. Artificial Molecular Shuttles: From Concepts to Devices. J. Mater. Chem 2009, 19, 2279–2294.
41. Fang, L; Olson, MA; Benitez, D; Tkatchouk, E; Goddard, WA, III; Stoddart, JF. Mechanically Bonded Macromolecules. Chem. Soc. Rev 2010, 39, 17–29.
42. Balzani, V; Credi, A; Venturi, M. Light Powered Molecular Machines. Chem. Soc. Rev 2009, 38, 1542–1550.
43. Collin, J-P; Durola, F; Lux, J; Sauvage, J-P. A Copper-based Shuttling [2]rotaxane with Two Bidentate Chelates in the Axis: Steric Control of the Motion. New J. Chem 2010, 34, 34–43.
44. Collin, J-P; Dietrich-Buchecker, C; Gavina, P; Jimenez-Molero, MC; Sauvage, J-P. Shuttles and Muscles: Linear Molecular Machines Based on Transition Metals. Acc. Chem. Res 2001, 34, 477–487.
45. Colasson, BX; Dietrich-Buchecker, C; Jimenez-Molero, MC; Sauvage, J-P. Towards Molecular Machines and Motors Based on Transition Metal Complexes. J. Phys. Org. Chem 2002, 15, 476–

483.

46. Jimenez-Molero, MC; Dietrich-Buchecker, C; Sauvage, J-P. Towards Artificial Muscles at the Nanometric Level. Chem. Commun 2003, 1613–1616.
47. Sauvage, J-P. Transition Metal-Complexed Catenanes and Rotaxanes as Molecular Machine Prototypes. Chem. Commun 2005, 1507–1510.
48. Champin, B; Mobian, P; Sauvage, J-P. Transition metal complexes as Molecular Machine Prototypes. Chem. Soc. Rev 2007, 36, 358–366.
49. Jimenez, MC; Dietrich-Buchecker, C; Sauvage, J-P. Towards Synthetic Molecular Muscles: Contraction and Stretching of a Linear Rotaxane Dimer. Angew. Chem. Int. Ed 2000, 39, 3284–3287.
50. Chuang, C-J; Li, W-S; Lai, C-C; Liu, Y-H; Peng, S-M; Chao, I; Chiu, S-H. A Molecular Cage-Based [2]Rotaxane That Behaves as a Molecular Muscle. Org. Lett 2009, 11, 385–388.
51. Badjic, JD; Balzani, V; Credi, A; Selvi, S; Stoddart, JF. A Molecular Elevator. Science 2004, 303, 1845–1849.
52. Goldup, SM; Leigh, DA; McGonigal, PR; Ronaldson, VE; Slawin, AMZ. Two Axles Threaded Using a Single Template Site: Active Metal Template Macrobicyclic [3]Rotaxanes. J. Am. Chem. Soc2010, 132, 315–320.
53. Fernandes, A; Viterisi, A; Coutrot, F; Potok, S; Leigh, DA; Aucagne, V; Papot, S. Rotaxane-Based Propeptides: Protection and Enzymatic Release of a Bioactive Pentapeptide. Angew. Chem. Int. Ed2009, 48, 6443–6447.
54. Moretto, A; Menegazzo, I; Crisma, M; Shotton, EJ; Nowell, H; Mammi, S; Toniolo, C. A Rigid Helical Peptide Axle for a [2] Rotaxane Molecular Machine. Angew. Chem. Int. Ed 2009, 48, 8986–8989.
55. Spruell, JM; Paxton, WF; Olsen, J-C; Benítez, D; Tkatchouk, E; Stern, CL; Trabolsi, A; Friedman, DC; Goddard, WA, III; Stoddart, JF. A Push-Button Molecular Switch. J. Am. Chem. Soc 2009, 131, 11571–11580.
56. Andreasson, J; Pischel, U. Smart Molecules at Work-Mimicking

Advanced Logic Operations. Chem. Soc. Rev 2010, 39, 174–188.

57. Credi, A; Balzani, V; Langford, SJ; Stoddart, JF. Logic Operations at the Molecular Level. An XOR Gate Based on a Molecular Machine. J. Am. Chem. Soc 1997, 119, 2679–2681.
58. Spruell, JM; Paxton, WF; Olsen, J; Bentez, D; Tkatchouk, E; Stern, CL; Trabolsi, A; Friedman, DC; Goddard, WA, III; Stoddart, JF. A Push-Button Molecular Switch. J. Am. Chem. Soc 2009, 131, 11571–11580.
59. Hsueh, S-Y; Lai, C-C; Chiu, S-H. Squaraine-Based [2]Rotaxanes that Function as Visibly Active Molecular Switches. Chem. Eur. J 2010, 16, 2997–3000.
60. Browne, WR; Feringa, BL. Making Molecular Machines Work. Nat. Nanotechnol 2006, 1, 25–35.
61. Pollard, MM; Klok, M; Pijper, D; Feringa, BL. Rate Acceleration of Light-Driven Rotary Molecular Motors. Adv. Funct. Mater 2007, 17, 718–729.
62. Goel, A; Vogel, V. Harnessing Biological Motors to Engineer Systems for Nanoscale Transport and Assembly. Nat. Nanotechnol 2008, 3, 465–475.
63. Fletcher, SP; Dumur, F; Pollard, MM; Feringa, BL. A Reversible, Unidirectional Molecular Rotary Motor Driven by Chemical Energy. Science 2005, 310, 80–82.
64. Klok, M; Boyle, N; Pryce, MT; Meetsma, A; Browne, WR; Feringa, BL. MHz Unidirectional Rotation of Molecular Rotary Motors. J. Am. Chem. Soc 2008, 130, 10484–10485.
65. Pijper, TC; Pijper, D; Pollard, MM; Dumur, F; Davey, SG; Meetsma, A; Feringa, BL. An Enantioselective Synthetic Route toward Second-Generation Light-Driven Rotary Molecular Motors. J. Org. Chem 2010, 75, 825–838.
66. Kulago, AA; Mes, EM; Klok, M; Meetsma, A; Brouwer, AM; Feringa, BL. Ultrafast Light Driven Nanomotors Based on an Acridine Stator. J. Org. Chem 2010, 75, 666–679.
67. Qu, D-H; Ferina, BL. Controlling Molecular Rotary Motion with a Self-Complexing Lock. Angew. Chem. Int. Ed 2010, 49, 1107–1110.
68. Shirai, Y; Osgood, AJ; Zhao, Y; Yao, Y; Saudan, L; Yang, H; Yu-

Hung, C; Alemany, LB; Sasaki, T; Morin, J-F; Guerrero, JM; Kelly, KF; Tour, JM. Surface-Rolling Molecules. J. Am. Chem. Soc 2006, 128, 4854–4864.

69. Shirai, Y; Morin, J-F; Sasaki, T; Guerrero, JM; Tour, JM. Recent Progress on Nanovehicles. Chem. Soc. Rev 2006, 35, 1043–1055.
70. Vives, G; Tour, JM. Synthesis of Single-Molecular Nanocars. Acc. Chem. Res 2009, 42, 473–487.
71. Shirai, Y; Osgood, AJ; Zhao, Y; Kelly, KF; Tour, JM. Synthesis of Single-Molecule Nanocars. Nano Lett 2005, 5, 2330–2334.
72. Sasaki, T; Morin, J-F; Lu, M; Tour, JM. Synthesis of a Single-Molecule Nanotruck. Tet. Lett 2007, 48, 5817–5820.
73. Vives, G; Tour, JM. Synthesis of a Nanocar with Organometallic Wheels. Tet. Lett 2009, 50, 1427–1430.
74. Morin, J-F; Shirai, Y; Tour, JM. En Route to a Motorized Nanocar. Org. Lett 2006, 8, 1713–1716.
75. Sasaki, T; Osgood, AJ; Klappes, JL; Kelly, KF; Tour, JM. Synthesis of a Porphyrin-Fullerene Pinwheel. Org. Lett 2008, 10, 1377–1380.
76. Sasaki, T; Guerrero, JM; Tour, JM. The Assembly Line: Self-Assembling Nanocars. Tetrahedron2008, 64, 8522–8529.
77. Sasaki, T; Guerrero, JM; Leonard, AD; Tour, JM. Nanotrains and Self-Assembled Two-Dimensional Arrays Built from Carboranes Linked by Hydrogen Bonding of Dipyridones. Nano Res 2008, 1, 412–419.
78. Von Delius, M; Geertsema, EM; Leigh, DA. A Synthetic Small Molecular That Can Walk Down a Track. Nat. Chem 2010, 2, 96–101.

PART II: MOLECULAR MOTORS

CHAPTER

3

MOLECULAR MOTOR PROTEINS AND AMYOTROPHIC LATERAL SCLEROSIS

Kai Y Soo [1], Manal Farg [1] and Julie D. Atkin [1,2,3]

[1]Department of Biochemistry, La Trobe Institute of Molecular Science, La Trobe University, VIC 3086, Australia

[2]Centre for Neuroscience, University of Melbourne, Parkville, VIC 3010, Australia

[3]Florey Neuroscience Institute, University of Melbourne, Parkville, VIC 3010, Australia

ABSTRACT

Amyotrophic lateral sclerosis (ALS) is a neurodegenerative disorder affecting motor neurons in the brain, brainstem and spinal cord, which is characterized by motor dysfunction, muscle dystrophy and progressive paralysis. Both inherited and sporadic forms of ALS share common pathological features, however, the initial trigger of neurodegeneration remains unknown. Motor neurons are uniquely targeted by ubiquitously expressed proteins in ALS but the reason for this selectively vulnerability is unclear. However motor neurons have unique characteristics such as very long axons, large cell bodies and high energetic metabolism, therefore placing high demands on cellular transport processes. Defects in cellular trafficking are now widely reported in ALS, including dysfunction to the molecular motors dynein and kinesin. Abnormalities to dynein in particular are linked to ALS, and defects in dynein-mediated axonal transport processes have been reported as one of the earliest pathologies in transgenic SOD1 mice. Furthermore, dynein is very highly expressed in neurons and neurons are particularly sensitive to dynein dysfunction. Hence, unravelling cellular transport processes mediated by molecular motor proteins may help shed light on motor neuron loss in ALS.

Keywords: Amyotrophic lateral sclerosis; axonal transport; kinesins; dynein; myosin

INTRODUCTION

Amyotrophic lateral sclerosis (ALS), also known as Charcot's sclerosis or Lou Gehrig's disease, is the most common form of motor neuron disease. It is characterized by progressive degeneration of motor neurons in the primary cortex, brainstem, and spinal cord, which results in muscle dystrophy, paralysis, and death. The fatal event is usually failure of the respiratory muscles. At least 10% of all ALS cases are inherited (familial ALS); however, most cases have no discernable genetic component (sporadic ALS). Several proteins are linked to sporadic and familial forms of ALS, notably superoxide dismutase 1 (SOD1) [1], TAR DNA binding protein 43 (TDP-43) [2], and fused in sarcoma (FUS) [3]. Both sporadic and familial forms of ALS are clinically and pathologically identical.

Of the cases of familial ALS, approximately 20% are linked to dominantly inherited mutations in SOD1. Mutant SOD1 has been extensively studied in both cellular and animal models of disease. As a multifactorial disorder, several mechanisms have been implicated in the neurodegenerative process in ALS. These include protein misfolding and aggregation, excitotoxicity [4], neuro-inflammation [5,6], endoplasmic reticulum (ER) stress, oxidative stress [7] and mitochondrial dysfunction [8], cytoskeleton abnormalities and defective axonal transport [9,10]. However, the initiating factors which trigger neurodegeneration still remain unknown. Recently, increasing evidence implicates dysfunction of the cellular transport machinery in ALS, including dysfunction to the molecular motor proteins, dynein and kinesin.

A perplexing characteristic of ALS is why motor neurons are targeted by ubiquitously expressed proteins. Motor neurons are large, highly differentiated, polarized cells, with extremely long axons, up to 1 metre in length in an adult human. Hence, they have high synthetic and energy requirements, which therefore places heavy demands on cellular transport processes [11]. Proteins, lipids, mRNA and organelles need to rapidly move from the cell body over large distances along the axon to the synaptic terminals or neuromuscular junctions, where they are required for axoplasmic membrane remodeling, energy production, neurotransmission and local protein synthesis [12]. Axonal transport is also required to collect neurotrophins, survival factors or potentially toxic factors from distal axons, peripheral synapses or muscle cells, back to the soma. The complexity and fine regulation of this system is highly sensitive to perturbation, and minor alterations of cellular and vesicular transport processes may result in motor neuron dysfunction [13].

Axonal transport is a key mechanism required for cellular viability in neuronal cells. Most proteins required in the axon and in synaptic terminals must be transported along the axon after synthesis in the cell body. Similarly RNA and organelles also need to be transported over long distances, and these transport processes require molecular motors, such as kinesins, dyneins and myosins, that operate along the cellular cytoskeleton. Dysfunction of axonal transport has been described in several neurodegenerative diseases, and is well documented in ALS [9,14,15]. In this review, we describe the functions of the molecular motors, including kinesins and dyneins, as well as their putative roles in ALS.

GENETICS OF ALS

Although most cases of ALS are sporadic, approximately 10% of all cases are inherited, and investigations into the genetic causes of ALS may prove to be important for uncovering common disease mechanisms [16,17]. Mutations in SOD1 were the first identified and remain the best characterized dominantly inherited causative genetic mutations in ALS [1]. However, in recent years additional genes have been linked with disease and most of these have arisen from studies of small familial pedigrees with dominantly-inherited ALS. However, mutations have also been identified in subsets of sporadic ALS cases, indicating that de novo mutations can cause disease, and suggesting that genetic changes are important in both familial and sporadic ALS.

Mutations in SOD1 are scattered amongst all five exons of the SOD1 gene. SOD1 is a cytoplasmic protein, and mutant SOD1 proteins are prone to aggregation and intracellular inclusion formation. The most widely used and well-characterized experimental model of human ALS is the $SOD1^{G93A}$transgenic mice, which develops progressive muscular weakness and paralysis similar to human ALS [18]. Structural abnormalities of motor neurons are evident in $SOD1^{G93A}$ transgenic mice, such as dilated ER in axons, dendrites, and cell bodies, as well as vacuoles associated with swollen degenerating mitochondria [19].

The second most common gene linked to ALS after SOD1 is fused in sarcoma (FUS) [3,20] and FUS mutations are found in up to 5% of familial ALS patients [3,20]. Mutations in FUS are found in both familial and sporadic ALS, as well as in patients with ALS and frontotemporal dementia (FTD), and in frontotemporal lobar degeneration (FTLD) patients without motor impairment [3,20–23], suggesting a pathogenic overlap between ALS and other neurodegenerative diseases. Most identified mutations cluster in the C-terminal portion of the protein, around the nuclear localization sequence.

TDP-43 is the major constituent of cytoplasmic and intracellular inclusions in neurons and glia of patients with sporadic and familial ALS [2,24]. More than 30 different mutations in the gene encoding TDP-43, TARDBP, have now been reported in 3 to 4% of cases of familial ALS [25,26]. The functions of both FUS and TDP-43 are largely uncharacterized, although the structural and functional similarities between these two proteins suggest that similar disease mechanisms could operate for both proteins. Transgenic mice overexpressing either wildtype or mutant TDP-43 produces similar phenotypes [27]. However, the mechanisms causing pathogenesis in the mutant TDP-43 transgenic mice remain unknown, although abnormal cytoplasmic mitochondrial aggregates and early lethality have been identified [27].

Several forms of juvenile-onset ALS have been characterized, such as ALS2, with a very slow disease progression. Mutations in ALS2, which are autosomal recessive, are thought to involve vesicle transport and membrane trafficking processes [28]. Another juvenile-onset ALS is autosomal dominant ALS4 linked to mutations in the SETX gene encoding senataxin, which is thought to possess DNA/RNA helicase acitivty [29]. A single mutation, P56S, in vesicle associated membrane protein/synaptobrevin-associated membrane protein B (VAPB) are also described in a form of autosomal-dominant late-onset ALS (ALS8) [30]. Mutations in tau and the p150 dynactin subunit are also described in ALS-dementia and lower ALS respectively, and cause progressive, autosomal dominant forms of diseases and without sensory symptoms [31,32]. In 2010, two further ALS-causative genes, optineurin [33] and valosin containing protein (VCP) [34], were identified. Mutations in optineurin are found in both sporadic and SOD1-linked ALS [33,35]; but VCP mutations have only been identified in familial ALS cases [34]. Mutations in the gene encoding ubiquitin-like protein ubiquilin 2, UBQLN2, which cause dominantly inherited, were also recently linked to chromosome-X-linked ALS and ALS/dementia [36]. Ubiquilin 2 is a member of the ubiquilin family, which regulates the degradation of ubiquitinated proteins [36], and mutations in UBQLN2 lead to an impairment of protein degradation [36]. A more recent report identifies a noncoding expanded GGGGCC hexanucleotide repeat in C9ORF72 as the cause of chromosome 9p-linked FTD/ALS and shows that this genetic defect is the most common cause of ALS and FTD identified to date [37].

POSSIBLE PATHOGENIC MECHANISMS IN ALS

Despite the fact that a number of genes have now been linked to ALS, the exact pathogenic mechanisms are still largely unclear. Recent studies have suggested that both sporadic and familial forms of ALS share common pathologic mechanisms [38,39].

Protein Misfolding and Aggregation

Protein misfolding and aggregation are prominent features of ALS, but the relationship to disease pathogenesis remains unclear. A number of different protein inclusions are found in human ALS, and animal models of ALS, including ubiquitinated skein-like inclusions, bunina bodies, and hyaline inclusions rich in neurofilaments proteins [40]. Many proteins have been found within the inclusions in ALS, including neurofilaments proteins and peripherin [41], SOD1, p38MAPK and Cdf4 kinases [42], cystatin C [43], FUS [44] and TDP-43 [2]. SOD1 inclusions have been detected in lower motor neurons of familial ALS patients, mutant SOD1 transgenic mice, and in cultured cells expressing mutant SOD1 [45]. Wildtype SOD1 does not form large intracellular inclusions but may form smaller aggregates under oxidative stress [46]. It is unclear how protein aggregation is toxic to motor neurons, either by sequestration of proteins critical for cell viability, or by interfering with vital cellular mechanisms, such as chaperone activity, inhibiting the ubiquitin-proteasome system and disrupting of cellular transport [16].

Excitotoxicity

Glutamate excitotoxicity results from the excessive influx of calcium cations through the over-stimulation of post-synaptic glutamate receptors, including NMDA and AMPA-type receptors [47]. This increase of calcium can activate enzymes such as phosphatases, proteases, lipases and endonucleases, causing protein and lipid alterations in cell membranes, generation of toxic reactive oxygen species (ROS), and mitochondrial damage and dysfunction [47]. Decreased levels of the excitatory amino acid transporter EAAT2, which is present on astrocytes and is responsible for rapid removal of glutamate from the synapse, are found in both human patients and in mutant SOD1 transgenic rodents [48]. Furthermore, increased glutamate is found in CSF of ALS patients [49]. Riluzole, which is the only approved treatment for ALS at present, is thought to inhibit glutamate release from synapses [50]; however, riluzole only moderately improves survival in ALS patients [51]. Overexpressing EAAT2 in mutant SOD1 transgenic mice decreases excitotoxicity, delays disease onset, slows disease progression and prolongs life-span [52], but only partially increases neuroprotection [53]. Thus, exicitotoxicity may be involved in modulation of disease progression.

Neuroinflammation

Neuroinflammation is characterized in ALS by the appearance of reactive microglial and astroglial cells [5,6], suggesting a non-cell autonomous process [54]. In ALS, reactive astrocytes produce nitric oxide and peroxynitrite, and trigger mitochondrial damage and apoptosis in motor neurons [55]. Astrocytes may also contribute to excitotoxic damage to motor neurons as discussed above. Furthermore, microglial cells are reported to be activated in the brain and spinal cord of patients with ALS, as well as mutant SOD1 transgenic mice [56] and activated microglia were detected before motor neuron loss [56]. Damage within motor neurons is enhanced by injury from microglial cells via an inflammatory response that accelerates disease progression [57]. Furthermore, treatment of ALS with minocycline, a tetracycline derivative that inhibits microglial activation, increases survival and delays disease onset in $SOD1^{G93A}$ transgenic mice [58]. Hence, understanding the cross-talk between motor neurons and their non-neuronal neighbours may lead to understanding how these factors act in concert to drive disease progression.

Mitochondrial Disturbances and Oxidative Stress

Changes in mitochondrial structure and function, and activation of mitochondrial dependent apoptosis, have been described in animal and cellular models of ALS, as well as in patients with ALS [59,60]. In $SOD1^{G93A}$ transgenic mice, mitochondrial degeneration and vacuolation are reported very early, at 2 weeks of age [8,61], suggesting that mitochondrial abnormalities are an early event in disease pathogenesis. Furthermore, mutant SOD1 associates with mitochondrial compartment, such as intermembrane space of mitochondria, from where it may trigger apoptosis [62]. Mitochondrial-dependent apoptosis, including redistribution of cytochrome c from mitochondria to cytosol and recruitment of proapoptotic Bax to mitochondria are associated with mutant SOD1 inclusion formation [60]. Dysfunction of components of the mitochondrial respiratory chain are also evident in the spinal cord of $SOD1^{G93A}$ transgenic mice at disease end stage [63]. Expression of mutant SOD1 in cell culture induces a significant loss of mitochondrial membrane potential and increases mitochondrial ROS production, suggesting that respiratory chain dysfunction and mitochondrial depolarization are underway [7]. Impaired axonal transport of mitochondria has been reported in ALS [64], although a recent study demonstrates that impairment of mitochondrial transport in $SOD1^{G93A}$ transgenic mice plays a minimal role in ALS pathology [65]. Furthermore, mutant SOD1 targeting to the mitochondria impaires mitochondrial dynamics and also is associated with impaired maintenance of neuritic processes [66]. Axonal transport of mitochondria may be essential to neurons due to their extensive processes between the cell body and the synapse at the cell periphery [67].

ER Stress

We and other groups have demonstrated that ER stress is an important pathway to cell death in ALS [68,69], and is triggered very early in $SOD1^{G93A}$ transgenic mice at p5 [70]. ER stress is triggered when misfolded proteins accumulate within the ER lumen, inducing the unfolded protein response (UPR). Although the initial phases of the UPR aim to promote cell survival, prolonged or severe ER stress triggers the apoptotic phase of the UPR. Upregulation of the three UPR sensor proteins, PERK, ATF6 and IRE1, have been observed both at the symptom onset and at disease end stage of $SOD1^{G93A}$ transgenic mice, implying the involvement of ER stress in disease mechanisms [68,71]. The ER chaperone, protein disulphide isomerase (PDI), was found to co-localize with mutant SOD1 inclusions in both cellular and animal models of ALS [72] and overexpression of PDI decreased mutant SOD1 aggregation, ER stress, and apoptosis [72]. However, it remains unclear how ER stress is triggered in ALS because SOD1 and the other proteins linked to ALS are not associated with the ER.

AXONAL TRANSPORT IN NEURONS

The major roles of axonal transport are to move newly synthesized proteins or lipids to the distal axon or synapse to maintain axonal and synaptic activity, and to clear proteins and organelles from the axon or synapse to the cell soma for efficient degradation. Axonal transport also plays a role in the communication of intracellular signals from the distal axon to the soma, allowing the neuron to respond to changes in the environment. Another important function is the transport of mitochondria to provide energy in the axon or synapse. Defects in either supply or clearance of material within an axon can lead to neuronal death.

Molecular motors from the kinesin, dynein, and myosin superfamilies transport cargo within neurons and other cell types [73,74]. Both microtubules and neurofilaments are the major longitudinal cytoskeletal filaments within the axon and dendrites and the molecular motors move along microtubules during transport. In the synaptic regions, such as pre-synaptic terminals and post-synaptic spines, actin filaments form the main cytoskeletal architecture [75], and myosins are the main molecular motor which conveys cargos along actin filaments [76]. Axonal transport occurs bi-directionally, from the cell body to the periphery (anterograde transport) and from the periphery to the cell body (retrograde transport). The direction of transport depends on the polarity of the rails. In the axon, microtubules have unipolar orientation: the fast growing "plus" end points towards the synapse, and the "minus" end points towards the soma [77]. However, in the dendrites, the microtubule polarity is mixed. Actin filaments also have polarity: the barbed end (the growing end) points to the plasma membrane in pre-synaptic and post-synaptic regions.

Classically and functionally, axonal transport is divided into fast and slow axonal transport based on the bulk speed of cargo movement. Vesicles and mitochondria move by fast axonal transport at speeds of ~1 μm/s, whereas cargoes such as cytoskeleton components move slowly at speeds of ~1 mm/day [78]. Movement is also intermittent in slow axonal transport with individual cargoes pausing during their transit within the axon [15,79].

MOLECULAR MOTOR PROTEINS: KINESIN, DYNEIN, MYOSIN

Kinesin superfamily proteins (KIFs) are composed of a motor domain, a stalk domain and a tail region [80]. The conserved globular motor domain consists of an ATP-binding site and a microtubule-binding sequence. The motor domain hydrolyzes ATP and transfers chemical energy for the movement of each KIF along microtubules with intrinsic directionality. The tail regions and the filamentous stalk regions recognize and bind to the cargo(s) [81]. While motor domains show high amino acid sequence homologies (30–60%) among the KIFs, the other regions exhibit significant variability [82,83].

KIFs can be broadly grouped into three types depending on the position of the motor domain within the molecule. The motor domain is found in the NH_2-terminal region in N-kinesins, in the middle of the protein in

M-kinesins, and at the COOH terminus in the C-kinesins. N-kinesins generally move toward microtubule plus ends, while C-kinesins move toward minus ends, and M-kinesins depolymerise microtubules [84,85]. The kinesin superfamily comprises at least 45 members in mammals such as human and mouse [86] and the KIFs are classified into 14 families based on head domain homology [86–89]. These include one M-kinesin family, one C-kinesin family, and 12 N-kinesin families. KIFs usually drive anterograde transport while most retrograde transport is mediated by cytoplasmic dynein, although a few kinesins also power retrograde transport [90–92].

In striking contrast to the kinesin family with its diverse members, cytoplasmic dynein utilizes its single structural form to carry out a wide variety of cellular functions, many of which are similar to those produced by the kinesins. Dyneins are members of the AAA+ superfamily (AAA), ATPase-associated proteins with various cellular activities [93]. Dynein motor proteins couple ATP hydrolysis to movement along microtubules towards the minus end and they can carry a large variety of cargoes. Dynein superfamily proteins comprise two major groups, cytoplasmic dynein and axonemal dyneins, which are also called ciliary or flagellar dyneins. The complete dynein motor is a large, multi-protein complex (1.5 MDa) composed of two identical heavy chains, two intermediate chains, four light intermediate chains and several light chains. Cytoplasmic dynein has an important associated protein complex called dynactin, containing $p150^{Glued}$, p62, dynamitin, actin related protein (Arp) 1, CAPZα and CAPZβ, p27, and p24. Dynactin regulates dynein activity and the binding capacity of dynein for its cargos [94]. The homodimer heavy chain of dynein binds to microtubules and enables dynein to move in an ATP-dependent manner [95]. The other subunits of dynein are thought to maintain the stability of the complex, to modulate its activity and to interact with accessory and cargo proteins [96–98]. Cytoplasmic dynein may also perform tasks other than the transport of cargos; for example, endosomes depend on dynein not just for their motility, but also for their maturation, morphology and receptor sorting [99].

Dynein and kinesins function in an inter-dependent manner, and the disruption of movement in one direction can also affect movement in the opposite direction [100]. The intermediate chain of dynein can bind directly to kinesin light chains 1 and 2 [101], and dynactin is able to interact not only with dynein but also with several kinesins [102]. Hence, the regulation of axonal transport processes extremely complex.

Myosin superfamily motor proteins bind to actin and hydrolyse ATP to generate force and movement along actin filaments. They are classified into 18 classes, mostly based on comparisons and phylogenetic analysis of the conserved motor domain [103]. They play significant roles in cell movement, muscle contraction, cytokinesis, membrane trafficking, and signal transduction. Most myosins form a dimer and consist of a motor domain, a neck region, and a tail region.

MOLECULAR MOTORS IN SOD1-ASSOCIATED ALS

There is increasing evidence for dysfunction of axonal transport in the pathogenesis of ALS. In $SOD1^{G93A}$ transgenic mice, defects in both anterograde and retrograde axonal transport are observed [104]. There is also evidence that impairment of axonal retrograde transport is one of the earliest axonal pathologies in $SOD1^{G93A}$ transgenic mice, suggesting deficits in axonal transport are a key pathogenic event in ALS [10]. Both fast and slow axonal transport were also impaired in the ventral roots of transgenic mice with low copy number of mutant SOD1 [105]. Swollen axon segments or spheroids are also present in spinal cords of human ALS and $SOD1^{G93A}$ transgenic mice [106]. Also, axonal disorganization and the accumulation of neurofilament proteins are hallmarks of both sporadic and inherited forms of ALS [107–109]. Mice overexpressing human neurofilament heavy-subunit gene (NF-H) posses dramatic defects in axonal transport, not only of neurofilament proteins but also of other proteins, including tubulin and actin [107]. Point mutation of the p150 subunit of dynactin has been detected in both sporadic and familial ALS patients [110]. Mutations in dynactin lead to a reduction in retrograde transport but disease progression is less severe than ALS in transgenic mice [31]. In contrast, mutations in cytoplasmic dynein can either result in pure sensory neuropathy or in a sensory neuropathy with motor neuron involvement [111]. A recent study suggested that alterations in retrograde signalling contribute to neurodegeneration in ALS, by shifting from the transport of survival-promoting to death-promoting signalling cargo molecules [112]. Mutant SOD1 also inhibits anterograde fast axonal transport of mitochondria enhances the retrograde movement resulting in a depletion of axonal mitochondria content [113]. Mislocalization of mutant SOD1 is observed in axonal mitochondria within motor neurons, accompanied by intra-axonal accumulation of misfolded SOD1 [114]. Considering the very long axon of the motor neuron and the relative abundance of mitochondria within these axons, mitochondrial transport disturbances are expected to have a significant impact on the overall function of the motor neuron [114]. Evidence suggests that axonal transport alterations are an early event in ALS, thus implicating axonal transport in the pathogenesis of ALS.

Kinesins

Kinesins mediate anterograde axonal transport, which is essential for synapse generation and for maintaining synaptic transmission [80]. Several studies have highlighted the importance of kinesin in ALS, particularly kinesin heavy chain genes, KIF5A and KIF1Bβ, which transport mitochondria, synaptic vesicles and macromolecular complexes. A recent study demonstrated that oxidized wildtype SOD1, and wildtype SOD1 immunopurified from sporadic ALS patient tissues, inhibited kinesin-based fast axonal transport in a manner similar to mutant SOD1 in familial ALS, suggesting common pathogenic mechanisms in both sporadic and familial ALS [38]. Disruption of KIF5A in neurons postnatally does not affect fast axonal transport; however, it induces a reduction in slow anterograde axonal transport, resulting in an accumulation of neurofilament proteins in the cell bodies of peripheral sensory neurons and a reduction in sensory axon calibre [115]. Similarly, the expression of KIF1Bβ mutants in mice induces defects in the anterograde transport of synaptic vesicle precursors, resulting in a late-onset axonopathy [116]. These pathological effects closely mimic those observed in individuals bearing mutations in the KIF5A and KIF1Bβ genes, which have been identified in families with hereditary spastic paraplegia [117] and Charcot-Marie Tooth type 2A neuropathy [116]. Furthermore, not only mutations, but lower expression of KIF has been associated with neurodegeneration. For example, reduction in the levels of kinesin associated protein 3 (KIFAP3) has been linked to increased survival in ALS patients [118], which may be due to changes in axonal transport of choline acetyltransferase (ChAT) [119]. Hence, KIFAP3 can be considered as a potential modifier of the ALS phenotype [120].

KIFAP3 is part of the trimeric motor kinesin II complex (KIF3) that is involved in multiple functions including intracellular transport [121]. Furthermore, a specific downregulation of kinesin-related proteins, KIF1Bβ and KIF3Aβ, has been detected in motor cortex specimens of sporadic ALS patients [122]. Nevertheless, interaction between mutant SOD1 and members of the kinesin-1 family (KIF5A, 5B, or 5C) was not observed in spinal cords of $SOD1^{G93A}$ transgenic mice at 60 days of age, suggesting that impairment of anterograde transport occurs on the verge of developing clinical symptoms [123]. In comparison to dynein, relatively few degenerative diseases have been directly linked to kinesin, possibly due to functional redundancy in the large kinesin superfamily.

Dyneins

Dynein is very highly expressed in neurons [124] and neurons are particularly sensitive to defects in dynein/dynactin function [78]. A dynein subunit was identified as a component of mutant SOD1-containing high molecular weight complexes prior to the onset of symptoms in ALS animal models and increased during disease progression [125]. Furthermore, mutations in the dynactin 1 gene (DCTN1), encoding the p150Glued subunit of dynactin, cause human distal hereditary motor neuropathy (HMN7B) [31,126]. Knock-in and transgenic mice expressing these mutations subsequently develop motor neuron degeneration [127,128]. Over-expression of the dynactin subunit dynamitin in mice leads to impairment of dynein-dynactin interaction and a late-onset progressive disease reminiscent of ALS [129]. Mutation of the dynein heavy chain 1 gene causes motor neuron degeneration and inhibition of axonal transport [130]. Loa (legs at odd angles) and Cra1 (Cramping1) mice which carry separate N-ethyl-N-nitrosourea-induced mutations of the dynein heavy chain 1 gene, show selective impairment of axonal retrograde transport as well as progressive locomotor disorders associated with spinal motor neuron degeneration [130]. Also, a third mouse model (Swl) with another dynein heavy chain gene mutation displays an early-onset sensory neuropathy with muscle spindle deficiency [131]. Hence, the relative contributions of sensory and motor neuron degeneration to the phenotypes of Loa, Cra1, and Swlmice are still under debate.

Surprisingly, crossing the $SOD1^{G93A}$ transgenic mouse with Loa or Cra1 animals delays disease progression and significantly increases lifespan [132]. This amelioration correlates with a complete recovery of the axonal transport deficits in motor neurons of these mice [132,133]. However, this observation was partially explained recently by a proteomics study which revealed that the SOD1 mutation augments retrograde transport of stress factors (p-JNK, caspase-8 and p75NTR cleavage fragment) and simultaneously impairs retrograde transport of survival factors (p-Trk and p-Erk1/2) [134]. Dynein also plays a role in the transport of nerve injury signals, phosphorylated Erk [135], and phosphorylated JNK [136]. Many types of protein cargoes transported by dynein are implicated in ALS [134]. Impaired retrograde transport of activated neurotrophin receptors by direct association of the dynein light chain with Trk neurotrophin receptors may also trigger neurodegeneration [137,138]. Dynein is involved in the autophagic clearance of misfolded proteins [139] and removal of damaged organelles and proteins from the axonal compartment [140]. Therefore, clearance of misfolded proteins in ALS is affected by dysfunction of dynein.

It is clear that dynein-mediated retrograde axonal transport is affected in motor neurons in ALS [123]; however, the underlying mechanisms are still unclear. Mutations in the dynein-dynactin machinery cause phenotypes that may influence multiple processes including neurotrophic factor delivery, transport, and homeostasis of mitochondria and protein aggregation or degradation. Disruption of dynein function could therefore link several proposed pathways relevant to ALS pathophysiology, including axonal transport, mitochondrial dysfunction, and mutant SOD1 aggregation.

DYNEIN-MEDIATED ER-GOLGI TRANSPORT IN ALS

The transport of secretory proteins between the ER and Golgi apparatus is also dynein mediated, and dysfunction of the ER-Golgi is also implicated in diseases of motor neurons, suggesting a link between ER-Golgi and axonal transport processes [129]. Also, dynein dysfunction can result in fragmentation of the Golgi apparatus [141], which occurs in spinal and cortical motor neurons in ALS patients, in cell lines over-expressing mutant SOD1 [142], and is also one of the first pathological events in $SOD1^{G93A}$transgenic mice [143].

The ER and Golgi apparatus regulate protein trafficking and secretion. The ER is responsible for the sorting, post-translational modification and trafficking of transmembrane and secretory proteins [144,145]. Vesicles containing secretory protein cargo bud from the ER, fuse with the cis-Golgi network and subsequently progress to the trans-Golgi network, from where they are transported to further cell was locations. Approximately a third of all cellular proteins are secreted via this classical ER-Golgi pathway. Hence, inhibition of ER-Golgi function and transport could severely impact on cellular function. The relationship between dynein-mediated ER-Golgi transport and axonal transport mechanisms is poorly understood, but the two processes are clearly linked (recently reviewed in [146]).

ER-Golgi and Diseases of Motor Neurons

Protein folding in the ER is required to maintain cellular function. Protein misfolding within the ER induces ER stress, triggering the UPR. The UPR mechanisms aim to restore homeostasis, but if homeostasis is not restored, cell death is triggered via apoptosis [147]. Cellular insults which lead to increased protein misfolding in the ER include changes in intracellular calcium concentration, alterations in the redox state of the ER, nutrient deprivation, failure of post-translational modifications and increases in secretory protein synthesis [148]. ER stress also induces ER-associated degradation (ERAD), a process by which proteins are removed from the ER and retrotranslocated to the cytosol, where they are degraded by the 26S proteasome. It has been reported that UPR induction is due to the interaction of cytoplasmic mutant SOD1 with Derlin-1, a key protein involved in ERAD [149], but this interaction was only detected after the onset of symptoms in transgenic SOD1 mice, suggesting that this mechanism is not the primary trigger of ER stress. However, it is now recognised that other cellular processes can cause secondary ER stress, such as disturbances of the ER, Golgi, endosomal and vesicular transport systems [150–152].

ER stress has been reported as a key role in mutant SOD1-linked ALS pathogenesis; however, the exact mechanisms that trigger ER stress in ALS remain elusive. Recent evidence also implicates the ER-Golgi in disease induced by TDP-43 and FUS in ALS. Recently, TDP-43 was detected in the rough ER in spinal cord motor neurons of sporadic

ALS patients [153], suggesting that TDP-43 could be redistributed to the ER in disease. Furthermore, TDP-43 has been detected in mouse brainstem microsomes which contain ER and other vesicles [154], and pharmacological induction of ER stress increases the accumulation of TDP-43 in cell culture [155]. Similarly, FUS has also been linked to the ER: electron microscopy of spinal motor neurons of patients with juvenile ALS demonstrated FUS-positive inclusions associated with disorganised rough ER [156]. A key UPR protein, BiP, is also found in FUS-immunoreactive inclusions in ALS patient tissue, providing further evidence that ER stress could be involved in mutant FUS-linked ALS [157].

ER-Golgi Vesicle Trafficking Defects

Vesicle trafficking defects have been linked to human motor neuron diseases [158]. Dynamin 2, a vesicular traffic regulator modulates the actin cytoskeleton, interacts with actin-binding proteins such as profilin and Abp1 [159] and is involved in the fusion and fission of vesicles and other membranous organelles [160]. Atlastin 1 (SPG3A) and REEP1 mutations (SPG31) also result in marked ER morphological defects and altered synaptic vesicle recycling. Several familial forms of ALS are also linked to genes encoding proteins involved in the regulation and control of vesicle transport: VAPB, VCP and ALS2 [34,161,162]. Mouse mutants with motor phenotypes have been linked to defects in intracellular trafficking, such as the progressive motor neuronopathy (pmn) mouse, which accumulates the mutant tubulin-specific chaperone TBCE in the cis-Golgi compartment [163]. Also, the wobbler mouse, which displays motor neuron dysfunction, contains mutations in vacuolar-vesicular protein sorting factor Vps54, which is involved in Golgi-associated vesicular trafficking [164].

(a) **Vesicle-associated membrane protein-associated protein B (VAPB).** Missense mutations of VAPB, a widely expressed ER transmembrane protein [165], cause autosomal dominant typical ALS, slowly progressing atypical ALS or late-onset spinal muscular atrophy (SMA). VAPB is a highly conserved protein, which is localized in the ER and associated with microtubules [166]. VAPB has important function in ER and Golgi maintenance and neuronal transmission [165,167,168].

It is also involved in the regulation of ER–Golgi vesicle transport [169] and in modulating ER stress [170]. The two VAPB mutations described in ALS, P56S and T46I, perturb ER and Golgi trafficking [171] and cause morphological disruption to the ER [168,172,173]. These mutations also appear to increase motor neuron vulnerability to ER stress-induced death, by sequestering wild-type VAPB [174]. Mutant VAPB also induces the co-aggregation of wild-type VAPB [170,173], suggesting a dominant-negative mode of pathogenesis. Mutant P56S VAPB inclusions are ER derived [165,168], and co-locate with a subset of ER proteins [172]. VAPB proteins are also involved in axonal guidance [175], which may lead to axonal dysfunctions and contribute to ALS [172].

(b) **Optineurin.** Optineurin is also one of the disease-causing genes in ALS although it was earlier reported to be a causative gene in primary open-angle glaucoma. Optineurin is a Golgi-localized protein [176,177]; however, mutations in optineurin induce the formation of cytoplasmic inclusions in both sporadic and SOD1 cases of ALS [33,35]. Several interacting partners of optineurin have been recently identified, including GTPase molecular Rab8 [178], myosin VI [179], and transferrin receptor [180]. The redistribution of optineurin from the Golgi apparatus to the cytoplasm leads to Golgi fragmentation [181]. Together these data indicate a possible role for optineurin in protein trafficking and axonal transport dysfunction [182].

(c) **Valosin containing protein (VCP).** Recently using exome sequencing, ALS-causative mutations in the gene encoding valosin containing protein (VCP/p97) were identified in familial ALS patients [34]. VCP is a hexameric type II ATPase of the AAA family, with many putative roles, including transcription, cell division, Golgi assembly, autophagy, and in particular, aiding proteasome function [183–186]. Knockdown of VCP in HeLa cells leads to an increase in ubiquitinated proteins derived from the ER [187], and reduced levels of cellular VCP induce ER stress, perhaps as a consequence of reduced ERAD and/or by disturbing the fusion of ER membranes

[188]. Motor neurons of an ALS patient with a VCP mutation were recently found to contain TDP-43-positive inclusions [34], similar to VCP mutations linked to frontotemporal lobar dementia (FTLD) [189]. VCP mutations in FTLD were also found to induce ER stress and cell death in cell culture [189]. Interestingly, VCP is involved in ERAD by binding to the ER transmembrane protein Derlin-1 [190,191]. The association of Derlin-1 with activation of the UPR in mutant SOD1 cellular and mouse models of ALS [192] suggests overlap in the pathogenic disease mechanisms triggered by VCP, TDP-43 and SOD1 related to ER stress and ERAD.

DO MUTANT TDP-43 AND MUTANT FUS ALSO INDUCE MOLECULAR MOTOR DYSFUNCTION?

Most of the evidence demonstrating the involvement of molecular motor proteins in ALS comes from the study of SOD1 disease models. A role for the other major proteins recently linked to ALS, TDP-43 and FUS, in molecular motor dysfunction remains yet to be established. Translocation of both TDP-43 and FUS from the nucleus to cytoplasm induces the formation of inclusions and is linked to pathology [193–195]. Previous studies have shown that axonal ligation induces transient redistribution of TDP-43 to the cytoplasm and peripheral axonal accumulation of TDP-43 in brainstem motor neurons [154,193]. Moreover, nuclear-excluded TDP-43 co-localized with RNA transport markers and stress granules, indicating that TDP-43 plays a role in sequestering and regulating mRNA levels in response to axonal injury [154]. Both TDP-43 and FUS also play a role in mRNA transport to dendrites in an activity-dependent manner [196]. These data together suggest that accumulation of TDP-43 or FUS in the cytoplasm could lead to transport impairment and axonal injury. Due to the functional similarities between TDP-43 and FUS, the pathological mechanisms associated with these two ALS-linked proteins may be overlapping.

CONCLUSIONS

Motor neurons are highly polarized cells, and hence cellular transport is a fundamental process for the survival and maintenance of motor neurons. Increasing evidence implicates transport defects in the etiology of ALS.

Studies using SOD1 transgenic mice clearly show that axonal transport impairment contributes to motor neuron degeneration. Moreover, mutations in genes encoding motor proteins provide direct evidence for a role in the pathogenesis of ALS. Although dysfunction of motor proteins in ALS directly triggers defects in transport, it is possible that other ALS-associated proteins are directly or indirectly linked to an impairment of transport. However, this requires further studies. Hence, identifying the common pathophysiological mechanisms in ALS is crucial. A powerful combination of genetic, cellular, and biochemical approaches have revealed new and often unexpected insights, and the further development of our understanding of molecular motors and their role in ALS will hopefully lead to the development of improved therapies for ALS.

ACKNOWLEDGMENTS

This study was supported by the National Health and Medical Research Council of Australia (NHMRC, Project grant # 454749 and # 1006141), Bethlehem Griffiths Research Foundation and Motor Neuron Disease Research Institute of Australia Henry Roth Foundation Grant to JDA.

REFERENCES

1. Rosen, D.R. Mutations in Cu/Zn superoxide dismutase gene are associated with familial amyotrophic lateral sclerosis. Nature **1993**, 364, 362. [Google Scholar]
2. Arai, T.; Hasegawa, M.; Akiyama, H.; Ikeda, K.; Nonaka, T.; Mori, H.; Mann, D.; Tsuchiya, K.; Yoshida, M.; Hashizume, Y.; et al. TDP-43 is a component of ubiquitin-positive tau-negative inclusions in frontotemporal lobar degeneration and amyotrophic lateral sclerosis. Biochem. Biophys. Res. Commun **2006**, 351, 602–611. [Google Scholar]
3. Kwiatkowski, T.J., Jr; Bosco, D.A.; Leclerc, A.L.; Tamrazian, E.; Vanderburg, C.R.; Russ, C.; Davis, A.; Gilchrist, J.; Kasarskis, E.J.; Munsat, T.; et al. Mutations in the FUS/TLS gene on chromosome 16 cause familial amyotrophic lateral sclerosis. Science **2009**, 323, 1205–1208. [Google Scholar]
4. Rothstein, J.D.; Tsai, G.; Kuncl, R.W.; Clawson, L.; Cornblath, D.R.; Drachman, D.B.; Pestronk, A.; Stauch, B.L.; Coyle, J.T. Abnormal

excitatory amino acid metabolism in amyotrophic lateral sclerosis. Ann. Neurol **1990**, 28, 18–25. [Google Scholar]

5. Van Den Bosch, L.; Robberecht, W. Crosstalk between astrocytes and motor neurons: what is the message? Exp. Neurol **2008**, 211, 1–6. [Google Scholar]
6. Neusch, C.; Bahr, M.; Schneider-Gold, C. Glia cells in amyotrophic lateral sclerosis: new clues to understanding an old disease? Muscle Nerve **2007**, 35, 712–724. [Google Scholar]
7. Rizzardini, M.; Mangolini, A.; Lupi, M.; Ubezio, P.; Bendotti, C.; Cantoni, L. Low levels of ALS-linked Cu/Zn superoxide dismutase increase the production of reactive oxygen species and cause mitochondrial damage and death in motor neuron-like cells. J. Neurol. Sci **2005**, 232, 95–103. [Google Scholar]
8. Bendotti, C.; Calvaresi, N.; Chiveri, L.; Prelle, A.; Moggio, M.; Braga, M.; Silani, V.; De Biasi, S. Early vacuolization and mitochondrial damage in motor neurons of FALS mice are not associated with apoptosis or with changes in cytochrome oxidase histochemical reactivity. J. Neurol. Sci **2001**, 191, 25–33. [Google Scholar]
9. Williamson, T.L.; Cleveland, D.W. Slowing of axonal transport is a very early event in the toxicity of ALS-linked SOD1 mutants to motor neurons. Nat. Neurosci **1999**, 2, 50–56. [Google Scholar]
10. Bilsland, L.G.; Sahai, E.; Kelly, G.; Golding, M.; Greensmith, L.; Schiavo, G. Deficits in axonal transport precede ALS symptoms in vivo. Proc. Natl. Acad. Sci. USA **2010**, 107, 20523–20528. [Google Scholar]
11. Morfini, G.; Pigino, G.; Brady, S.T. Polyglutamine expansion diseases: Failing to deliver. Trends Mol. Med **2005**, 11, 64–70. [Google Scholar]
12. Chevalier-Larsen, E.; Holzbaur, E.L. Axonal transport and neurodegenerative disease. Biochim. Biophys. Acta **2006**, 1762, 1094–1108. [Google Scholar]
13. Sau, D.; Rusmini, P.; Crippa, V.; Onesto, E.; Bolzoni, E.; Ratti, A.; Poletti, A. Dysregulation of axonal transport and motorneuron diseases. Biol. Cell **2011**, 103, 87–107. [Google Scholar]

14. Kieran, D.; Hafezparast, M.; Bohnert, S.; Dick, J.R.; Martin, J.; Schiavo, G.; Fisher, E.M.; Greensmith, L. A mutation in dynein rescues axonal transport defects and extends the life span of ALS mice. J. Cell Biol **2005**, 169, 561–567. [Google Scholar]
15. Roy, S.; Coffee, P.; Smith, G.; Liem, R.K.; Brady, S.T.; Black, M.M. Neurofilaments are transported rapidly but intermittently in axons: Implications for slow axonal transport. J. Neurosci **2000**, 20, 6849–6861. [Google Scholar]
16. Rothstein, J.D. Current hypotheses for the underlying biology of amyotrophic lateral sclerosis. Ann. Neurol **2009**, 65, S3–S9. [Google Scholar]
17. Dion, P.A.; Daoud, H.; Rouleau, G.A. Genetics of motor neuron disorders: new insights into pathogenic mechanisms. Nat. Rev. Genet **2009**, 10, 769–782. [Google Scholar]
18. Gurney, M.E.; Pu, H.; Chiu, A.Y.; Dal Canto, M.C.; Polchow, C.Y.; Alexander, D.D.; Caliendo, J.; Hentati, A.; Kwon, Y.W.; Deng, H.X.; et al. Motor neuron degeneration in mice that express a human Cu,Zn superoxide dismutase mutation. Science **1994**, 264, 1772–1775. [Google Scholar]
19. Jaarsma, D.; Haasdijk, E.D.; Grashorn, J.A.; Hawkins, R.; van Duijn, W.; Verspaget, H.W.; London, J.; Holstege, J.C. Human Cu/Zn superoxide dismutase (SOD1) overexpression in mice causes mitochondrial vacuolization, axonal degeneration, and premature motoneuron death and accelerates motoneuron disease in mice expressing a familial amyotrophic lateral sclerosis mutant SOD1. Neurobiol. Dis **2000**, 7, 623–643. [Google Scholar]
20. Blair, I.P.; Williams, K.L.; Warraich, S.T.; Durnall, J.C.; Thoeng, A.D.; Manavis, J.; Blumbergs, P.C.; Vucic, S.; Kiernan, M.C.; Nicholson, G.A. FUS mutations in amyotrophic lateral sclerosis: clinical, pathological, neurophysiological and genetic analysis. J. Neurol. Neurosurg. Psychiatry **2010**, 81, 639–645. [Google Scholar]
21. Vance, C.; Rogelj, B.; Hortobagyi, T.; De Vos, K.J.; Nishimura, A.L.; Sreedharan, J.; Hu, X.; Smith, B.; Ruddy, D.; Wright, P.; et al. Mutations in FUS, an RNA processing protein, cause familial amyotrophic lateral sclerosis type 6. Science **2009**, 323, 1208–1211. [Google Scholar]

22. Belzil, V.V.; Valdmanis, P.N.; Dion, P.A.; Daoud, H.; Kabashi, E.; Noreau, A.; Gauthier, J.; Hince, P.; Desjarlais, A.; Bouchard, J.P.; et al. Mutations in FUS cause FALS and SALS in French and French Canadian populations. Neurology **2009**, 73, 1176–1179. [Google Scholar]
23. Van Langenhove, T.; van der Zee, J.; Sleegers, K.; Engelborghs, S.; Vandenberghe, R.; Gijselinck, I.; Van den Broeck, M.; Mattheijssens, M.; Peeters, K.; De Deyn, P.P.; et al. Genetic contribution of FUS to frontotemporal lobar degeneration. Neurology **2010**, 74, 366–371. [Google Scholar]
24. Neumann, M.; Sampathu, D.M.; Kwong, L.K.; Truax, A.C.; Micsenyi, M.C.; Chou, T.T.; Bruce, J.; Schuck, T.; Grossman, M.; Clark, C.M.; et al. Ubiquitinated TDP-43 in frontotemporal lobar degeneration and amyotrophic lateral sclerosis. Science **2006**, 314, 130–133. [Google Scholar]
25. Sreedharan, J.; Blair, I.P.; Tripathi, V.B.; Hu, X.; Vance, C.; Rogelj, B.; Ackerley, S.; Durnall, J.C.; Williams, K.L.; Buratti, E.; et al. TDP-43 mutations in familial and sporadic amyotrophic lateral sclerosis. Science **2008**, 319, 1668–1672. [Google Scholar]
26. Yokoseki, A.; Shiga, A.; Tan, C.F.; Tagawa, A.; Kaneko, H.; Koyama, A.; Eguchi, H.; Tsujino, A.; Ikeuchi, T.; Kakita, A.; et al. TDP-43 mutation in familial amyotrophic lateral sclerosis. Ann. Neurol**2008**, 63, 538–542. [Google Scholar]
27. Xu, Y.F.; Zhang, Y.J.; Lin, W.L.; Cao, X.; Stetler, C.; Dickson, D.W.; Lewis, J.; Petrucelli, L. Expression of mutant TDP-43 induces neuronal dysfunction in transgenic mice. Mol. Neurodegener**2011**, 6, 73. [Google Scholar]
28. Chandran, J.; Ding, J.; Cai, H. Alsin and the molecular pathways of amyotrophic lateral sclerosis. Mol. Neurobiol **2007**, 36, 224–231. [Google Scholar]
29. Chen, Y.Z.; Bennett, C.L.; Huynh, H.M.; Blair, I.P.; Puls, I.; Irobi, J.; Dierick, I.; Abel, A.; Kennerson, M.L.; Rabin, B.A.; et al. DNA/RNA helicase gene mutations in a form of juvenile amyotrophic lateral sclerosis (ALS4). Am. J. Hum. Genet **2004**, 74, 1128–1135. [Google Scholar]
30. Nishimura, A.L.; Mitne-Neto, M.; Silva, H.C.; Richieri-Costa, A.;

Middleton, S.; Cascio, D.; Kok, F.; Oliveira, J.R.; Gillingwater, T.; Webb, J.; et al. A mutation in the vesicle-trafficking protein VAPB causes late-onset spinal muscular atrophy and amyotrophic lateral sclerosis. Am. J. Hum. Genet**2004**, 75, 822–831. [Google Scholar]

31. Puls, I.; Jonnakuty, C.; LaMonte, B.H.; Holzbaur, E.L.; Tokito, M.; Mann, E.; Floeter, M.K.; Bidus, K.; Drayna, D.; Oh, S.J.; et al. Mutant dynactin in motor neuron disease. Nat. Genet **2003**, 33, 455–456. [Google Scholar]
32. Strong, M.J. The syndromes of frontotemporal dysfunction in amyotrophic lateral sclerosis. Amyotroph. Lateral Scler **2008**, 9, 323–338. [Google Scholar]
33. Maruyama, H.; Morino, H.; Ito, H.; Izumi, Y.; Kato, H.; Watanabe, Y.; Kinoshita, Y.; Kamada, M.; Nodera, H.; Suzuki, H.; et al. Mutations of optineurin in amyotrophic lateral sclerosis. Nature**2010**, 465, 223–226. [Google Scholar]
34. Johnson, J.O.; Mandrioli, J.; Benatar, M.; Abramzon, Y.; Van Deerlin, V.M.; Trojanowski, J.Q.; Gibbs, J.R.; Brunetti, M.; Gronka, S.; Wuu, J.; et al. Exome sequencing reveals VCP mutations as a cause of familial ALS. Neuron **2010**, 68, 857–864. [Google Scholar]
35. Osawa, T.; Mizuno, Y.; Fujita, Y.; Takatama, M.; Nakazato, Y.; Okamoto, K. Optineurin in neurodegenerative diseases. Neuropathology **2011**, 31, 569–574. [Google Scholar]
36. Deng, H.X.; Chen, W.; Hong, S.T.; Boycott, K.M.; Gorrie, G.H.; Siddique, N.; Yang, Y.; Fecto, F.; Shi, Y.; Zhai, H.; et al. Mutations in UBQLN2 cause dominant X-linked juvenile and adult-onset ALS and ALS/dementia. Nature **2011**, 477, 211–215. [Google Scholar]
37. Dejesus-Hernandez, M.; Mackenzie, I.R.; Boeve, B.F.; Boxer, A.L.; Baker, M.; Rutherford, N.J.; Nicholson, A.M.; Finch, N.A.; Flynn, H.; Adamson, J.; et al. Expanded GGGGCC hexanucleotide repeat in noncoding region of C9ORF72 causes chromosome 9p-Linked FTD and ALS. Neuron**2011**, 72, 245–256. [Google Scholar]
38. Bosco, D.A.; Morfini, G.; Karabacak, N.M.; Song, Y.; Gros-Louis, F.; Pasinelli, P.; Goolsby, H.; Fontaine, B.A.; Lemay, N.; McKenna-Yasek, D.; et al. Wild-type and mutant SOD1 share an aberrant conformation and a common pathogenic pathway in ALS. Nat. Neurosci **2010**, 13, 1396–1403. [Google Scholar]

39. Forsberg, K.; Jonsson, P.A.; Andersen, P.M.; Bergemalm, D.; Graffmo, K.S.; Hultdin, M.; Jacobsson, J.; Rosquist, R.; Marklund, S.L.; Brannstrom, T. Novel antibodies reveal inclusions containing non-native SOD1 in sporadic ALS patients. PLoS One **2010**, 5, e11552. [Google Scholar]
40. Wood, J.D.; Beaujeux, T.P.; Shaw, P.J. Protein aggregation in motor neurone disorders. Neuropathol. Appl. Neurobiol **2003**, 29, 529–545. [Google Scholar]
41. Lariviere, R.C.; Julien, J.P. Functions of intermediate filaments in neuronal development and disease. J. Neurobiol **2004**, 58, 131–148. [Google Scholar]
42. Nakamura, S.; Kawamoto, Y.; Nakano, S.; Ikemoto, A.; Akiguchi, I.; Kimura, J. Cyclindependent kinase 5 in Lewy body-like inclusions in anterior horn cells of a patient with sporadic amyotrophic lateral sclerosis. Neurology **1997**, 48, 267–270. [Google Scholar]
43. Okamoto, K.; Hirai, S.; Amari, M.; Watanabe, M.; Sakurai, A. Bunina bodies in amyotrophic lateral sclerosis immunostained with rabbit anti-cystatin C serum. Neurosci. Lett **1993**, 162, 125–128. [Google Scholar]
44. Deng, H.X.; Zhai, H.; Bigio, E.H.; Yan, J.; Fecto, F.; Ajroud, K.; Mishra, M.; Ajroud-Driss, S.; Heller, S.; Sufit, R.; et al. FUS-immunoreactive inclusions are a common feature in sporadic and non-SOD1 familial amyotrophic lateral sclerosis. Ann. Neurol **2010**, 67, 739–748. [Google Scholar]
45. Johnston, J.A.; Dalton, M.J.; Gurney, M.E.; Kopito, R.R. Formation of high molecular weight complexes of mutant Cu, Zn-superoxide dismutase in a mouse model for familial amyotrophic lateral sclerosis. Proc. Natl. Acad. Sci. USA **2000**, 97, 12571–12576. [Google Scholar]
46. Ezzi, S.A.; Urushitani, M.; Julien, J.P. Wild-type superoxide dismutase acquires binding and toxic properties of ALS-linked mutant forms through oxidation. J. Neurochem **2007**, 102, 170–178. [Google Scholar]
47. Corona, J.C.; Tovar-y-Romo, L.B.; Tapia, R. Glutamate excitotoxicity and therapeutic targets for amyotrophic lateral sclerosis. Expert Opin. Ther. Targets **2007**, 11, 1415–1428. [Google Scholar]

48. Rothstein, J.D.; Van Kammen, M.; Levey, A.I.; Martin, L.J.; Kuncl, R.W. Selective loss of glial glutamate transporter GLT-1 in amyotrophic lateral sclerosis. Ann. Neurol **1995**, 38, 73–84. [Google Scholar]

49. Spreux-Varoquaux, O.; Bensimon, G.; Lacomblez, L.; Salachas, F.; Pradat, P.F.; Le Forestier, N.; Marouan, A.; Dib, M.; Meininger, V. Glutamate levels in cerebrospinal fluid in amyotrophic lateral sclerosis: a reappraisal using a new HPLC method with coulometric detection in a large cohort of patients. J. Neurol. Sci **2002**, 193, 73–78. [Google Scholar]

50. Rothstein, J.D.; Kuncl, R.W. Neuroprotective strategies in a model of chronic glutamatemediated motor neuron toxicity. J. Neurochem **1995**, 65, 643–651. [Google Scholar]

51. Lacomblez, L.; Bensimon, G.; Leigh, P.N.; Guillet, P.; Meininger, V. Dose-ranging study of riluzole in amyotrophic lateral sclerosis. Amyotrophic Lateral Sclerosis/Riluzole Study Group II. Lancet**1996**, 347, 1425–1431. [Google Scholar]

52. Rothstein, J.D.; Patel, S.; Regan, M.R.; Haenggeli, C.; Huang, Y.H.; Bergles, D.E.; Jin, L.; Dykes Hoberg, M.; Vidensky, S.; Chung, D.S.; et al. Beta-lactam antibiotics offer neuroprotection by increasing glutamate transporter expression. Nature **2005**, 433, 73–77. [Google Scholar]

53. Guo, H.; Lai, L.; Butchbach, M.E.; Stockinger, M.P.; Shan, X.; Bishop, G.A.; Lin, C.L. Increased expression of the glial glutamate transporter EAAT2 modulates excitotoxicity and delays the onset but not the outcome of ALS in mice. Hum. Mol. Genet **2003**, 12, 2519–2532. [Google Scholar]

54. Clement, A.M.; Nguyen, M.D.; Roberts, E.A.; Garcia, M.L.; Boillee, S.; Rule, M.; McMahon, A.P.; Doucette, W.; Siwek, D.; Ferrante, R.J.; et al. Wild-type nonneuronal cells extend survival of SOD1 mutant motor neurons in ALS mice. Science **2003**, 302, 113–117. [Google Scholar]

55. Cassina, P.; Cassina, A.; Pehar, M.; Castellanos, R.; Gandelman, M.; de Leon, A.; Robinson, K.M.; Mason, R.P.; Beckman, J.S.; Barbeito, L.; et al. Mitochondrial dysfunction in SOD1G93Abearing astrocytes promotes motor neuron degeneration: prevention by

mitochondrial-targeted antioxidants. J. Neurosci **2008**, 28, 4115–4122. [Google Scholar]

56. Henkel, J.S.; Beers, D.R.; Siklos, L.; Appel, S.H. The chemokine MCP-1 and the dendritic and myeloid cells it attracts are increased in the mSOD1 mouse model of ALS. Mol. Cell. Neurosci **2006**, 31, 427–437. [Google Scholar]
57. Barbeito, L.H.; Pehar, M.; Cassina, P.; Vargas, M.R.; Peluffo, H.; Viera, L.; Estevez, A.G.; Beckman, J.S. A role for astrocytes in motor neuron loss in amyotrophic lateral sclerosis. Brain Res. Brain Res. Rev **2004**, 47, 263–274. [Google Scholar]
58. Zhu, S.; Stavrovskaya, I.G.; Drozda, M.; Kim, B.Y.; Ona, V.; Li, M.; Sarang, S.; Liu, A.S.; Hartley, D.M.; Wu, D.C.; et al. Minocycline inhibits cytochrome c release and delays progression of amyotrophic lateral sclerosis in mice. Nature **2002**, 417, 74–78. [Google Scholar]
59. Guegan, C.; Vila, M.; Rosoklija, G.; Hays, A.P.; Przedborski, S. Recruitment of the mitochondrialdependent apoptotic pathway in amyotrophic lateral sclerosis. J. Neurosci **2001**, 21, 6569–6576. [Google Scholar]
60. Soo, K.Y.; Atkin, J.D.; Horne, M.K.; Nagley, P. Recruitment of mitochondria into apoptotic signaling correlates with the presence of inclusions formed by amyotrophic lateral sclerosisassociated SOD1 mutations. J. Neurochem **2009**, 108, 578–590. [Google Scholar]
61. Kong, J.; Xu, Z. Massive mitochondrial degeneration in motor neurons triggers the onset of amyotrophic lateral sclerosis in mice expressing a mutant SOD1. J. Neurosci **1998**, 18, 3241–3250. [Google Scholar]
62. Okado-Matsumoto, A.; Fridovich, I. Amyotrophic lateral sclerosis: a proposed mechanism. Proc. Natl. Acad. Sci. USA **2002**, 99, 9010–9014. [Google Scholar]
63. Mattiazzi, M.; D'Aurelio, M.; Gajewski, C.D.; Martushova, K.; Kiaei, M.; Beal, M.F.; Manfredi, G. Mutated human SOD1 causes dysfunction of oxidative phosphorylation in mitochondria of transgenic mice. J. Biol. Chem **2002**, 277, 29626–29633. [Google Scholar]
64. Shi, P.; Wei, Y.; Zhang, J.; Gal, J.; Zhu, H. Mitochondrial dysfunction is a converging point of multiple pathological pathways

in amyotrophic lateral sclerosis. J. Alzheimers Dis **2010**, 20, S311–S324. [Google Scholar]

65. Zhu, Y.B.; Sheng, Z.H. Increased axonal mitochondrial mobility does not slow amyotrophic lateral sclerosis (ALS)-like disease in mutant SOD1 mice. J. Biol. Chem **2011**, 286, 23432–23440. [Google Scholar]
66. Magrane, J.; Hervias, I.; Henning, M.S.; Damiano, M.; Kawamata, H.; Manfredi, G. Mutant SOD1 in neuronal mitochondria causes toxicity and mitochondrial dynamics abnormalities. Hum. Mol. Genet **2009**, 18, 4552–4564. [Google Scholar]
67. Hollenbeck, P.J.; Saxton, W.M. The axonal transport of mitochondria. J. Cell Sci **2005**, 118, 5411–5419. [Google Scholar]
68. Atkin, J.D.; Farg, M.A.; Turner, B.J.; Tomas, D.; Lysaght, J.A.; Nunan, J.; Rembach, A.; Nagley, P.; Beart, P.M.; Cheema, S.S.; et al. Induction of the unfolded protein response in familial amyotrophic lateral sclerosis and association of protein-disulphide isomerase with superoxide dismutase 1. J. Biol. Chem **2006**, 281, 30152–30165. [Google Scholar]
69. Atkin, J.D.; Farg, M.A.; Walker, A.K.; McLean, D.; Tomas, D.; Horne, M.K. Endoplasmic reticulum stress and induction of the unfolded protein response in human sporadic amyotrophic lateral sclerosis. Neurobiol. Dis **2008**, 30, 400–407. [Google Scholar]
70. Saxena, S.; Cabuy, E.; Caroni, P. A role for motoneuron subtype-selective ER stress in disease manifestations of FALS mice. Nat. Neurosci **2009**, 12, 627–636. [Google Scholar]
71. Kikuchi, H.; Almer, G.; Yamashita, S.; Guegan, C.; Nagai, M.; Xu, Z.; Sosunov, A.A.; McKhann, G.M., 2nd; Przedborski, S. Spinal cord endoplasmic reticulum stress associated with a microsomal accumulation of mutant superoxide dismutase-1 in an ALS model. Proc. Natl. Acad. Sci. USA **2006**, 103, 6025–6030. [Google Scholar]
72. Walker, A.K.; Farg, M.A.; Bye, C.R.; McLean, C.A.; Horne, M.K.; Atkin, J.D. Protein disulphide isomerase protects against protein aggregation and is S-nitrosylated in amyotrophic lateral sclerosis. Brain **2010**, 133, 105–116. [Google Scholar]
73. Vale, R.D. The molecular motor toolbox for intracellular transport. Cell **2003**, 112, 467–480. [Google Scholar]

74. Hirokawa, N.; Niwa, S.; Tanaka, Y. Molecular motors in neurons: Transport mechanisms and roles in brain function, development, and disease. Neuron **2010**, 68, 610–638. [Google Scholar]
75. Nagle, B.W.; Okamoto, C.; Taggart, B.; Burnside, B. The teleost cone cytoskeleton. Localization of actin, microtubules, and intermediate filaments. Invest. Ophthalmol. Vis. Sci **1986**, 27, 689–701. [Google Scholar]
76. Bridgman, P.C.; Dailey, M.E. The organization of myosin and actin in rapid frozen nerve growth cones. J. Cell Biol **1989**, 108, 95–109. [Google Scholar]
77. Smith, D.A.; Simmons, R.M. Models of motor-assisted transport of intracellular particles. Biophys. J**2001**, 80, 45–68. [Google Scholar]
78. De Vos, K.J.; Grierson, A.J.; Ackerley, S.; Miller, C.C. Role of axonal transport in neurodegenerative diseases. Annu. Rev. Neurosci **2008**, 31, 151–173. [Google Scholar]
79. Wang, L.; Ho, C.L.; Sun, D.; Liem, R.K.; Brown, A. Rapid movement of axonal neurofilaments interrupted by prolonged pauses. Nat. Cell Biol **2000**, 2, 137–141. [Google Scholar]
80. Hirokawa, N.; Noda, Y.; Tanaka, Y.; Niwa, S. Kinesin superfamily motor proteins and intracellular transport. Nat. Rev. Mol. Cell Biol **2009**, 10, 682–696. [Google Scholar]
81. Hirokawa, N.; Noda, Y. Intracellular transport and kinesin superfamily proteins, KIFs: structure, function, and dynamics. Physiol. Rev **2008**, 88, 1089–1118. [Google Scholar]
82. Hirokawa, N.; Noda, Y.; Okada, Y. Kinesin and dynein superfamily proteins in organelle transport and cell division. Curr. Opin. Cell Biol **1998**, 10, 60–73. [Google Scholar]
83. Hirokawa, N. Organelle transport along microtubules - the role of KIFs. Trends Cell Biol **1996**, 6, 135–141. [Google Scholar]
84. Hirokawa, N. Kinesin and dynein superfamily proteins and the mechanism of organelle transport. Science **1998**, 279, 519–526. [Google Scholar]
85. Dagenbach, E.M.; Endow, S.A. A new kinesin tree. J. Cell Sci **2004**, 117, 3–7. [Google Scholar]

86. Miki, H.; Okada, Y.; Hirokawa, N. Analysis of the kinesin superfamily: insights into structure and function. Trends Cell Biol **2005**, 15, 467–476. [Google Scholar]

87. Lawrence, C.J.; Dawe, R.K.; Christie, K.R.; Cleveland, D.W.; Dawson, S.C.; Endow, S.A.; Goldstein, L.S.; Goodson, H.V.; Hirokawa, N.; Howard, J.; et al. A standardized kinesin nomenclature. J. Cell Biol **2004**, 167, 19–22. [Google Scholar]

88. Lawrence, C.J.; Malmberg, R.L.; Muszynski, M.G.; Dawe, R.K. Maximum likelihood methods reveal conservation of function among closely related kinesin families. J. Mol. Evol **2002**, 54, 42–53. [Google Scholar]

89. Miki, H.; Setou, M.; Kaneshiro, K.; Hirokawa, N. All kinesin superfamily protein, KIF, genes in mouse and human. Proc. Natl. Acad. Sci. USA **2001**, 98, 7004–7011. [Google Scholar]

90. Brady, S.T. A novel brain ATPase with properties expected for the fast axonal transport motor. Nature **1985**, 317, 73–75. [Google Scholar]

91. Vale, R.D.; Schnapp, B.J.; Reese, T.S.; Sheetz, M.P. Movement of organelles along filaments dissociated from the axoplasm of the squid giant axon. Cell **1985**, 40, 449–454. [Google Scholar]

92. Paschal, B.M.; King, S.M.; Moss, A.G.; Collins, C.A.; Vallee, R.B.; Witman, G.B. Isolated flagellar outer arm dynein translocates brain microtubules in vitro. Nature **1987**, 330, 672–674. [Google Scholar]

93. Neuwald, A.F.; Aravind, L.; Spouge, J.L.; Koonin, E.V. AAA+: A class of chaperone-like ATPases associated with the assembly, operation, and disassembly of protein complexes. Genome Res **1999**, 9, 27–43. [Google Scholar]

94. Schroer, T.A. Dynactin. Annu. Rev. Cell Dev. Biol **2004**, 20, 759–779. [Google Scholar]

95. Gennerich, A.; Carter, A.P.; Reck-Peterson, S.L.; Vale, R.D. Force-induced bidirectional stepping of cytoplasmic dynein. Cell **2007**, 131, 952–965. [Google Scholar]

96. Karki, S.; Holzbaur, E.L. Cytoplasmic dynein and dynactin in cell division and intracellular transport. Curr. Opin. Cell Biol **1999**, 11, 45–53. [Google Scholar]

97. Naisbitt, S.; Valtschanoff, J.; Allison, D.W.; Sala, C.; Kim, E.; Craig, A.M.; Weinberg, R.J.; Sheng, M. Interaction of the postsynaptic density-95/guanylate kinase domain-associated protein complex with a light chain of myosin-V and dynein. J. Neurosci **2000**, 20, 4524–4534. [Google Scholar]
98. Lee, K.H.; Lee, S.; Kim, B.; Chang, S.; Kim, S.W.; Paick, J.S.; Rhee, K. Dazl can bind to dynein motor complex and may play a role in transport of specific mRNAs. EMBO J **2006**, 25, 4263–4270. [Google Scholar]
99. Driskell, O.J.; Mironov, A.; Allan, V.J.; Woodman, P.G. Dynein is required for receptor sorting and the morphogenesis of early endosomes. Nat. Cell Biol **2007**, 9, 113–120. [Google Scholar]
100. Waterman-Storer, C.M.; Karki, S.B.; Kuznetsov, S.A.; Tabb, J.S.; Weiss, D.G.; Langford, G.M.; Holzbaur, E.L. The interaction between cytoplasmic dynein and dynactin is required for fast axonal transport. Proc. Natl. Acad. Sci. USA **1997**, 94, 12180–12185. [Google Scholar]
101. Ligon, L.A.; Tokito, M.; Finklestein, J.M.; Grossman, F.E.; Holzbaur, E.L. A direct interaction between cytoplasmic dynein and kinesin I may coordinate motor activity. J. Biol. Chem **2004**, 279, 19201–19208. [Google Scholar]
102. Deacon, S.W.; Serpinskaya, A.S.; Vaughan, P.S.; Lopez Fanarraga, M.; Vernos, I.; Vaughan, K.T.; Gelfand, V.I. Dynactin is required for bidirectional organelle transport. J. Cell Biol **2003**, 160, 297–301. [Google Scholar]
103. Foth, B.J.; Goedecke, M.C.; Soldati, D. New insights into myosin evolution and classification. Proc. Natl. Acad. Sci. USA **2006**, 103, 3681–3686. [Google Scholar]
104. Morfini, G.A.; Burns, M.; Binder, L.I.; Kanaan, N.M.; LaPointe, N.; Bosco, D.A.; Brown, R.H., Jr; Brown, H.; Tiwari, A.; Hayward, L.; et al. Axonal transport defects in neurodegenerative diseases. J. Neurosci **2009**, 29, 12776–12786. [Google Scholar]
105. Zhang, B.; Tu, P.; Abtahian, F.; Trojanowski, J.Q.; Lee, V.M. Neurofilaments and orthograde transport are reduced in ventral root axons of transgenic mice that express human SOD1 with a G93A mutation. J. Cell Biol **1997**, 139, 1307–1315. [Google Scholar]

106. Hirano, A. Neuropathology of ALS: An overview. Neurology **1996**, 47, S63–S66. [Google Scholar]

107. Collard, J.F.; Cote, F.; Julien, J.P. Defective axonal transport in a transgenic mouse model of amyotrophic lateral sclerosis. Nature **1995**, 375, 61–64. [Google Scholar]

108. Hirano, A.; Nakano, I.; Kurland, L.T.; Mulder, D.W.; Holley, P.W.; Saccomanno, G. Fine structural study of neurofibrillary changes in a family with amyotrophic lateral sclerosis. J. Neuropathol. Exp. Neurol **1984**, 43, 471–480. [Google Scholar]

109. Rouleau, G.A.; Clark, A.W.; Rooke, K.; Pramatarova, A.; Krizus, A.; Suchowersky, O.; Julien, J.P.; Figlewicz, D. SOD1 mutation is associated with accumulation of neurofilaments in amyotrophic lateral sclerosis. Ann. Neurol **1996**, 39, 128–131. [Google Scholar]

110. Munch, C.; Sedlmeier, R.; Meyer, T.; Homberg, V.; Sperfeld, A.D.; Kurt, A.; Prudlo, J.; Peraus, G.; Hanemann, C.O.; Stumm, G.; et al. Point mutations of the p150 subunit of dynactin (DCTN1) gene in ALS. Neurology **2004**, 63, 724–726. [Google Scholar]

111. Chen, X.J.; Levedakou, E.N.; Millen, K.J.; Wollmann, R.L.; Soliven, B.; Popko, B. Proprioceptive sensory neuropathy in mice with a mutation in the cytoplasmic Dynein heavy chain 1 gene. J. Neurosci **2007**, 27, 14515–14524. [Google Scholar]

112. Perlson, E.; Jeong, G.B.; Ross, J.L.; Dixit, R.; Wallace, K.E.; Kalb, R.G.; Holzbaur, E.L. A switch in retrograde signaling from survival to stress in rapid-onset neurodegeneration. J. Neurosci **2009**, 29, 9903–9917. [Google Scholar]

113. De Vos, K.J.; Chapman, A.L.; Tennant, M.E.; Manser, C.; Tudor, E.L.; Lau, K.F.; Brownlees, J.; Ackerley, S.; Shaw, P.J.; McLoughlin, D.M.; et al. Familial amyotrophic lateral sclerosis-linked SOD1 mutants perturb fast axonal transport to reduce axonal mitochondria content. Hum. Mol. Genet **2007**, 16, 2720–2728. [Google Scholar]

114. Vande Velde, C.; McDonald, K.K.; Boukhedimi, Y.; McAlonis-Downes, M.; Lobsiger, C.S.; Bel Hadj, S.; Zandona, A.; Julien, J.P.; Shah, S.B.; Cleveland, D.W. Misfolded SOD1 associated with motor neuron mitochondria alters mitochondrial shape and distribution prior to clinical onset. PLoS One**2011**, 6, e22031. [Google Scholar]

115. Xia, C.H.; Roberts, E.A.; Her, L.S.; Liu, X.; Williams, D.S.;

Cleveland, D.W.; Goldstein, L.S. Abnormal neurofilament transport caused by targeted disruption of neuronal kinesin heavy chain KIF5A. J. Cell Biol **2003**, 161, 55–66. [Google Scholar]

116. Zhao, C.; Takita, J.; Tanaka, Y.; Setou, M.; Nakagawa, T.; Takeda, S.; Yang, H.W.; Terada, S.; Nakata, T.; Takei, Y.; et al. Charcot-Marie-Tooth disease type 2A caused by mutation in a microtubule motor KIF1Bbeta. Cell **2001**, 105, 587–597. [Google Scholar]
117. Reid, E.; Kloos, M.; Ashley-Koch, A.; Hughes, L.; Bevan, S.; Svenson, I.K.; Graham, F.L.; Gaskell, P.C.; Dearlove, A.; Pericak-Vance, M.A.; et al. A kinesin heavy chain (KIF5A) mutation in hereditary spastic paraplegia (SPG10). Am. J. Hum. Genet **2002**, 71, 1189–1194. [Google Scholar]
118. Landers, J.E.; Melki, J.; Meininger, V.; Glass, J.D.; van den Berg, L.H.; van Es, M.A.; Sapp, P.C.; van Vught, P.W.; McKenna-Yasek, D.M.; Blauw, H.M.; et al. Reduced expression of the Kinesin-Associated Protein 3 (KIFAP3) gene increases survival in sporadic amyotrophic lateral sclerosis. Proc. Natl. Acad. Sci. USA **2009**, 106, 9004–9009. [Google Scholar]
119. Tateno, M.; Kato, S.; Sakurai, T.; Nukina, N.; Takahashi, R.; Araki, T. Mutant SOD1 impairs axonal transport of choline acetyltransferase and acetylcholine release by sequestering KAP3. Hum. Mol. Genet **2009**, 18, 942–955. [Google Scholar]
120. Orsetti, V.; Pegoraro, E.; Cima, V.; D'Ascenzo, C.; Palmieri, A.; Querin, G.; Volpe, M.; Ermani, M.; Angelini, C.; Soraru, G. Genetic variation in KIFAP3 is associated with an upper motor neuronpredominant phenotype in amyotrophic lateral sclerosis. Neurodegener. Dis **2011**, 8, 491–495. [Google Scholar]
121. Manning, B.D.; Snyder, M. Drivers and passengers wanted! The role of kinesin-associated proteins. Trends Cell Biol **2000**, 10, 281–289. [Google Scholar]
122. Pantelidou, M.; Zographos, S.E.; Lederer, C.W.; Kyriakides, T.; Pfaffl, M.W.; Santama, N. Differential expression of molecular motors in the motor cortex of sporadic ALS. Neurobiol. Dis**2007**, 26, 577–589. [Google Scholar]
123. Shi, P.; Strom, A.L.; Gal, J.; Zhu, H. Effects of ALS-related SOD1 mutants on dynein- and KIF5-mediated retrograde and anterograde

axonal transport. Biochim. Biophys. Acta **2010**, 1802, 707–716. [Google Scholar]

124. Melloni, R.H., Jr; Tokito, M.K.; Holzbaur, E.L. Expression of the p150Glued component of the dynactin complex in developing and adult rat brain. J. Comp. Neurol. **1995**, 357, 15–24. [Google Scholar]
125. Zhang, F.J.; Strom, A.L.; Fukada, K.; Lee, S.; Hayward, L.J.; Zhu, H.N. Interaction between familial amyotrophic lateral sclerosis (ALS)-linked SOD1 mutants and the dynein complex. J. Biol. Chem**2007**, 282, 16691–16699. [Google Scholar]
126. Munch, C.; Prechter, F.; Xu, R.; Linke, P.; Prudlo, J.; Kuzma, M.; Kwiecinski, H.; Ludolph, A.C.; Meyer, T. Frequency of a tau genotype in amyotrophic lateral sclerosis. J. Neurol. Sci **2005**, 236, 13–16. [Google Scholar]
127. Lai, C.; Lin, X.; Chandran, J.; Shim, H.; Yang, W.J.; Cai, H. The G59S mutation in p150(glued) causes dysfunction of dynactin in mice. J. Neurosci **2007**, 27, 13982–13990. [Google Scholar]
128. Laird, F.M.; Farah, M.H.; Ackerley, S.; Hoke, A.; Maragakis, N.; Rothstein, J.D.; Griffin, J.; Price, D.L.; Martin, L.J.; Wong, P.C. Motor neuron disease occurring in a mutant dynactin mouse model is characterized by defects in vesicular trafficking. J. Neurosci **2008**, 28, 1997–2005. [Google Scholar]
129. LaMonte, B.H.; Wallace, K.E.; Holloway, B.A.; Shelly, S.S.; Ascano, J.; Tokito, M.; Van Winkle, T.; Howland, D.S.; Holzbaur, E.L. Disruption of dynein/dynactin inhibits axonal transport in motor neurons causing late-onset progressive degeneration. Neuron **2002**, 34, 715–727. [Google Scholar]
130. Hafezparast, M.; Klocke, R.; Ruhrberg, C.; Marquardt, A.; Ahmad-Annuar, A.; Bowen, S.; Lalli, G.; Witherden, A.S.; Hummerich, H.; Nicholson, S.; et al. Mutations in dynein link motor neuron degeneration to defects in retrograde transport. Science **2003**, 300, 808–812. [Google Scholar]
131. Chen, X.J.; Levedakou, E.N.; Millen, K.J.; Wollmann, R.L.; Soliven, B.; Popko, B. Proprioceptive sensory neuropathy in mice with a mutation in the cytoplasmic Dynein heavy chain 1 gene. J. Neurosci **2007**, 27, 14515–14524. [Google Scholar]
132. Kieran, D.; Hafezparast, M.; Bohnert, S.; Dick, J.R.; Martin, J.;

Schiavo, G.; Fisher, E.M.; Greensmith, L. A mutation in dynein rescues axonal transport defects and extends the life span of ALS mice. J. Cell Biol **2005**, 169, 561–567. [Google Scholar]

133. Teuchert, M.; Fischer, D.; Schwalenstoecker, B.; Habisch, H.J.; Bockers, T.M.; Ludolph, A.C. A dynein mutation attenuates motor neuron degeneration in SOD1(G93A) mice. Exp. Neurol **2006**, 198, 271–274. [Google Scholar]
134. Perlson, E.; Jeong, G.B.; Ross, J.L.; Dixit, R.; Wallace, K.E.; Kalb, R.G.; Holzbaur, E.L. A switch in retrograde signaling from survival to stress in rapid-onset neurodegeneration. J. Neurosci **2009**, 29, 9903–9917. [Google Scholar]
135. Perlson, E.; Hanz, S.; Ben-Yaakov, K.; Segal-Ruder, Y.; Seger, R.; Fainzilber, M. Vimentindependent spatial translocation of an activated MAP kinase in injured nerve. Neuron **2005**, 45, 715–726. [Google Scholar]
136. Cavalli, V.; Kujala, P.; Klumperman, J.; Goldstein, L.S. Sunday Driver links axonal transport to damage signaling. J. Cell Biol **2005**, 168, 775–787. [Google Scholar]
137. Heerssen, H.M.; Pazyra, M.F.; Segal, R.A. Dynein motors transport activated Trks to promote survival of target-dependent neurons. Nat. Neurosci **2004**, 7, 596–604. [Google Scholar]
138. Yano, H.; Lee, F.S.; Kong, H.; Chuang, J.; Arevalo, J.; Perez, P.; Sung, C.; Chao, M.V. Association of Trk neurotrophin receptors with components of the cytoplasmic dynein motor. J. Neurosci **2001**, 21, RC125. [Google Scholar]
139. Komatsu, M.; Waguri, S.; Koike, M.; Sou, Y.S.; Ueno, T.; Hara, T.; Mizushima, N.; Iwata, J.; Ezaki, J.; Murata, S.; et al. Homeostatic levels of p62 control cytoplasmic inclusion body formation in autophagy-deficient mice. Cell **2007**, 131, 1149–1163. [Google Scholar]
140. Strom, A.L.; Shi, P.; Zhang, F.; Gal, J.; Kilty, R.; Hayward, L.J.; Zhu, H. Interaction of amyotrophic lateral sclerosis (ALS)-related mutant copper-zinc superoxide dismutase with the dynein-dynactin complex contributes to inclusion formation. J. Biol. Chem **2008**, 283, 22795–227805. [Google Scholar]
141. Gonatas, N.K.; Stieber, A.; Gonatas, J.O. Fragmentation of the

Golgi apparatus in neurodegenerative diseases and cell death. J. Neurol. Sci **2006**, 246, 21–30. [Google Scholar]

142. Turner, B.J.; Atkin, J.D. ER stress and UPR in familial amyotrophic lateral sclerosis. Curr. Mol. Med**2006**, 6, 79–86. [Google Scholar]
143. Mourelatos, Z.; Gonatas, N.K.; Stieber, A.; Gurney, M.E.; Dal Canto, M.C. The Golgi apparatus of spinal cord motor neurons in transgenic mice expressing mutant Cu,Zn superoxide dismutase becomes fragmented in early, preclinical stages of the disease. Proc. Natl. Acad. Sci. USA **1996**, 93, 5472–5477. [Google Scholar]
144. Schroder, M. Endoplasmic reticulum stress responses. Cell. Mol. Life Sci **2008**, 65, 862–894. [Google Scholar]
145. Routledge, K.E.; Gupta, V.; Balch, W.E. Emergent properties of proteostasis-COPII coupled systems in human health and disease. Mol. Membr. Biol **2010**, 27, 385–397. [Google Scholar]
146. Ramirez, O.A.; Couve, A. The endoplasmic reticulum and protein trafficking in dendrites and axons. Trends Cell Biol **2011**, 21, 219–227. [Google Scholar]
147. Szegezdi, E.; Logue, S.E.; Gorman, A.M.; Samali, A. Mediators of endoplasmic reticulum stress-induced apoptosis. EMBO Rep **2006**, 7, 880–885. [Google Scholar]
148. Rutkowski, D.T.; Wu, J.; Back, S.H.; Callaghan, M.U.; Ferris, S.P.; Iqbal, J.; Clark, R.; Miao, H.; Hassler, J.R.; Fornek, J.; et al. UPR pathways combine to prevent hepatic steatosis caused by ER stress-mediated suppression of transcriptional master regulators. Dev. Cell **2008**, 15, 829–840. [Google Scholar]
149. Nishitoh, H.; Kadowaki, H.; Nagai, A.; Maruyama, T.; Yokota, T.; Fukutomi, H.; Noguchi, T.; Matsuzawa, A.; Takeda, K.; Ichijo, H. ALS-linked mutant SOD1 induces ER stress- and ASK1-dependent motor neuron death by targeting Derlin-1. Genes Dev **2008**, 22, 1451–1464. [Google Scholar]
150. Short, D.M.; Heron, I.D.; Birse-Archbold, J.L.; Kerr, L.E.; Sharkey, J.; McCulloch, J. Apoptosis induced by staurosporine alters chaperone and endoplasmic reticulum proteins: Identification by quantitative proteomics. Proteomics **2007**, 7, 3085–3096. [Google Scholar]
151. Preston, A.M.; Gurisik, E.; Bartley, C.; Laybutt, D.R.; Biden, T.J.

Reduced endoplasmic reticulum (ER)-to-Golgi protein trafficking contributes to ER stress in lipotoxic mouse beta cells by promoting protein overload. Diabetologia **2009**, 52, 2369–2373. [Google Scholar]

152. Jonikas, M.C.; Collins, S.R.; Denic, V.; Oh, E.; Quan, E.M.; Schmid, V.; Weibezahn, J.; Schwappach, B.; Walter, P.; Weissman, J.S.; Schuldiner, M. Comprehensive characterization of genes required for protein folding in the endoplasmic reticulum. Science **2009**, 323, 1693–1697. [Google Scholar]
153. Sasaki, S. Endoplasmic reticulum stress in motor neurons of the spinal cord in sporadic amyotrophic lateral sclerosis. J. Neuropathol. Exp. Neurol **2010**, 69, 346–355. [Google Scholar]
154. Sato, T.; Takeuchi, S.; Saito, A.; Ding, W.; Bamba, H.; Matsuura, H.; Hisa, Y.; Tooyama, I.; Urushitani, M. Axonal ligation induces transient redistribution of TDP-43 in brainstem motor neurons. Neuroscience **2009**, 164, 1565–1578. [Google Scholar]
155. Suzuki, H.; Lee, K.; Matsuoka, M. TDP-43-induced death is associated with altered regulation of BIM and Bcl-xL and attenuated by caspase-mediated TDP-43 cleavage. J. Biol. Chem **2011**, 286, 13171–13183. [Google Scholar]
156. Huang, E.J.; Zhang, J.; Geser, F.; Trojanowski, J.Q.; Strober, J.B.; Dickson, D.W.; Brown, R.H., Jr; Shapiro, B.E.; Lomen-Hoerth, C. Extensive FUS-immunoreactive pathology in juvenile amyotrophic lateral sclerosis with basophilic inclusions. Brain Pathol. **2010**, 20, 1069–1076. [Google Scholar]
157. Tateishi, T.; Hokonohara, T.; Yamasaki, R.; Miura, S.; Kikuchi, H.; Iwaki, A.; Tashiro, H.; Furuya, H.; Nagara, Y.; Ohyagi, Y.; et al. Multiple system degeneration with basophilic inclusions in Japanese ALS patients with FUS mutation. Acta Neuropathol **2010**, 119, 355–364. [Google Scholar]
158. Wishart, M.J.; Dixon, J.E. PTEN and myotubularin phosphatases: from 3-phosphoinositide dephosphorylation to disease. Trends Cell Biol **2002**, 12, 579–585. [Google Scholar]
159. Orth, J.D.; McNiven, M.A. Dynamin at the actin-membrane interface. Curr. Opin. Cell Biol **2003**, 15, 31–39. [Google Scholar]
160. McNiven, M.A. Dynamin in disease. Nat. Genet **2005**, 37, 215–216.

[Google Scholar]

161. Yang, Y.; Hentati, A.; Deng, H.X.; Dabbagh, O.; Sasaki, T.; Hirano, M.; Hung, W.Y.; Ouahchi, K.; Yan, J.; Azim, A.C.; et al. The gene encoding alsin, a protein with three guanine-nucleotide exchange factor domains, is mutated in a form of recessive amyotrophic lateral sclerosis. Nat. Genet **2001**, 29, 160–165. [Google Scholar]
162. Otomo, A.; Hadano, S.; Okada, T.; Mizumura, H.; Kunita, R.; Nishijima, H.; Showguchi-Miyata, J.; Yanagisawa, Y.; Kohiki, E.; Suga, E.; et al. ALS2, a novel guanine nucleotide exchange factor for the small GTPase Rab5, is implicated in endosomal dynamics. Hum. Mol. Genet **2003**, 12, 1671–1687. [Google Scholar]
163. Schaefer, M.K.; Schmalbruch, H.; Buhler, E.; Lopez, C.; Martin, N.; Guenet, J.L.; Haase, G. Progressive motor neuronopathy: A critical role of the tubulin chaperone TBCE in axonal tubulin routing from the Golgi apparatus. J. Neurosci **2007**, 27, 8779–8789. [Google Scholar]
164. Schmitt-John, T.; Drepper, C.; Mussmann, A.; Hahn, P.; Kuhlmann, M.; Thiel, C.; Hafner, M.; Lengeling, A.; Heimann, P.; Jones, J.M.; et al. Mutation of Vps54 causes motor neuron disease and defective spermiogenesis in the wobbler mouse. Nat. Genet **2005**, 37, 1213–1215. [Google Scholar]
165. Teuling, E.; Ahmed, S.; Haasdijk, E.; Demmers, J.; Steinmetz, M.O.; Akhmanova, A.; Jaarsma, D.; Hoogenraad, C.C. Motor neuron disease-associated mutant vesicle-associated membrane protein-associated protein (VAP) B recruits wild-type VAPs into endoplasmic reticulum-derived tubular aggregates. J. Neurosci **2007**, 27, 9801–9815. [Google Scholar]
166. Skehel, P.A.; Fabian-Fine, R.; Kandel, E.R. Mouse VAP33 is associated with the endoplasmic reticulum and microtubules. Proc. Natl. Acad. Sci. USA **2000**, 97, 1101–1106. [Google Scholar]
167. Peretti, D.; Dahan, N.; Shimoni, E.; Hirschberg, K.; Lev, S. Coordinated lipid transfer between the endoplasmic reticulum and the Golgi complex requires the VAP proteins and is essential for Golgi-mediated transport. Mol. Biol. Cell **2008**, 19, 3871–3884. [Google Scholar]
168. Fasana, E.; Fossati, M.; Ruggiano, A.; Brambillasca, S.; Hoogenraad,

C.C.; Navone, F.; Francolini, M.; Borgese, N. A VAPB mutant linked to amyotrophic lateral sclerosis generates a novel form of organized smooth endoplasmic reticulum. FASEB J **2010**, 24, 1419–1430. [Google Scholar]

169. Foster, L.J.; Weir, M.L.; Lim, D.Y.; Liu, Z.; Trimble, W.S.; Klip, A. A functional role for VAP-33 in insulin-stimulated GLUT4 traffic. Traffic **2000**, 1, 512–521. [Google Scholar]
170. Kanekura, K.; Nishimoto, I.; Aiso, S.; Matsuoka, M. Characterization of amyotrophic lateral sclerosis-linked P56S mutation of vesicle-associated membrane protein-associated protein B (VAPB/ALS8). J. Biol. Chem **2006**, 281, 30223–30233. [Google Scholar]
171. Nishimura, A.L.; Mitne-Neto, M.; Silva, H.C.; Richieri-Costa, A.; Middleton, S.; Cascio, D.; Kok, F.; Oliveira, J.R.; Gillingwater, T.; Webb, J.; et al. A mutation in the vesicle-trafficking protein VAPB causes late-onset spinal muscular atrophy and amyotrophic lateral sclerosis. Am. J. Hum. Genet**2004**, 75, 822–831. [Google Scholar]
172. Tsuda, H.; Han, S.M.; Yang, Y.; Tong, C.; Lin, Y.Q.; Mohan, K.; Haueter, C.; Zoghbi, A.; Harati, Y.; Kwan, J.; et al. The amyotrophic lateral sclerosis 8 protein VAPB is cleaved, secreted, and acts as a ligand for Eph receptors. Cell **2008**, 133, 963–977. [Google Scholar]
173. Chen, H.J.; Anagnostou, G.; Chai, A.; Withers, J.; Morris, A.; Adhikaree, J.; Pennetta, G.; de Belleroche, J.S. Characterization of the properties of a novel mutation in VAPB in familial amyotrophic lateral sclerosis. J. Biol. Chem **2010**, 285, 40266–40281. [Google Scholar]
174. Suzuki, H.; Kanekura, K.; Levine, T.P.; Kohno, K.; Olkkonen, V.M.; Aiso, S.; Matsuoka, M. ALS-linked P56S-VAPB, an aggregated loss-of-function mutant of VAPB, predisposes motor neurons to ER stress-related death by inducing aggregation of co-expressed wild-type VAPB. J. Neurochem**2009**, 108, 973–985. [Google Scholar]
175. Flanagan, J.G.; Vanderhaeghen, P. The ephrins and Eph receptors in neural development. Annu. Rev. Neurosci **1998**, 21, 309–345. [Google Scholar]
176. Chalasani, M.L.; Swarup, G.; Balasubramanian, D. Optineurin and its mutants: molecules associated with some forms of glaucoma. Ophthalmic Res **2009**, 42, 176–184. [Google Scholar]

177. Ray, K.; Mookherjee, S. Molecular complexity of primary open angle glaucoma: Current concepts. J. Genet **2009**, 88, 451–467. [Google Scholar]

178. Chibalina, M.V.; Roberts, R.C.; Arden, S.D.; Kendrick-Jones, J.; Buss, F. Rab8-optineurin-myosin VI: Analysis of interactions and functions in the secretory pathway. Methods Enzymol **2008**, 438, 11–24. [Google Scholar]

179. Sahlender, D.A.; Roberts, R.C.; Arden, S.D.; Spudich, G.; Taylor, M.J.; Luzio, J.P.; Kendrick-Jones, J.; Buss, F. Optineurin links myosin VI to the Golgi complex and is involved in Golgi organization and exocytosis. J. Cell Biol **2005**, 169, 285–295. [Google Scholar]

180. Nagabhushana, A.; Chalasani, M.L.; Jain, N.; Radha, V.; Rangaraj, N.; Balasubramanian, D.; Swarup, G. Regulation of endocytic trafficking of transferrin receptor by optineurin and its impairment by a glaucoma-associated mutant. BMC Cell Biol **2010**, 11, 4. [Google Scholar]

181. Fujita, Y.; Okamoto, K.; Sakurai, A.; Gonatas, N.K.; Hirano, A. Fragmentation of the Golgi apparatus of the anterior horn cells in patients with familial amyotrophic lateral sclerosis with SOD1 mutations and posterior column involvement. J. Neurol. Sci. **2000**, 174, 137–140. [Google Scholar]

182. Hammerschlag, R.; Stone, G.C.; Bolen, F.A.; Lindsey, J.D.; Ellisman, M.H. Evidence that all newly synthesized proteins destined for fast axonal transport pass through the Golgi apparatus. J. Cell Biol **1982**, 93, 568–575. [Google Scholar]

183. Woodman, P.G. p97, a protein coping with multiple identities. J. Cell Sci **2003**, 116, 4283–4290. [Google Scholar]

184. Dreveny, I.; Pye, V.E.; Beuron, F.; Briggs, L.C.; Isaacson, R.L.; Matthews, S.J.; McKeown, C.; Yuan, X.; Zhang, X.; Freemont, P.S. p97 and close encounters of every kind: A brief review. Biochem. Soc. Trans **2004**, 32, 715–720. [Google Scholar]

185. Halawani, D.; Latterich, M. p97: The cell's molecular purgatory? Mol. Cell **2006**, 22, 713–717. [Google Scholar]

186. Patel, S.; Latterich, M. The AAA team: Related ATPases with diverse functions. Trends Cell Biol**1998**, 8, 65–71. [Google Scholar]

187. Wojcik, C.; Yano, M.; DeMartino, G.N. RNA interference of valosin-containing protein (VCP/p97) reveals multiple cellular roles linked to ubiquitin/proteasome-dependent proteolysis. J. Cell Sci**2004**, 117, 281–292. [Google Scholar]

188. Nowis, D.; McConnell, E.; Wojcik, C. Destabilization of the VCP-Ufd1-Npl4 complex is associated with decreased levels of ERAD substrates. Exp. Cell Res **2006**, 312, 2921–2932. [Google Scholar]

189. Gitcho, M.A.; Strider, J.; Carter, D.; Taylor-Reinwald, L.; Forman, M.S.; Goate, A.M.; Cairns, N.J. VCP mutations causing frontotemporal lobar degeneration disrupt localization of TDP-43 and induce cell death. J. Biol. Chem **2009**, 284, 12384–12398. [Google Scholar]

190. Rabinovich, E.; Kerem, A.; Frohlich, K.U.; Diamant, N.; Bar-Nun, S. AAA-ATPase p97/Cdc48p, a cytosolic chaperone required for endoplasmic reticulum-associated protein degradation. Mol. Cell. Biol **2002**, 22, 626–634. [Google Scholar]

191. Ye, Y.; Shibata, Y.; Yun, C.; Ron, D.; Rapoport, T.A. A membrane protein complex mediates retro-translocation from the ER lumen into the cytosol. Nature **2004**, 429, 841–847. [Google Scholar]

192. Nishitoh, H.; Kadowaki, H.; Nagai, A.; Maruyama, T.; Yokota, T.; Fukutomi, H.; Noguchi, T.; Matsuzawa, A.; Takeda, K.; Ichijo, H. ALS-linked mutant SOD1 induces ER stress- and ASK1-dependent motor neuron death by targeting Derlin-1. Genes Dev **2008**, 22, 1451–1464. [Google Scholar]

193. Moisse, K.; Mepham, J.; Volkening, K.; Welch, I.; Hill, T.; Strong, M.J. Cytosolic TDP-43 expression following axotomy is associated with caspase 3 activation in NFL−/− mice: support for a role for TDP-43 in the physiological response to neuronal injury. Brain Res **2009**, 1296, 176–186. [Google Scholar]

194. Shan, X.; Vocadlo, D.; Krieger, C. Mislocalization of TDP-43 in the G93A mutant SOD1 transgenic mouse model of ALS. Neurosci. Lett **2009**, 458, 70–74. [Google Scholar]

195. Ju, S.; Tardiff, D.F.; Han, H.; Divya, K.; Zhong, Q.; Maquat, L.E.; Bosco, D.A.; Hayward, L.J.; Brown, R.H., Jr; Lindquist, S.; et al. A yeast model of FUS/TLS-dependent cytotoxicity. PLoS Biol**2011**, 9, e1001052. [Google Scholar]

196. Fujii, R.; Takumi, T. TLS facilitates transport of mRNA encoding an actin-stabilizing protein to dendritic spines. J. Cell Sci **2005**, 118, 5755–5765. [Google Scholar]

CHAPTER

4

HOW MOLECULAR MOTORS ARE ARRANGED ON A CARGO IS IMPORTANT FOR VESICULAR TRANSPORT

Robert P. Erickson[1], Zhiyuan Jia[1], Steven P. Gross[1,2], Clare C. Yu[1]

[1] Department of Physics and Astronomy, University of California, Irvine, Irvine, California, United States of America,

[2] Department of Developmental and Cell Biology, University of California, Irvine, Irvine, California, United States of America

ABSTRACT

The spatial organization of the cell depends upon intracellular trafficking of cargos hauled along microtubules and actin filaments by the molecular motor proteins kinesin, dynein, and myosin. Although much is known about how single motors function, there is significant evidence that cargos in vivo are carried by multiple motors. While some aspects of multiple motor function have received attention, how the cargo itself —and motor organization on the cargo—affects transport has not been considered. To address this, we have developed a three-dimensional Monte Carlo simulation of motors transporting a spherical cargo, subject to thermal fluctuations that produce both rotational and translational diffusion. We found that these fluctuations could exert a load on the motor(s), significantly decreasing the mean travel distance and velocity of large cargos, especially at large viscosities. In addition, the presence of the cargo could dramatically help the motor to bind productively to the microtubule: the relatively slow translational and rotational diffusion of moderately sized cargos gave the motors ample opportunity to bind to a microtubule before the motor/cargo ensemble diffuses out of range of that microtubule. For rapidly diffusing cargos, the probability of their binding to a microtubule was high if there were nearby microtubules that they could easily reach by translational diffusion. Our simulations found that one reason why motors may be approximately 100 nm long is to improve their 'on' rates when attached to comparably sized cargos. Finally, our results suggested that to efficiently regulate the number of active motors, motors should be clustered together rather than spread randomly over the surface of the cargo. While our simulation uses the specific parameters for kinesin, these effects result from generic properties of the motors, cargos, and filaments, so they should apply to other motors as well.

AUTHOR SUMMARY

The spatial organization of living cells depends upon a transportation system consisting of molecular motor proteins that act like porters carrying cargos along filaments that are analogous to roads. The breakdown of this transportation system has been associated with neurodegenerative diseases such as Alzheimer's and Huntington's disease. In living cells, cargos are typically carried by multiple motors. While some aspects of multiple motor function have received attention, how the cargo itself affects transport has not been considered. To address this, we developed a three-dimensional computer simulation of motors transporting a spherical cargo subject to fluctuations produced when small molecules in the intracellular environment buffet the cargo. These fluctuations can cause the cargo to pull on the motors, slowing them down and making them detach from the filament (road). This effect increases as the cargo size and viscosity of the medium increase. We also found that the presence of the cargo helped the motors to bind to a filament before it drifted away.

If other filaments were present, then the cargo could bind to one of them. Our results also indicated that it is better to group the motors on the cargo rather than spread them randomly over the surface.

INTRODUCTION

Cells are highly organized, and much of this organization results from motors that move cargos along microtubules. The single-molecule properties of molecular motors are relatively well understood both experimentally and theoretically. With this as a starting point, we investigated how the presence of the cargo itself alters transport. Aside from exerting viscous drag, the cargo could in principle alter single-motor based transport both by changing the motors' diffusion and ability to contact the filament (a free motor diffuses very differently from a cargo-bound one), and also by exposing the motor to the random forces resulting from thermal fluctuations of the cargo which depend on the size of the cargo and the viscosity of the environment. Whether such effects are significant are investigated here.

Recent studies show that cargos in vivo are frequently moved by more than one microtubule-based motor [1], [2], [3], [4]. This raises the question of how multiple motors function together, the subject of recent theoretical and experimental work [1], [5], [6], [7]. In vitro, when more than one motor is actively hauling a cargo, the run length, i.e., the distance that the cargo travels along the microtubule before detaching, increases with the number of active motors. However, the presence of the cargo itself may be important when there are multiple motors. In addition to possibly changing the single-molecule's function, the cargo's size may alter the relationship between the total number of motors present and the number of motors actively engaged in transporting the cargo (assuming random motor organization on the cargo's surface). If motors are not randomly organized, details of this organization will also be important. How each of these factors contributes to overall transport is unknown.

To approach these problems requires a new theoretical framework: past studies simplified the problem using essentially one-dimensional models [5], [6], [8], [9] that had the motors attached to the cargo at a single point, with the cargo represented by a single point (though potentially experiencing viscous drag proportional to a specific diameter). Here we have developed a bone-fide three dimensional Monte Carlo simulation that allows us to directly investigate how the presence of the cargo itself affects single-motor driven transport and motor-microtubule attachment, as well as how the relationship between cargo size and the arrangement of motors on the cargo affects ultimate cargo motion, all within the context of a cargo experiencing random Brownian translational and rotational motion.

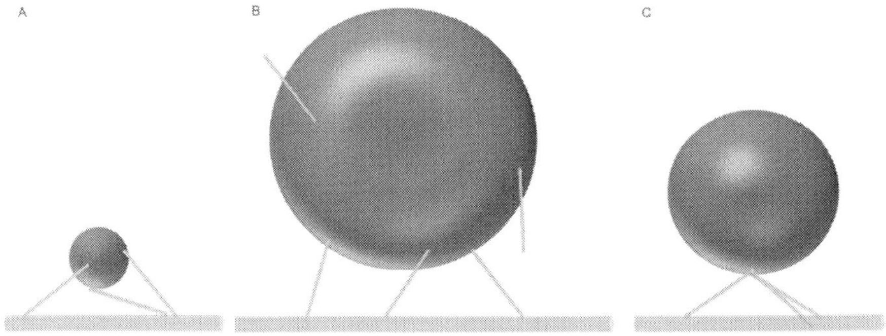

Figure 1. Kinesin motors (sticks) with a length of 110 nm are attached to cargos in various arrangements. (A) Motors on a cargo with diameter of 50 nm can easily reach the microtubule. (B) Motors on a cargo with a diameter of 500 nm have difficulty reaching the microtubule from most places on the cargo. (C) Motors attached to the cargo at the same point (South Pole) can easily reach the microtubule.

The attachment of motors to a cargo of finite size, rather than an idealized point mass, has a number of ramifications. First, the function of the motor(s) might be altered by the translational and rotational diffusion of the cargo; the larger the cargo, the more effect it has on the motors› diffusion, and thus, potentially, on the motors› ability to contact/interact with a microtubule. Second, when a motor is attached to both the microtubule and the cargo, it will feel instantaneous forces due to the cargo›s thermal motion. These forces will depend on the cargo›s size; and the random thermal 'tugs' from the cargo could slow the rate of travel of a motor and, in principle, induce the motor to detach from the filament. Third, there is a relationship between the cargo size, the total number of motors present, how they are arranged, and how many can be engaged. To illustrate this, imagine one cargo that is 50 nm in diameter, and another that is 500 nm in diameter. In the first case, even if the motors are randomly distributed on the cargo, because the length of an individual motor is more than 100 nm, all of those on the lower half of the cargo, and some on the upper half, will be able to reach a nearby microtubule (Figure 1A). In contrast for the 500 nm cargo, most motors will be unable to reach if they are randomly distributed on the cargo (Figure 1B). However, if all the motors were clumped at a single point, the size of the cargo essentially becomes irrelevant, because if one motor can reach, they all can (Figure 1C).

We thus set out to answer the following questions:

A. How does a cargo affect the rate at which the motor(s) on the cargo bind to the microtubule?

   i. For a single motor on a cargo, how does the presence of the cargo affect the motor's effective 'on' rate, i.e., the rate at which the motor binds to the microtubule?

   ii. How does the cargo's size and viscosity affect the probability that the cargo will bind to the microtubule before diffusing away?

   iii. Further, how does the length of the motor compared to the cargo size contribute to these properties?

   iv. What about the binding probability of a cargo with multiple motors?

   v. How does the distance between microtubules affect the probability that a cargo with one or more motors will bind to a microtubule?

B. Does the cargo's Brownian motion affect the motor's function as measured by its travel distance?

   i. For a cargo with a single motor?

   ii. For a cargo with multiple motors?

C. Number of engaged motors.

i. For randomly distributed motors, does doubling, say, the total number of motors on the cargo double the number of motors engaged in hauling the cargo along the microtubule?

ii. What is the relationship between the cargo size, the viscosity, the number of motors present, the average number of motors actively engaged in transporting the cargo, and the cargos' mean travel distance?

D. For motors "clustered" on the cargo, what is the relationship between cargo size, the viscosity, the size of the cluster, the number of motors in the cluster, and the mean travel distance?

We organized the presentation of our results according to these questions.

METHODS

Monte Carlo simulations

To address these questions, we developed three-dimensional Monte Carlo simulations. Generally speaking, Monte Carlo is an approach to computer simulations in which an event A occurs with a certain probability P_A where $0 \leq P_A \leq 1$. In practice, during each time step, a random number x is generated with uniform probability between 0 and 1. If $x \leq P_A$, event A occurs; if $x > P_A$, event A does not occur.

Our simulations were carried out as follows. To this cargo, we attached kinesin motor(s) that are modeled as bungee cords, i.e., they behave as springs with a spring constant of 0.32 pN/nm [5], [10] when stretched beyond their relaxed length of 110 nm but produce no force when compressed. We started the simulation so that potentially one or more motors could bind to a cylindrical microtubule (25 nm diameter). The motors then moved the cargo along the microtubules, taking 8 nm steps. The general idea is that at each time step Δt, we consider all motors present, calculate all forces acting upon them, and then ask what each of them does.

We start by describing how we simulate transport of a cargo with motors attached. Our basic algorithm is as follows. Consider one or more motors attached at random points to the cargo surface. The cargo is then suspended above the microtubule, with a well-defined separation distance between the bottom of the cargo and the top of the microtubule, and the motors are each given an opportunity to attach to the microtubule. If none do (either because none can reach, or because although they can reach, they stochastically are not able to attach in the allotted time with the 'on' rate assumed to be ~2/sec [11], [12], [13]), we use one of two initial conditions. If we want to find the time it takes for a cargo with a single motor to attach, then the cargo is allowed to rotate consistent with Brownian diffusion, and the procedure is repeated. Eventually, the motor binds. The time between when the simulation is started and when the motor attaches is the 'on' rate for the cargo; since only one motor is present, it reflects how the presence of the cargo affects the motors' on-rate.

The other initial condition is used if there are multiple motors and we are more interested in transport along the microtubule after the motors attach to the filament. In this case, if none of the motors attaches after being given the opportunity to do so, the cargo is rotated so that at least one motor attaches to the microtubule.

Once some subset of the motors is attached, the cargo travels along the microtubule. At each time step of the simulation, each motor on the cargo is given the opportunity to detach from the MT if it is attached, or attach if it is detached (and geometrically can reach the MT). If a motor is attached to a MT, then there is some probability that it will bind and hydrolyze ATP, and subsequently take a step. Although kinesin is a two headed motor, we model each motor by a single kinesin head that hydrolyzes ATP in such a way that Michaelis-Menten kinetics is obeyed. The probabilities of a motor detaching from the MT, releasing ATP, and taking a step are all dependent on the load on the cargo because the cargo exerts force on the motors. This load has contributions from the externally applied force, the other motors which are pulling the cargo, and from thermal fluctuations. The thermal fluctuations randomly rotate and translate the cargo which, in turn, can stretch the motor linkage and exert a load on the motor. (See below for further details on thermal fluctuations.) Once all the motors have been given a chance to step, the cargo is translated and rotated according to the force and torque to which it is subjected. The cargo travels along the microtubule until all the motors detach from the microtubule, and the 'run' ends; this then determines the run length of the cargo. The velocity is calculated by dividing the distance the cargo moves by the travel time τ, where τ is typically 1 msec but may be as long as 10 msec. Averaging over these velocities gives the average velocity. To get good statistics, we simulate a specified number of runs with the same initial conditions to get a set of runs. We also simulate a number of sets with different initial conditions to obtain good statistics.

In our simulations, the spherical cargo is subjected to thermal fluctuations which we can divide into translational and rotational components. The equation of the cargo's translational motion is given by the Langevin equation:

$$m\frac{d\vec{v}(t)}{dt} = -\alpha_T \vec{v}(t) + \vec{f}(\vec{x},t) + \vec{F}_T(t) \tag{1.1}$$

where m is the cargo's mass and $\vec{v}$ is the cargo's velocity. The drag force on the cargo is proportional to its velocity with the drag coefficient $\alpha_T = 6\pi\eta R$,, where R is the cargo's radius and η is the coefficient of viscosity which is the kinematic viscosity multiplied by the specific gravity of the fluid. $\vec{f}(\vec{x},t)$ is the sum of the forces due to an external force of magnitude F_L and the force of the engaged motors pulling on the cargo. The solution here for the position of the cargo at time step t+Δt:

$$\vec{x}(t+\Delta t) = \vec{x}(t) + \frac{\Delta t}{\alpha_T}\left\langle \vec{f}(\vec{x},t)\right\rangle + \sigma_T\vec{\varepsilon} \tag{1.2}$$

where $\sigma_T = \sqrt{(2k_B T \Delta t)/\alpha_T}$ is the standard deviation of a normal distribution and $\vec{\varepsilon} = (\varepsilon_x, \varepsilon_y, \varepsilon_z)$ is a vector in Cartesian coordinates of the laboratory frame of reference that represents three independent random variates drawn on a normal distribution having zero mean and unit standard deviation.

For the cargo›s rotational motion, the corresponding Langevin equation is

$$I\frac{d\vec{\Omega}(t)}{dt} = -\alpha_R \vec{\Omega}(t) + \vec{\tau}(\vec{x},t) + \vec{N}_R(t) \tag{1.3}$$

where $I = 2mR^2/5$ is the moment of inertia of a solid spherical cargo, and $\alpha_R = 8\pi\eta R^3$ is the drag coefficient proportional to the angular velocity $\vec{\Omega}(t)$. $\vec{\tau}(\vec{x},t)$ is the torque on the cargo referenced from the center of mass due to the engaged motors. $\vec{N}_R(t)$ is the rapidly varying random torque due to the thermal fluctuations of the environment. These formulas are analogous to Eq. (1.2). As we shall see, rotational diffusion due to thermal fluctuations can play a significant role in limiting the distance that a cargo can travel.

After considering motors randomly attached anywhere on the cargo, we consider cases which have a restricted region of the cargo surface area where motors can attach. For these cases, we start each simulation with N motors randomly attached to the cargo›s surface within a region specified by the cone angle as shown in Figure 2. The area available for attachment can be described by a cone with its apex at the center of the sphere. A line extends from the apex to the base of the cone. The cluster angle φ is the angle between this line and the side of the cone. The intersection of the cone with the surface of the cargo defines the allowed region of motor attachment. The cluster angle can vary between 0 and 180 degrees. A cluster angle of 90 degrees defines the lower hemisphere of the cargo. A cluster angle of 180 degrees corresponds to the entire spherical surface, and means that the motors can attach anywhere on the sphere.

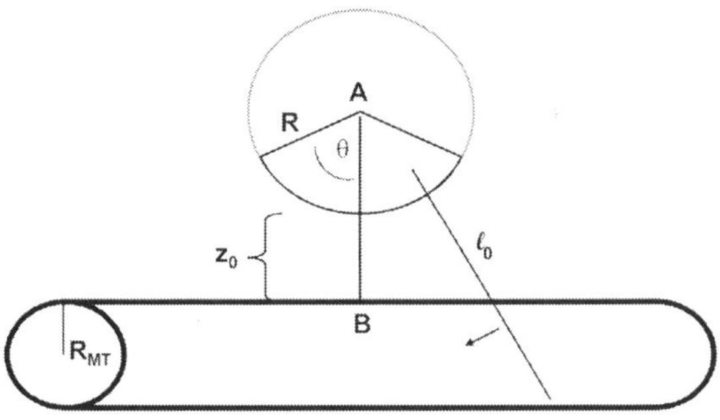

Figure 2. Cluster angle.

RESULTS

We organize our results according to the questions posed in the introduction.

A. How does a cargo affect the rate at which the motor(s) on the cargo bind to the microtubule?

A. i. For a single motor on a cargo, how does the presence of the cargo affect the motor's effective 'on' rate, i.e., the rate at which the motor binds to the microtubule?

The cargo affected a motor's 'on' rate. Let us start with our investigation of the effect of the presence of the cargo on motor attachment to a microtubule. Kinesin is estimated to have an on-rate of approximately 2 sec^{-1}, so that one expects a free motor close to a microtubule to take roughly 0.5 seconds to attach to the microtubule.

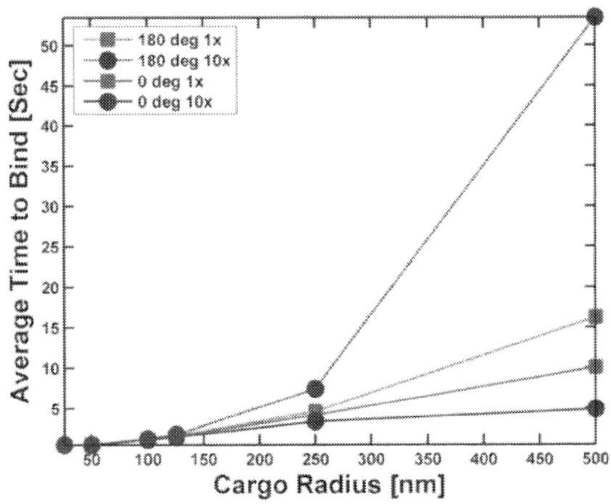

Figure 3. The average time for a motor (with an on-rate of 2/sec) to bind to a microtubule. The motor is initially attached to either the South Pole (cone angle=0, solid lines) of the cargo facing the microtubule or randomly attached somewhere on the cargo (cone angle=180 degrees, dashed lines). The cargo rests on the microtubule, and rotates randomly due to thermal effects but cannot diffuse away from the microtubule. The simulations were run long enough to allow at least 95% of the cargos to bind to the microtubules. The binding time was the averaged over the cargos that bound to the microtubule in this time. 1× represents a viscosity equal to that of water and 10× means the viscosity is 10 times greater than that of water. We took the viscosity of water to be 10^{-9} pN-s/nm^2throughout the paper.

If we attached that same motor to a small (25 or 50 nm radius) cargo, and held the cargo on the microtubule so that it could rotate randomly due to thermal motion but not diffuse away, we saw that the presence of the cargo had little effect on the typical time for the motor to bind (Figure 3). However, larger cargos started to have a significant effect on the time for a motor randomly positioned on the cargo to attach to the microtubule. Indeed, for a cargo with a radius of 250 nm, the typical time to attach increased to more than 4 seconds, and for a 500 nm cargo the time is up to 16 seconds. As might be expected, because viscosity slows rotational diffusion, increasing the viscosity significantly increased the typical binding time for motors attached to large beads—for a viscosity 10 times that of water, for a 500 nm cargo the typical attachment time is up to 53 seconds. Thus, we concluded that the presence of the cargo can indeed have a very significant effect on the typical time for a motor attached to the cargo to bind to a microtubule.

A. ii. *How does the cargo's size and viscosity affect the probability that the cargo will bind to the microtubule before diffusing away?*

The cargo affected the percentage of 'free' cargos that bound to the microtubule. In Figure 4 we plot the fraction of cargos with a single motor that bind to the microtubule as a function of the time allowed for different cargo sizes and different viscosities. In contrast to the study in Figure 3 where cargos were not allowed to diffuse away from the microtubule, in this study the cargos started by resting on the microtubule, and then could both rotate and diffuse away as a result of thermal fluctuations. A cargo was given up to 30 seconds for its motor to bind. If it had not bound in that time or if it diffused a distance greater than 50 times the length of the motor (5500 nm), then the trial was deemed a failure. The fraction of successful binding events as a function of time is plotted in Figure 4. When the motor was closest to the microtubule, the larger the cargo and the larger the viscosity, the greater was the probability of binding because these were the cases where there was the least amount of rotational and translational diffusion. When the motor was randomly attached somewhere on the cargo, larger cargos still had the greatest chance of binding at low viscosities and long times. However, for large viscosity, having the motor randomly attached to the cargo reduced its chances of binding since, if the motor could not reach the microtubule, it had a harder time to come within reach before the cargo diffused away from the microtubule. Nonetheless, it is clear overall that higher viscosity improves the cargo›s binding rate—rotational diffusion tends to bring the motor close to the MT before the cargo diffuses too far from the MT, so the presence of the cargo dramatically increases the fraction of successful binding events at high viscosity.

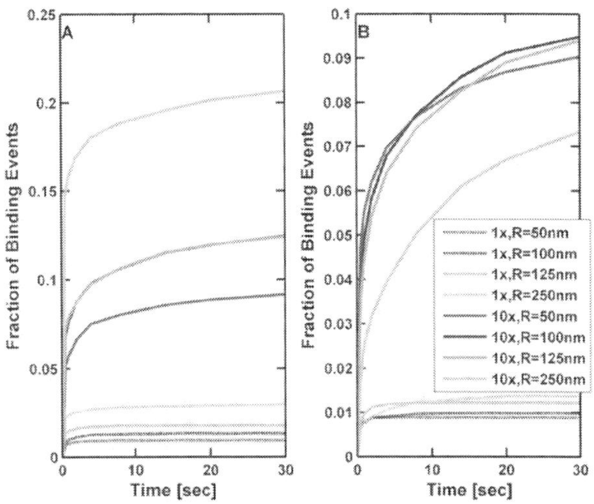

Figure 4. Fraction of cargos with a single motor that have bound to the microtubule versus time in an environment with the viscosity of water (1×) and 10 times the viscosity of water (10×). The cargos are initially resting on the microtubule, and can rotate and diffuse away due to thermal fluctuations. In the first plot (A), the simulation was started with the motor located at the South Pole facing the microtubule, and in the second plot (B), the motor was randomly placed anywhere on the cargo. Each curve represents the outcome of 6000 trials.

A. iii. Further, how does the length of the motor compared to the cargo size contribute to these properties?

The motor's length was an important contributor to its effective on-rate. When considering the system of the motor plus cargo, one might wonder how important the physical length of the motor is. We investigated this in the context of the on-rate. We varied the cargo radius from 25 to 250 nm, the motor length from 25 to 1000 nm, and the viscosity from that of water to 10 times that of water. As might be expected, the higher the viscosity, the longer it took for a single motor attached to a cargo to bind to a microtubule. In addition, motors with shorter stalks on average tended to take longer to bind to the microtubule. Interestingly, we discovered that the effective on-rate improved until the motor was approximately equal to the cargo's radius; for motors much longer than the cargo's radius there was little additional improvement (Figure 5). Biologically, the relevant time to bind should be at most a few seconds. Again we found that in order to have the time to bind to a microtubule be less than or equal to 5 seconds, the motor needed to be longer or comparable to the radius of the cargo. In order to maximize the on-rate for a cargo driven by one or two motors, our model thus suggests that one might therefore choose to have the motors be comparable to the cargo's radius. Axonal cargos driven by kinesin are rarely larger than 200 nm in diameter [14], [15], [16]; consistent with this, kinesin is 110 nm long. Below, we investigate this question more completely, with multiple motors.

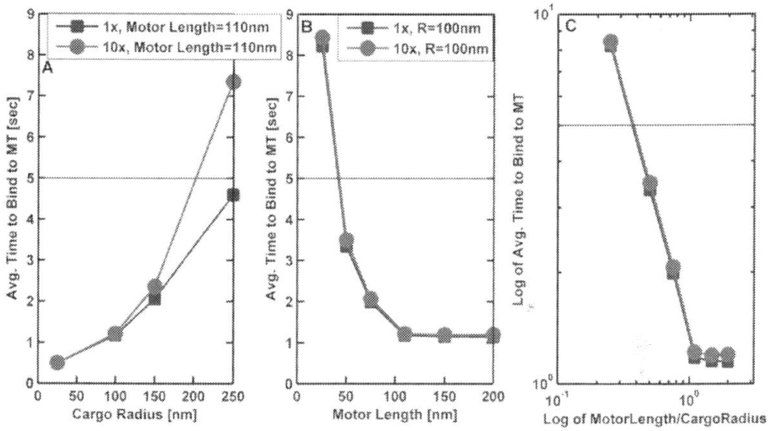

Figure 5. Average time for a cargo with a single motor to bind to a microtubule as a function of cargo radius and motor length. (A) Average time for a 110 nm motor to bind to a microtubule versus cargo radius. (B) Average time for a motor bound to a cargo with radius 100 nm to bind to a microtubule as a function of motor length. (C) Log-log plot of the average time to bind (in seconds) versus the ratio of motor length to the cargo radius. The solid horizontal line marks a time to bind of 5 seconds. The cargo was initially resting on the MT with the motor randomly placed on the cargo's surface. The cargo was allowed to rotate randomly due to thermal effects, but it could not diffuse away from the microtubule. The simulations were run long enough to allow at least 95% of the cargos to bind to the microtubules. The binding time was the averaged over the cargos

that bound to the microtubule in this time. The solid lines are guides to the eye. In all 3 plots the blue squares correspond to the viscosity of water (1×), and the red circles correspond to 10 times the viscosity of water.

A cargo moved by a single motor had unexpected oscillations. When we considered a cargo hauled along a microtubule by a single active motor without any thermal fluctuations, we found that the motor and cargo underwent an oscillating porpoise-like motion as the cargo traveled down the microtubule. This is illustrated in Figure 6. The upward and downward angular motion produced a load on the motor by tugging on the motor linkage. The oscillations, and hence the tugging, became more pronounced as the cargo radius increased. These oscillations were damped by viscosity and thermal effects as seen in Figure 7where we show the oscillations in the angle θ (measured with respect to the z-axis perpendicular to the microtubule) at which the cargo's motor detached from the microtubule in the different simulation runs. This implies that viscous damping in the cell will render these oscillations of no consequence.

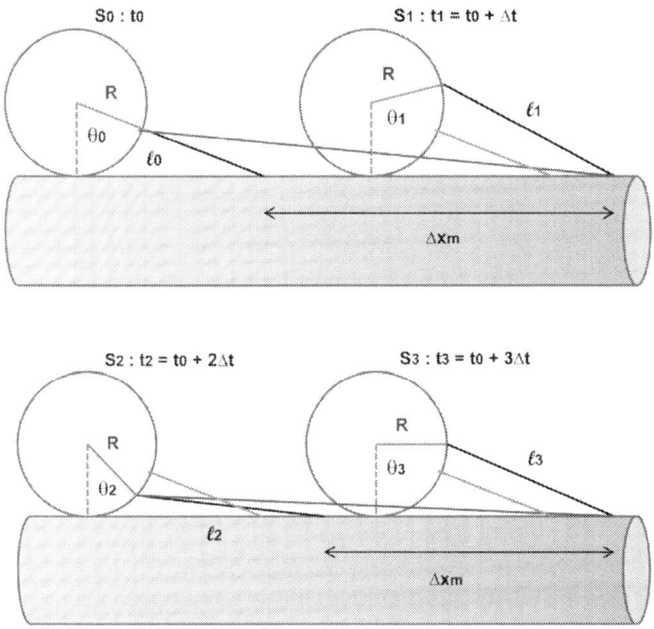

Figure 6. First few steps of a cargo pulled along a microtubule by a single motor. The figure illustrates porpoise-like oscillations about the equilibrium point.

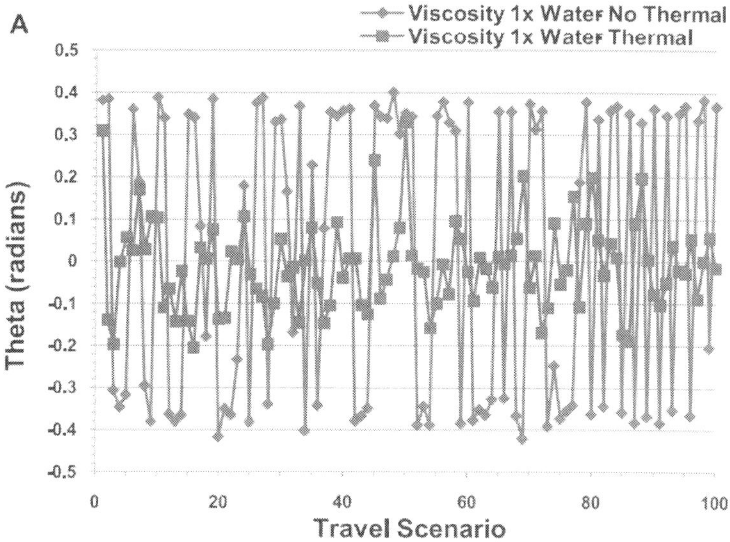

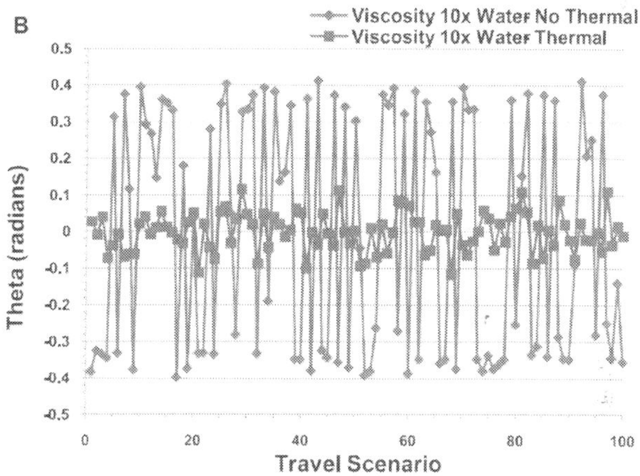

Figure 7. Rotation angle θ at which the motor detached during each of 100 Monte Carlo simulations of a cargo with 500 nm radius pulled by a single motor in a medium with the viscosity of water and 10 times the viscosity of water. The cluster angle is zero. Blue lines, labeled "no thermal," correspond to the case of no thermal rotational and no thermal translational diffusion. Red lines, labeled "thermal," correspond to the presence of both thermal rotational and thermal translational diffusion.

A. iv. What about the binding probability of a cargo with multiple motors?

How multiple motors functioned together to haul a cargo. Since our single motor simulations showed that the presence of a cargo can affect the on-rate of a motor, we wanted to see how the time it took for at least one motor to bind was affected by the total number of motors attached to the cargo as well as the size of the cargo. We considered the case of multiple motors randomly attached anywhere on the cargo surface. We placed motors randomly on the surface of a cargo, and measured the average time that it took to have at least one motor bind to the microtubule. An example of our results is shown in Figure 8. The binding time increases with cargo radius, and decreases as 1/N where N is the total number of motors on the cargo.

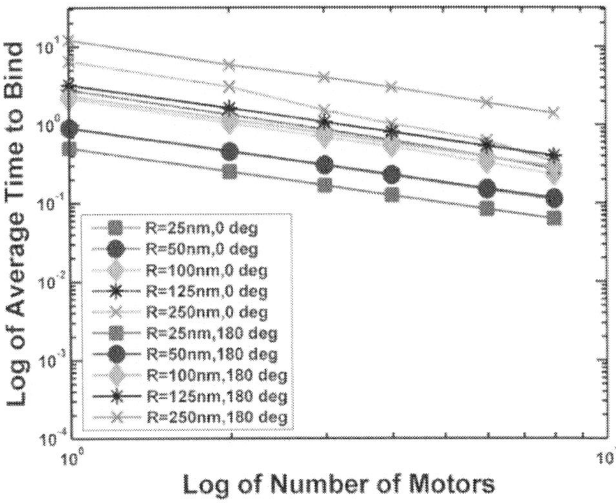

Figure 8. Log-log plot of the average binding time versus the number of motors on a cargo placed 50 nm above a microtubule. The different lines correspond to different cargo radii ranging from 25 nm to 250 nm. The cargo is allowed to rotate due to thermal diffusion but it cannot diffuse away. The medium has the viscosity of water. Having a cone angle of 180 degrees means that the motors are randomly placed on the cargo's surface. Having a cone angle of 0 degrees means that the motors are all located at the South Pole facing the microtubule.

Dependence of cargo binding time on motor length and the number of motors. Above, we investigated how the motor's length affected on-rate in the single-motor case. We extended these studies to determine the time it took for a motor to bind to a microtubule, as a function of the motor length (25 to 1000 nm), the cargo radius (25 to 250 nm), and the number of motors on the cargo. For these simulations the cargo rested on the microtubule and was not allowed to diffuse away, but did rotate randomly due to thermal effects, so that if the motor(s) were initially unable to reach the cargo, they soon came within reach of the microtubule. The motors were either clustered at one random point on the cargo's surface or else they were spread randomly over the surface. We tried motors of different lengths and cargos of different sizes. Our results are shown in Figure 9. We found that the average time τ_{bind} that it took for the first motor to bind to the microtubule had a power law dependence on the length L of the motor when the motor was shorter than or comparable to the cargo radius; namely, $\tau_{bind} \sim L^{-b}$ where the exponent b varied between 1.3 and 1.7. This meant that the longer the motor, the lower the average binding time. When the motor was much longer than the cargo radius, the average binding time became independent of the motor length. For motors short compared to the radius of the cargo, the binding time decreased as the number of motors increased. But as the motor length increased, the time it took to bind was less sensitive to the number of motors on the cargo. We can see that from Figure 9 that the binding time for kinesin, which is 110 nm long, is rather insensitive to the number of motors on the cargo. This may be one reason why molecular motors involved in intracellular transport are not shorter than about 60 nm which is the length of dynein [17]. If motors involved in transport were shorter, it would take too long for cargos to attach to filaments unless there were a large number of motors on the cargo. Intriguingly, some have suggested that more dyneins may be functioning on a cargo than kinesins [3], [4]. While this remains to be fully established, from an on-rate point of view, additional dynein motors present could compensate for dynein›s shorter overall length relative to kinesin.

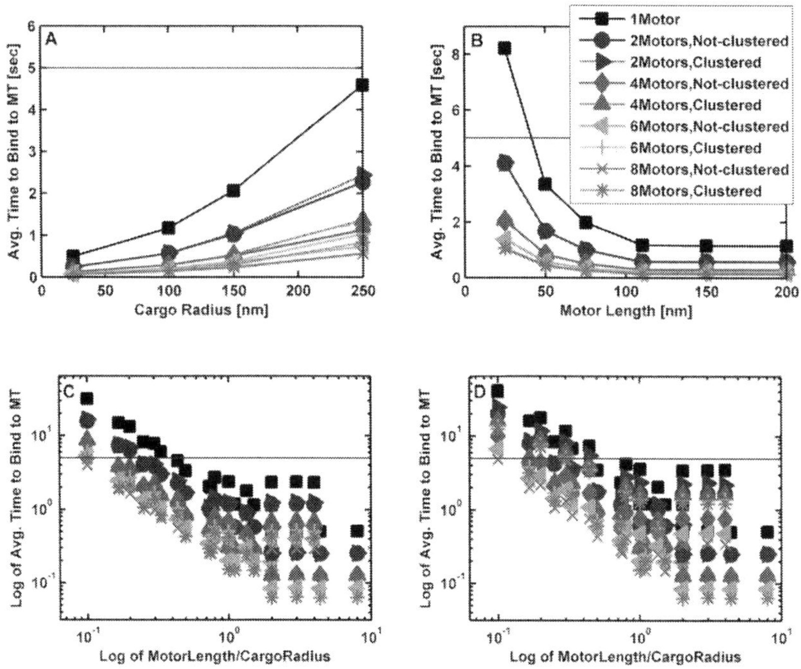

Figure 9. Average time for a cargo with up to 8 motors to bind to a microtubule as a function of cargo radius and motor length. (A) Average time for 110 nm motors to bind to a microtubule versus cargo radius at the viscosity of water. (B) Average time for motors bound to a cargo with radius 100 nm to bind to a microtubule as a function of motor length at the viscosity of water. (C) Log-log plot of the average time to bind (in seconds) versus the ratio of motor length to the cargo radius at the viscosity of water. (D) Log-log plot of the average time to bind (in seconds) versus the ratio of motor length to the cargo radius at 10 times the viscosity of water. The solid horizontal line marks a time to bind of 5 seconds. The motors are either clustered at one point on the surface of the cargo or they are randomly distributed over the surface of the cargo. The cargo was allowed to rotate randomly due to thermal effects, but it could not diffuse away from the microtubule. The log-log plots show that the time to bind goes as L^{-b} where L is the length of the motor and the exponent b varies between 1.3 and 1.7. The solid and dashed lines are guides to the eye.

A. v. How does the distance between microtubules affect the probability that a cargo with one or more motors will bind to a microtubule?

Rapidly diffusing cargos can easily reach and bind to nearby microtubules. So far we have only considered how long it takes a cargo that is sitting on a microtubule to actually bind to a single microtubule, but unattached cargos may not start on a microtubule. Rather they may be floating in the cytoplasm and may need to come within reach of a microtubule. In this case the time for a motor on a cargo to bind to a microtubule depends on the time to diffuse to a microtubule, and once it finds a microtubule, to bind to it before the cargo diffuses away. The time to diffuse between microtubules depends on the viscosity, the size of the cargo, and the distance between microtubules. The typical intracellular environment consists of multiple microtubules extending radially from the nucleus. In a flat 2D cell, the microtubule density decreases as 1/r where r is the distance from the microtubule organizing center (MTOC). At the periphery of Xenopus melanophores, the typical microtubule separation is about 800 nm [18].

We performed simulations to see how the binding fraction (the fraction of cargos where at least one motor binds to a microtubule) depends on the distance between microtubules, the radius of the cargo, and the viscosity. The geometry was a slab (extending from $z=-1$ micron to $z=+1$ micron) with either one microtubule or an infinite number of evenly spaced microtubules running parallel to the y-axis. For the case of one microtubule, it lays along the y-axis with the plus end in the positive y-direction, and slab extended to infinity in the x and y directions. For the case of microtubules evenly spaced by a distance x_{MT}, we placed one along the y-axis, and the next one a distance x_{MT} away. To obtain an infinite number of microtubules, we used periodic boundary conditions in the x direction such that if a cargo has a position $x > x_{MT}$ ($x < x_{MT}$), we mapped x to $x-x_{MT}$ ($x+x_{MT}$). Initially the cargo rested on a microtubule with motor(s) attached randomly on the surface of the cargo. The cargo was allowed 60 seconds to attach; if it failed to attach in that time, the trial was deemed a failure. The cargo was able to diffuse translationally and rotationally. The z-position of cargos that entered the floor or ceiling in the z direction was reset such that the cargo just touched the floor or ceiling. We considered microtubule spacings between 400 and 1200 nm, cargo radii from 25 nm to 250 nm, and viscosities from that of water to 1000 times that of water. We ran 1000 trials for a given set of values of the parameters.

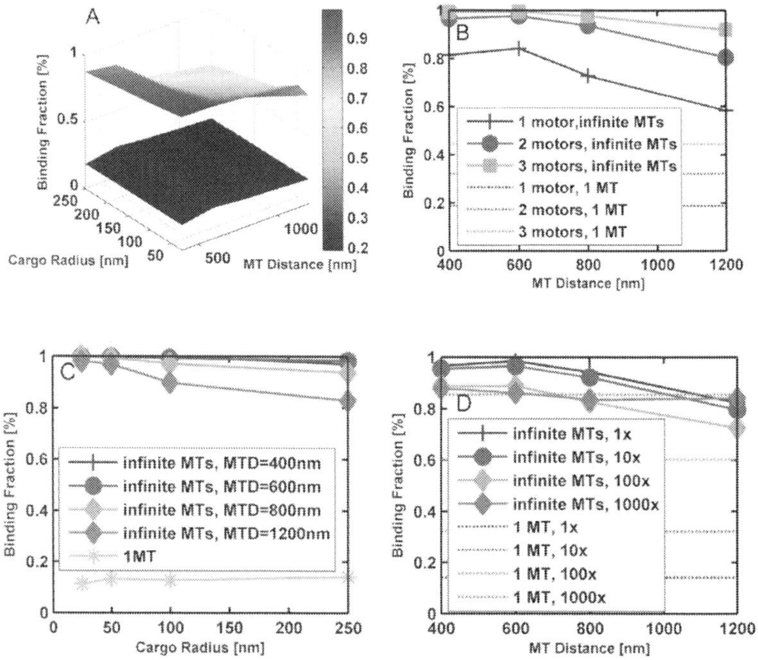

Figure 10. Binding fraction (fraction of cargos that bind to the microtubule in 60 seconds) vs. cargo radius and the distance between microtubules. (A) 3D plot of binding fraction vs. cargo radius and microtubule (MT) distance for a cargo with 1 motor randomly attached on its surface at 10 times the viscosity of water for both translational and rotational diffusion. The lower blue surface is for the case of 1 microtubule where there is obviously no dependence on microtubule distance. The upper surface is for the case of an infinite number of microtubules. (B) Binding fraction vs. microtubule distance for a cargo with radius 250 nm at 10 times the viscosity of water for both translational and rotational diffusion. The solid lines are for an infinite number of microtubules, while the single microtubule case is represented by the straight dot-dash lines which have no dependence on microtubule distance. Blue pluses, red circles, and green squares are for 1, 2, and 3 motors, respectively, placed randomly on the surface of the cargo. (C) Binding fraction vs. cargo radius for a cargo with 2 motors randomly attached to its surface at the viscosity of water for translational diffusion and 10 times the viscosity of water for rotational diffusion.

MTD stands for microtubule distance, i.e., the spacing between microtubules. (D) Binding fraction vs. microtubule (MT) distance for a cargo with a radius of 250 nm and 2 motors randomly attached to its surface at 10 times the viscosity of water for rotational diffusion. The solid lines are for the case of an infinite number of microtubules and the dot-dash lines are for one microtubule. Blue, red, green, and magenta correspond to 1, 10, 100, and 1000 times the viscosity of water for translational diffusion.

Our results for the binding fraction are shown in Figure 10. One can see that the binding fraction was much higher when an infinite number of microtubules were available compared to just one microtubule for viscosities up to 100 times that of water. This was because the time to diffuse between neighboring microtubules was relatively short. So if a cargo drifted away from a microtubule, it quickly found another one. However, if only one microtubule was available, the cargo tended to diffuse away before it had a chance to bind. When the viscosity was 1000 times that of water, the cargo was so immobile that it was localized near its initial microtubule and did not visit other microtubules. As a result, the binding fraction for an infinite number of microtubules was about the same as for 1 microtubule as seen in Figure 10D. As expected for the case with multiple microtubules, the binding fraction decreased somewhat with increasing distance between the microtubules (see Figure 10B and 10C), with increasing cargo size (see Figure 10C), and with increasing viscosity (see Figure 10D).

B. Effect of thermal fluctuations on cargo travel distance

B. i. Does the cargo's Brownian motion affect the motor's function as measured by its travel distance for a cargo with a single motor?

Run length decreased with increasing viscosity at large cargo sizes, due to diffusive rotation as well as viscous drag.

The presence of the cargo could also affect motor function by amplifying thermal noise effects, leading to increased random forces acting on the motor, and thus possibly affecting travel distance. We considered a cargo being hauled along a microtubule by a single active motor. The radius of the cargo ranged from 50 nm to 500 nm, and we studied environments with the viscosity of water and 10 times the viscosity of water. In Figure 11 we show the results where we plot the average cargo run length and velocity versus the cargo radius with and without rotational diffusion for different viscosities. Translational diffusion was present in all cases, and the run length was the distance traveled by the cargo before falling off the microtubule.

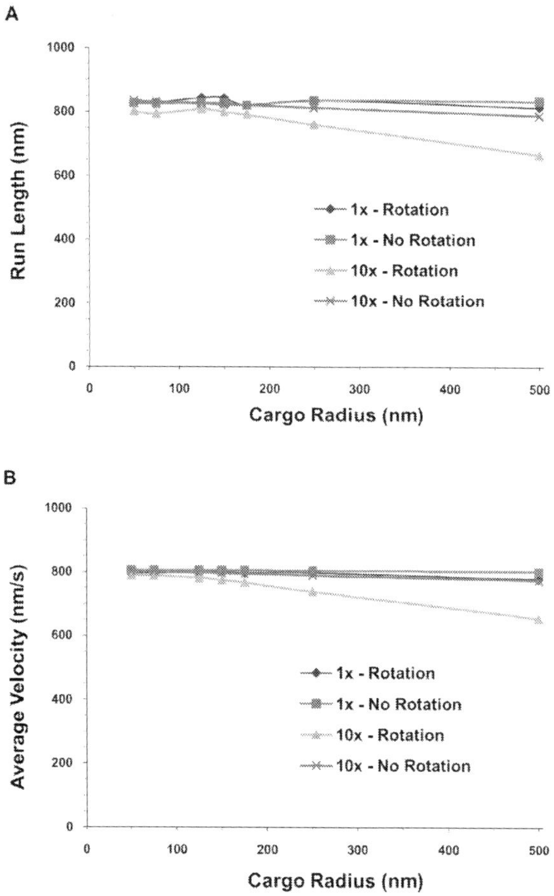

Figure 11. Average run length and average velocity of a cargo vs. cargo radius. (A) Cargo run length and (B) cargo velocity vs. cargo radius for a cargo hauled by a single motor with and without rotation at the viscosity of water and 10 times the viscosity of water.

Notice that the run length and velocity were approximately independent of cargo size when both rotational and translational diffusion were present for low viscosity. However, at 10 times the viscosity of water, when there was rotational diffusion, the run length and velocity decreased by about 20% for large cargo sizes, implying that rotational diffusion can play a significant role in decreasing run lengths under certain conditions. Looking at Equation (1.3), we see that the reason is related to the increased drag torque $-6\eta V\vec{\Omega}(t)$ which increases with viscosity $\eta$and cargo volume V. This increased drag torque reduced the rotations at each time step, but when there was a large random rotation that stretched the motor linkage, it lasted longer, increasing the time that there was load on the motor and the probability of detachment.

The contribution of rotational diffusion in limiting the run length continued out to higher viscosities, though not quite so strikingly. This is seen in Figure 12a where we show the run length versus viscosity for small (R=50 nm) and large (R=500 nm) cargos hauled by a single motor both in the presence and absence of rotational diffusion. Here we see that increasing the viscosity to 100 times that of water had little effect on the run length of the small cargo, but resulted in a decrease of the run length of the large cargo by about 40% when rotational diffusion was present and by about 30% when there was no rotational diffusion compared to the case when the viscosity was that of water.

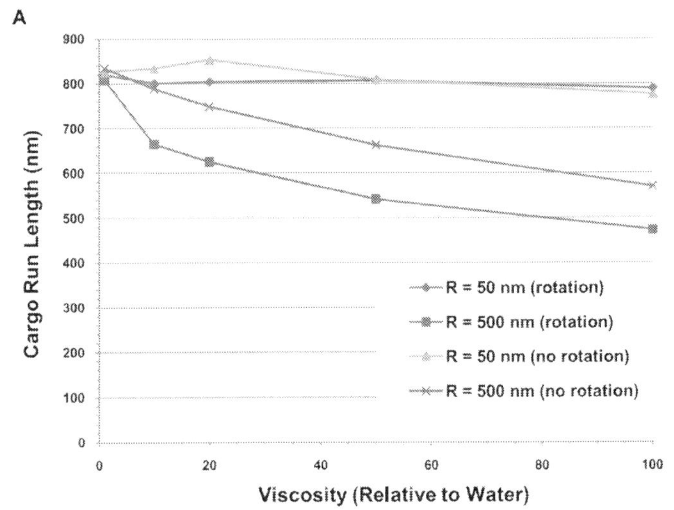

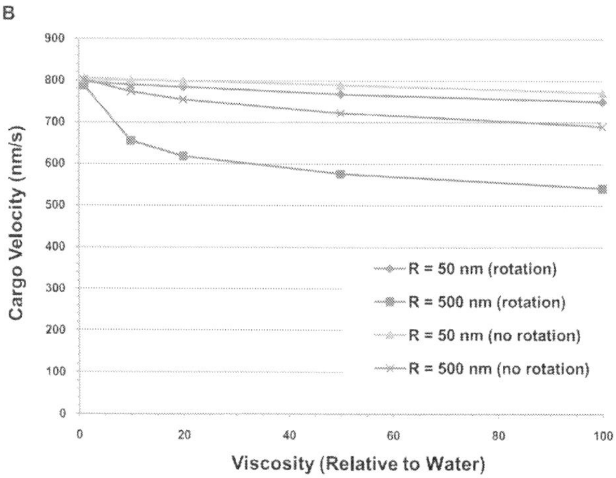

Figure 12. Average cargo run length and velocity vs. viscosity. (A) Run length and (B) average velocity of a cargo carried by a single motor vs. the viscosity relative to water for 50 nm and 500 nm radius cargos both with and without rotational diffusion.

Figures 11b and 12b show that viscosity can have a significant effect on the average velocity of large cargos. There was little effect of viscosity on the velocity of small cargos, but when the viscosity was increased by a factor of 100, the velocity of large cargos dropped by about 30% when rotational diffusion was present but only by about 14% when there was no rotational diffusion.

B. ii. Does the cargo's Brownian motion affect the motor's function as measured by its travel distance for a cargo with multiple motors?

Thermal motion that increased with cargo size decreased run lengths. As we discussed for the case of one motor on a cargo, a cargo that was tethered to a filament by one or more motors was subjected to thermal motion in the form of translational and rotational diffusion. These thermally generated forces and torques increased with the size of the cargo according to Stokes' law which says that the translational drag coefficient $\alpha_T=6\pi\eta R$, and the rotational drag coefficient $\alpha_R=8\pi\eta R^3$. As a result, the run lengths decreased with increasing viscosity and cargo radius for a fixed number of motors and cluster angle. We saw this for the case of one motor (Figures 8 and 9) where the run length decreased noticeably at large cargo sizes at the larger viscosity due to the increases in drag and thermal fluctuations.

We show the case of multiple motors in Figure 13 where the 20th percentile run lengths L_{80} are plotted versus cargo radius for 5 motors at different values of the viscosity and cluster angle. By 20th percentile run length, we mean the run length such that 80% of the motors traveled at least this far. We denote this run length by L_{80}. One can clearly see that L_{80} decreased with increasing cargo radius. This decrease was accentuated by increasing viscosity, especially in the simplest case of zero cluster angle.

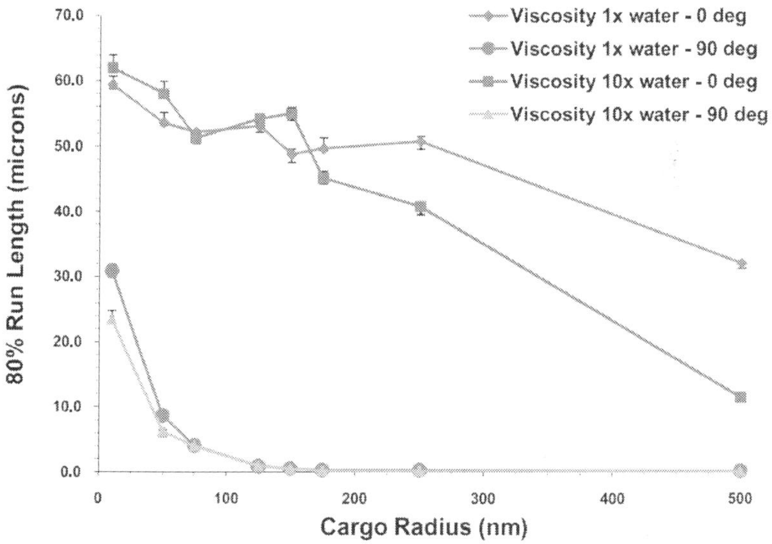

Figure 13. 20th **percentile run length vs. cargo radius for 5 motors at the viscosity** of water and 10 times the viscosity of water for cluster angles of 0 and 90°. The error bars underestimate the error and the curves would be made smoother by increasing the sample space. The roughness of the curves is not indicative of any physical behavior.

To further investigate the importance of rotational diffusion, we measured L_{80} for 5 motors for a sphere with R=500 nm at the viscosity of water, and R=250 nm at twice the viscosity of water. Note that the translational drag coefficient α_T was the same for both cases, but α_R was larger for the 500 nm sphere. We found that L_{80} was the same in both cases if we turned off rotational diffusion, but if we included rotational diffusion, then L_{80} (R=250 nm)=9.9 microns compared to L_{80} (R=500 nm)=7.7 microns. Clearly rotational diffusion of the cargo played an important role in increasing the load on the motors and decreasing the travel distance.

C. Number of engaged motors

C. i. *For randomly distributed motors, does doubling, say, the total number of motors on the cargo double the number of motors engaged in hauling the cargo along the microtubule?*

The number of engaged motors depended on the total number of motors randomly placed on the cargo and the cargo's size.

Since our simulations suggested that the presence of the cargo can affect both effective motor 'on' rates and the mean travel distance, we were also interested in how multiple motors attached to the cargo might function. We first considered motors randomly on the surface of a cargo, and during the simulation, calculated the average number of engaged motors as well as the distribution of the number of engaged motors as a function of the total number N of motors on the cargo. This was done as a function of the cargo radius R, for both the case where the cargo moved through a medium with the viscosity of water, and also for a medium with10 times the viscosity of water. The results in Figure 14 show that the average number of engaged motors increased linearly with the total number N of motors on the cargo of a fixed radius R. However, the slope of the line decreased significantly as the radius of the cargo increased, and for fixed N, the average number of engaged motors decreased quite rapidly with increasing R, because the active motors that can carry the cargo along a filament must be able to reach the filament. Thus, if motors are randomly distributed over the surface of the cargo (corresponding to a cluster angle of 180 degrees as described above), the number of motors that can reach the cargo depends on the diameter of the cargo compared to the length of the motor.

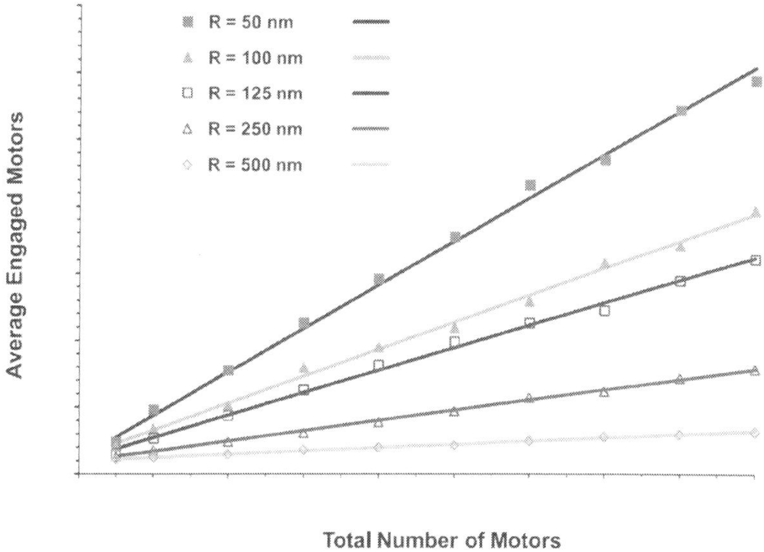

Figure 14. Average number of engaged motors vs. total number of motors at the viscosity of water for various cargo radii. Plot at 10× viscosity of water is similar.

From a biological point of view, the relationship between the number of motors engaged and the total number of motors present suggested that for most cargo sizes, doubling the total number of motors on the cargo did not double the number of engaged motors. Put differently, even though the number of actively engaged motors is linearly proportional to the total number N of motors on the cargo, it does not necessarily follow that the number of engaged motors will double if the total number of active motors on the cargo doubles because the constant of proportionality may be too small to correspond to doubling the number of active motors (Figure 15). (By active motors, we mean the number of motors that are able to haul the cargo even though they may not actually be doing so. In other words active motors have not been inactivated or incapacitated by some interfering protein or some conformational change.) For example, consider a cargo with a 100 nm diameter (R=50 nm). If N=5, the average number of engaged motors was 2.4. If the total number of motors is 10 times larger (N=50), the average number of engaged motors was 17.7 or approximately 6 times larger. In contrast, for a 500 nm diameter cargo (R=250 nm), if N=5, the average number of engaged motors was 1.3. But with10 times as many motors (N=50), the average number of engaged motors was merely tripled to 4.7.

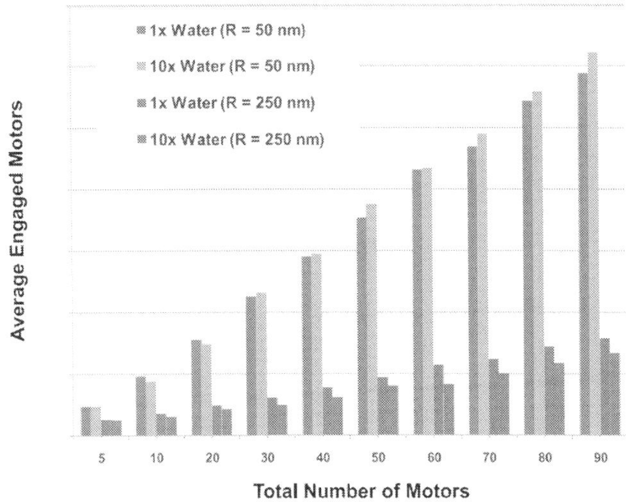

Figure 15. Average number of engaged motors vs. total number of motors for cargos with radii of 50 nm and 250 nm in media with a viscosity equal to that of water and 10 times that of water. The number of engaged motors grows more slowly for larger cargos.

What this means from a regulatory point of view is that, in the absence of some higher-order organization of motors (see below), to control motion by recruitment of motors to the cargo, different numbers of motors must be recruited depending on the cargo size. In the two cases above, if the number n of engaged motors is approximately 1, to recruit enough motors to end up with n~2 requires one to quadruple the total number of motors present in the R=50 nm case, but increase the total number of motors by a factor of 20 in the R=250 nm case.

Alternatively, one could hypothesize that motor recruitment could be controlled locally, so that the surface density of motors could be controlled. Interestingly, because of geometrical effects, fixed densities of motors did not equate to the same number of engaged motors on cargos of different sizes. For example, a density of 6.3×10^{-5} motors/nm^2 corresponded to 2 total motors (and 1.5 engaged motors) on a cargo with R=50 nm, but to 50 total motors (and 4.7 engaged motors) on a cargo with R=250 nm. The relationship between motor density and the number of engaged motors is investigated more fully below.

Use of Poisson statistics to estimate the number of engaged motors. Frequently, one uses Poisson statistics to estimate the number of motors that are engaged. We were able to directly test this assumption. By recording the number of engaged motors at each time step, we were able to calculate the distribution $P(N_{engaged})$ of engaged motors. We found that $P(N_{engaged})$ obeyed Poisson statistics to a good approximation that improved as the total number N of motors increased. For 30 motors or more, it was an excellent approximation. An example is shown in Figure 16. For a given total number of motors on the cargo, the only adjustable parameter is the Poisson mean for which we used the simulation results to obtain average number of engaged motors.

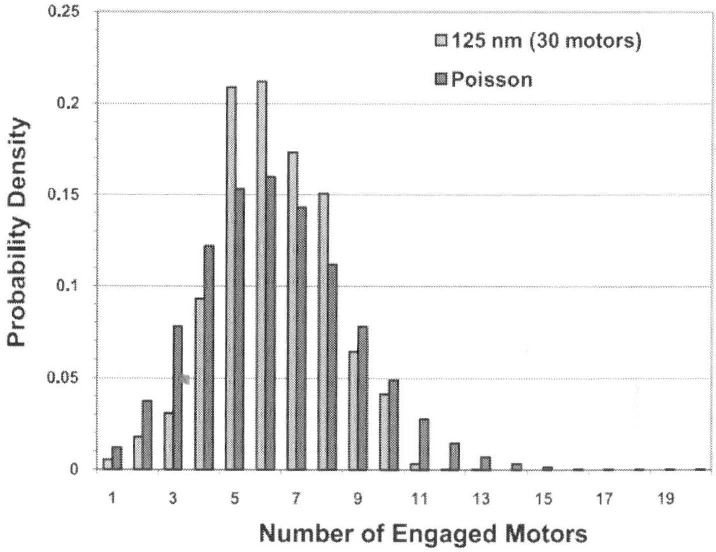

Figure 16. Distribution of the number of engaged motors at the viscosity of water. 30 total motors on a cargo with radius of 125 nm.

C. ii. What is the relationship between the cargo size, the viscosity, the number of motors present, the average number of motors actively engaged in transporting the cargo, and the cargos' mean travel distance?

Relationship between the cargo size, the motor density, and the number of engaged motors. If motors were randomly distributed on a cargo, a key quantity that helped to determine the number of engaged motors, and hence the run length, was the density of motors on the cargo surface. Figure 17 shows the average number of engaged motors as a function of the surface density of motors. In general, the larger the cargo, the lower the motor density required to achieve a specific mean number of engaged motors.

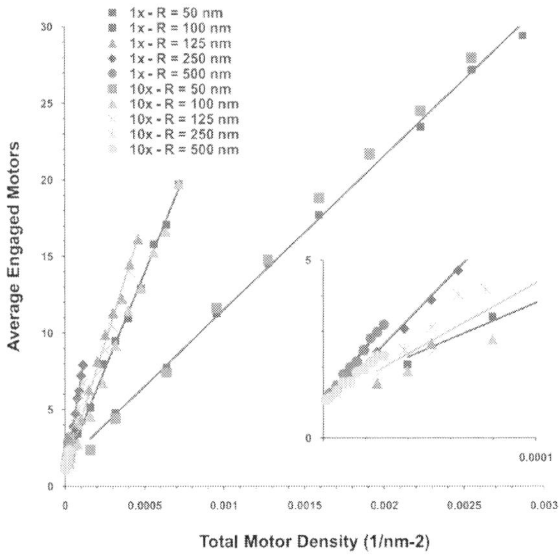

Figure 17. Average number of engaged motors as a function of the density of motors on the cargo surface for a variety of cargo radii. The lines are interpolated between points with the viscosity of water. The points at 10 times the viscosity of water have no lines. Inset is a blow-up of the region near the origin.

In our example above, 5 motors on a cargo with a diameter of 100 nm correspond to a density of $1.6\times10^{-4}/nm^2$, and 50 motors correspond to a density of $1.6\times10^{-3}/nm^2$. This is in contrast to a 500 nm diameter cargo where density of $6.4\times10^{-6}/nm^2$ corresponds to 5 motors, and a density of $6.4\times10^{-5}/nm^2$ corresponds to 50 motors. This is summarized in Table 1.

Table 1. Relationship between the cargo size, the motor density, and the number of engaged motors.

Radius [nm]	Total number of Motors	Density [nm^{-2}]	Number of Engaged Motors
50	5	1.6×10^{-4}	2.4
50	50	1.6×10^{-3}	17.7
250	5	6.4×10^{-6}	1.3
250	50	6.4×10^{-5}	4.7

In addition to motor density, cargo size is also important because motors have a harder time reaching the microtubule from a large cargo. Since the motors cannot pass through the cargo, there is more excluded volume for large cargos.

Non-Random Organization of Multiple Motors on Cargos

D. For motors "clustered" on the cargo, what is the relationship between cargo size, the viscosity, the size of the cluster, the number of motors in the cluster, and the mean travel distance?

The run length increased exponentially with the total number of motors on the cargo. The simplest non-random organization for a group of motors is to cluster them together. We investigated the effects of clustering by specifying the cluster size in terms of the cone angle subtended by the cluster (see Figure 2). We assumed the clusters to be cylindrically symmetric about the axis of the cone. A 0 degree cluster placed all motors at a single point on the cargo; a 90 degree cluster positioned them randomly on the lower hemisphere between −90 and +90 degrees. If only one motor was attached to the cargo, then the run length was about 800 nm, independent of the cargo radius and cluster angle, though run length did decrease slightly under high viscosity conditions for large cargos. For multiple motors, as the number of motors increased, the run length increased exponentially for a fixed cargo radius and cluster angle. We found this to be true for cargo radii of 50, 75, 125, 150, 175, 250 and 500 nm, cluster angles varying from 0 to 90 degrees, and solvent viscosities equal to that of water and ten times that of water. An example is shown in Figure 18. The increase of run length with the total number of motors makes sense since the cargo was able to continue traveling along the filament even if some of the motors detached from the filament. If some motors were detached and some were still attached to the filament, then the detached motors had a chance to reattach to the filament and help carry the cargo.

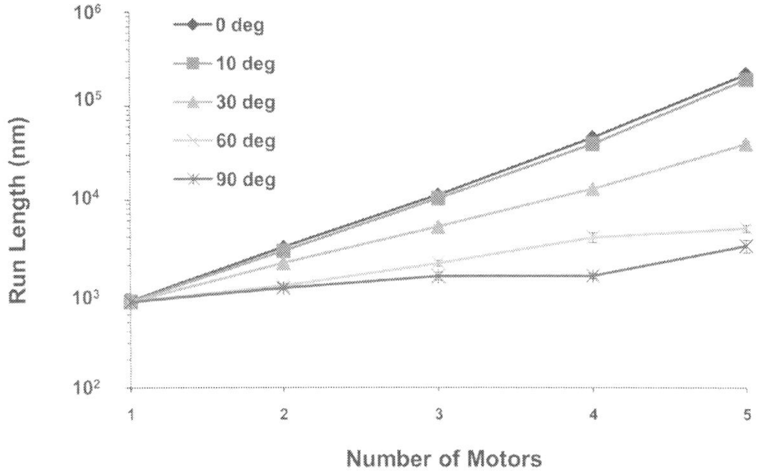

Figure 18. Semi-logarithmic plot of the run length for a 250 nm cargo vs. the total number of motors for different cluster angles at the viscosity of water. The run length increases exponentially with the number of motors. The error bars underestimate the error and the curves would be made smoother by increasing the sample space. The roughness of the curves is not indicative of any physical behavior.

Run length decreased with increasing cluster angle and increasing cargo radius. For two or more motors, the run length decreased as the cluster angle and cargo radius increased. An example is shown in Figure 19 for 5 motors. This decrease in run length was due to the fact that it was harder for the motors to attach to the filament with increasing cargo radius. Increasing the cluster angle increased the cargo surface area where the motors can attach to the cargo; and the farther motors were from the South Pole (nearest point to the filament), the harder it was for them to attach. There were also greater loads on the motors that were barely able to attach, making the probability of detaching from the filament more likely. It is obvious that for larger cargos, clustering is critical to achieve good performance from a limited number of motors.

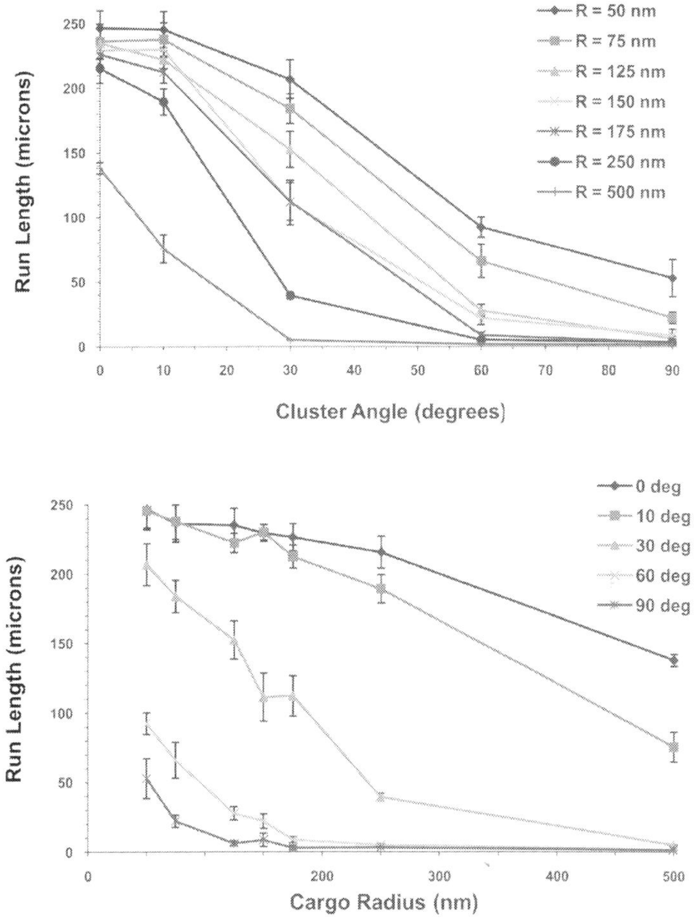

Figure 19. Run length for cargo with 5 motors as a function of cluster angle and cargo radius at the viscosity of water. (A) Run length vs. cluster angle. (B) Run length vs. cargo radius. The error bars underestimate the error and the curves would be made smoother by increasing the sample space. The roughness of the curves is not indicative of any physical behavior.

Another way to exhibit this data is shown in Figure 20 where we plot the 20th percentile run lengths L_{80} as a function of the cargo radius and cluster angle. We can see from the figures that L_{80}, the minimum distance traveled by 80% of the motors, decreased with increasing cluster angle and cargo radius. This decrease was more rapid for large cargos and for large cluster angles.

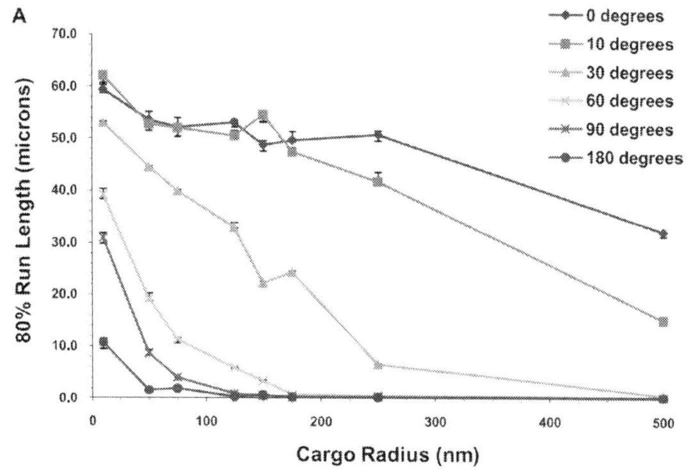

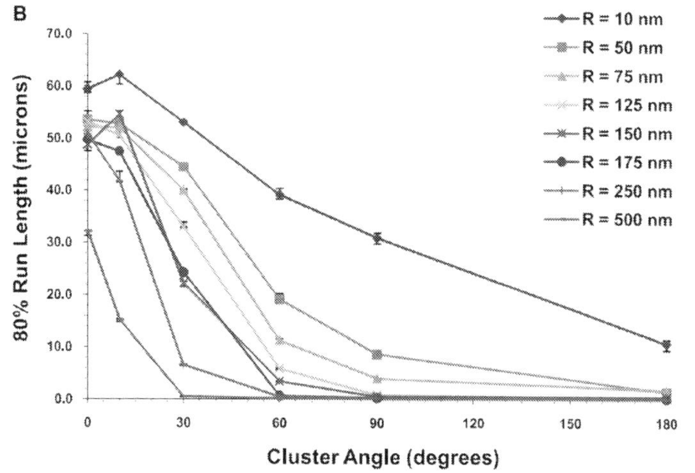

Figure 20. 20th **percentile run lengths at the viscosity** of water. 80% of the cargos traveled at least a distance that we call the 20th percentile run length. 20th percentile run lengths vs. (A) cargo radius and (B) cluster angle for 5 motors. The error bars underestimate the error and the curves would be made smoother by increasing the sample space. The roughness of the curves is not indicative of any physical behavior.

Run lengths decreased with increasing viscosity. In Figures 20 and 21, we show the 20th percentile run lengths at the viscosity of water and 10 times the viscosity of water as a function of the cargo radius and the cluster angle. At the higher viscosity, one can see that

L_{80} decreased faster with increasing cargo radius and cluster angle. This is to be expected since the higher viscosity medium produced a greater drag on the cargo, and hence a higher load on the motors. In our model the probability of a motor detaching from the microtubule increased exponentially with increasing load [5], leading to shorter run lengths.

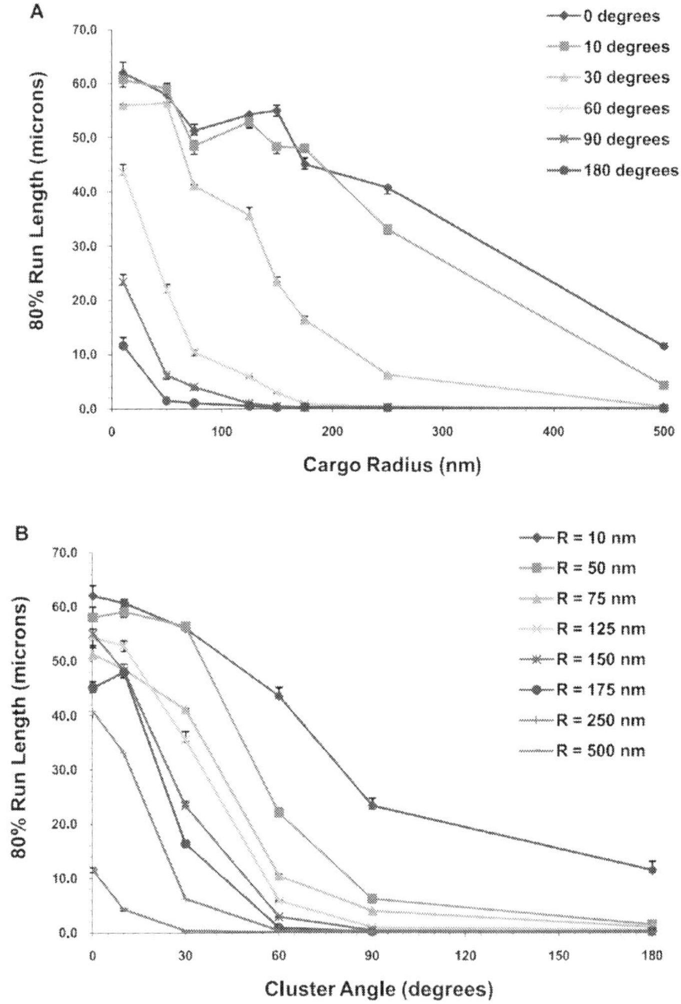

Figure 21. 20th **percentile run lengths at 10 times the viscosity** of water. 20th percentile run lengths vs. (A) cargo radius and (B) cluster angle for 5 motors on the cargo. The error bars underestimate the error and the curves would be made smoother by increasing the sample space. The roughness of the curves is not indicative of any physical behavior.

Surprisingly, clustering motors had little effect on the rate of attaching the cargo to the microtubule.

The results described above showed that run length increased if the motors were clustered rather than randomly spread over the surface of the cargo. We wondered if one cost of clustering might be to increase the time for the cargo to attach, compared to randomly placed motors. We performed simulations in which the cargo initially was resting on the microtubule and was subjected to thermal fluctuations that could cause it to diffuse away or to rotate randomly. In Figure 22 we compare the fraction of binding events (in which the motors attached to the microtubule) of cargos with motors randomly spread over the cargo surface to the case where the motors are clustered. One can see that clustering did not significantly affect the ability of the cargo to bind except for very large cargos at high viscosities (10 times that of water).

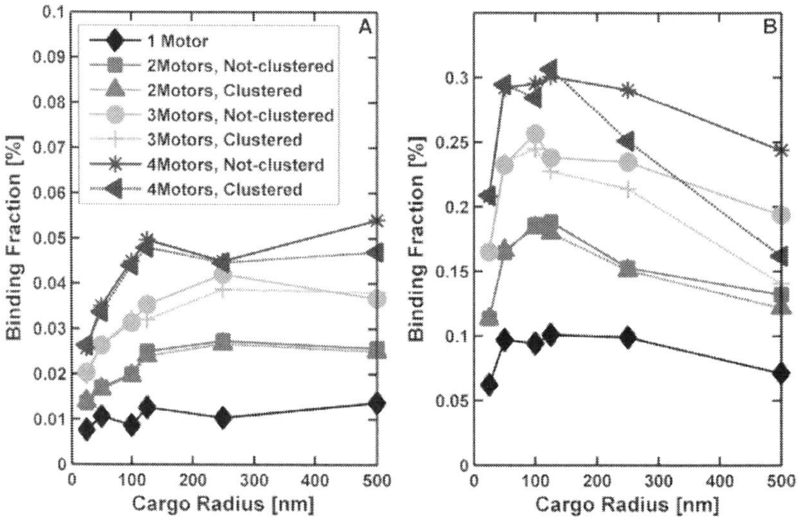

Figure 22. Binding fraction vs. cargo radius for various numbers of motors on the cargo. Each trial was terminated when the cargo bound to the microtubule, or if the cargo did not bind after 60 seconds, or if the cargo did not bind and diffused more than 5.5 microns away from the microtubule. (5.5 microns is 50 times the length of the motor.) The dashed lines correspond to the motors being clustered, i.e., attached to the cargo at one point. The solid lines correspond to the motors being randomly distributed over the surface of the cargo. The cargo initially was resting on the microtubule, and was allowed to randomly rotate and diffuse away due to thermal fluctuations. (A) Medium had the viscosity of

water. (B) Medium had ten times the viscosity of water. The curves would be made smoother by increasing the sample space. The roughness of the curves is not indicative of any physical behavior.

DISCUSSION

Our study of the effects of the cargo on transport has a number of 'take-home' messages. The first is that, at both the single-motor and multiple-motor levels, the presence of the cargo can significantly alter the effective 'on' rate/probability of successful binding of the motor(s) to the filament, because the center of mass of the cargo diffuses away from the microtubule relatively slowly, and while this is occurring, its rotational diffusion frequently brings the motor close enough to the microtubule to allow attachment. Thus, the cargo 'helps' the motor to attach, though the degree of assistance depends on cargo size and viscosity of the medium surrounding the cargo. Rapidly diffusing cargos might not linger long in the vicinity of a microtubule, but in a cell where there are multiple filaments available, these cargos could quickly find and bind to a filament.

Second, in order to for a motor to attach to the filament in a reasonable amount of time, the motor length needs to be longer or comparable to the radius of the cargo which may explain why motors are 60 to 110 nm in length.

Third, if motors are randomly arranged on the cargo's surface, the relationship between the number of motors present and the number of actually engaged motors depends strongly on the cargo size, so that different simple models of regulating cargo motion by recruiting motors to the cargo surface (either by a specified change in total number of motors, or by a specified change in local motor surface density) will have different effects on overall cargo motion as a function of cargo size. Thus, in order to have regulation affect a set of cargos equally, independent in variations in cargo size, it is best to have motors clustered in a small region on the cargo.

A further finding also supports the utility of motor clustering: for large cargos, if motors are randomly placed, achieving a reasonable number of engaged motors (n=3–6) would require a large number of motors (50–100) to be present on the cargo, which appears inconsistent with biochemical characterizations of cargo-bound microtubule motors [4], though it is consistent with biochemical characterizations of cargo-bound myosin motors [18] which are likely randomly arranged on cargos [18], [19]. Overall, our findings suggest that, in vivo, microtubule motors are likely organized into clusters when present on large cargos, but that such clustering is unnecessary for small cargos.

In addition, a reasonable number of engaged motors would be required for long travel distances of several microns but not for short run lengths. Since microtubules can be tens of microns long compared to actin filaments which have a typical decay length of 1.6 microns [18], we expect long travel distances along microtubules but relatively short run lengths along actin filaments. Thus we predict the microtubule motors kinesin and dynein to be clustered on cargos while we expect the actin motor myosin V to bind randomly to cargos. There is clear experimental evidence for the random arrangement of myosin on cargos in vivo, and weak experimental evidence for the clustering of kinesin and cytoplasmic dynein [19].

For the purposes of this paper, we have assumed that the points where motors are attached to the cargos are fixed on the cargo's surface. This is true in some cases, e.g., when motors bind to dynactin which in turn binds to spectrin which is a filament that coats some vesicles [20], [21]. However, in other cases, the attachment points can diffuse through the fluid membrane of the vesicle and cluster at one location. An example of this is an experiment showing that motors dynamically accumulate at the tip of membrane tubes growing out of a vesicle as a consequence of the fluidity of the membrane [13], [22].

Clustering does not seem to affect the rate at which the first motor of a cargo attaches to a microtubule unless the cargo is large (greater than 200 nm) and the viscosity is high. Motor proteins are sufficiently long (greater than 50 nm) and rotational diffusion sufficiently rapid that the number of motors on a cargo does not significantly affect the rate at which the cargo binds to the microtubule.

ACKNOWLEDGMENTS

CCY thanks the Aspen Center for Physics for its hospitality during the initial stages of this work.

AUTHOR CONTRIBUTIONS

Conceived and designed the experiments: ZJ SPG CCY. Performed the experiments: RPE ZJ. Analyzed the data: RPE ZJ SPG CCY. Contributed reagents/materials/analysis tools: RPE ZJ CCY. Wrote the paper: RPE ZJ SPG CCY.

REFERENCES

1. Shubeita GT, Tran SL, Xu J, Vershinin M, Cermelli S, et al. (2008) Consequences of motor copy number on the intracellular transport of kinesin-1-driven lipid droplets. Cell 135: 1098–1107.GT ShubeitaSL TranJ. XuM. VershininS. Cermelli2008Consequences of motor copy number on the intracellular transport of kinesin-1-driven lipid droplets.Cell13510981107
2. Mallik R, Gross SP (2009) Intracellular transport: how do motors work together? Curr Biol 19: R416–418.R. MallikSP Gross2009Intracellular transport: how do motors work together?Curr Biol19R416418
3. Soppina V, Rai AK, Ramaiya AJ, Barak P, Mallik R (2009) Tug-of-war between dissimilar teams of microtubule motors regulates transport and fission of endosomes. Proc Natl Acad Sci U S A 106: 19381–19386.V. SoppinaAK RaiAJ RamaiyaP. BarakR. Mallik2009Tug-of-war between dissimilar teams of microtubule motors regulates transport and fission of endosomes.Proc Natl Acad Sci U S A1061938119386
4. Hendricks AG, Perlson E, Ross JL, Schroeder HW 3rd, Tokito M, et al. (2010) Motor Coordination via a Tug-of-War Mechanism Drives Bidirectional Vesicle Transport. Curr Biol 20: 697–702.AG HendricksE. PerlsonJL RossHW Schroeder 3rdM. Tokito2010Motor Coordination via a Tug-of-War Mechanism Drives Bidirectional Vesicle Transport.Curr Biol20697702
5. Kunwar A, Vershinin M, Xu J, Gross SP (2008) Stepping, strain gating, and an unexpected force-velocity curve for multiple-motor-based transport. Curr Biol 18: 1173–1183.A. KunwarM. VershininJ. XuSP Gross2008Stepping, strain gating, and an unexpected force-velocity curve for multiple-motor-based transport.Curr Biol1811731183

6. Klumpp S, Lipowsky R (2005) Cooperative cargo transport by several molecular motors. Proc Natl Acad Sci U S A 102: 17284–17289.S. KlumppR. Lipowsky2005Cooperative cargo transport by several molecular motors.Proc Natl Acad Sci U S A1021728417289

7. Vershinin M, Carter BC, Razafsky DS, King SJ, Gross SP (2007) Multiple-motor based transport and its regulation by Tau. Proc Natl Acad Sci U S A 104: 87–92.M. VershininBC CarterDS RazafskySJ KingSP Gross2007Multiple-motor based transport and its regulation by Tau.Proc Natl Acad Sci U S A1048792

8. Aghababaie Y, Menon GI, Plischke M (1999) Universal properties of interacting Brownian motors. Physical Review E 59: 2578–2586.Y. AghababaieGI MenonM. Plischke1999Universal properties of interacting Brownian motors.Physical Review E5925782586

9. Muhuri S, Pagonabarraga I (2010) Lattice-gas model for active vesicle transport by molecular motors with opposite polarities. Phys Rev E 82: 021925-021921–021925-021913.S. MuhuriI. Pagonabarraga2010Lattice-gas model for active vesicle transport by molecular motors with opposite polarities.Phys Rev E82021925-021921021925-021913

10. Coppin CM, Pierce DW, Hsu L, Vale RD (1997) The load dependence of kinesin's mechanical cycle. Proc Natl Acad Sci U S A 94: 8539–8544.CM CoppinDW PierceL. HsuRD Vale1997The load dependence of kinesin's mechanical cycle.Proc Natl Acad Sci U S A9485398544

11. King SJ, Schroer TA (2000) Dynactin increases the processivity of the cytoplasmic dynein motor. Nat Cell Biol 2: 20–24.SJ KingTA Schroer2000Dynactin increases the processivity of the cytoplasmic dynein motor.Nat Cell Biol22024

12. Reck-Peterson SL, Yildiz A, Carter AP, Gennerich A, Zhang N, et al. (2006) Single-molecule analysis of dynein processivity and stepping behavior. Cell 126: 335–348.SL Reck-PetersonA. YildizAP CarterA. GennerichN. Zhang2006Single-molecule analysis of dynein processivity and stepping behavior.Cell126335348

13. Leduc C, Campas O, Zeldovich KB, Roux A, Jolimaitre P, et al. (2004) Cooperative extraction of membrane nanotubes by molecular motors. Proc Natl Acad Sci U S A 101: 17096–17101.C. LeducO.

CampasKB ZeldovichA. RouxP. Jolimaitre2004Cooperative extraction of membrane nanotubes by molecular motors.Proc Natl Acad Sci U S A1011709617101

14. Cook RD, Burnstock G (1976) The ultrastructure of Auerbach's plexus in the guinea-pig. II. Non-neuronal elements. J Neurocytol 5: 195–206.RD CookG. Burnstock1976The ultrastructure of Auerbach's plexus in the guinea-pig. II. Non-neuronal elements.J Neurocytol5195206
15. Eranko O, Klinge E, Sjostrand NO (1976) Different types of synaptic vesicles in axons of the retractor penis muscle of the bull. Experientia 32: 1335–1337.O. ErankoE. KlingeNO Sjostrand1976Different types of synaptic vesicles in axons of the retractor penis muscle of the bull.Experientia3213351337
16. Geffen LB, Ostberg A (1969) Distribution of granular vesicles in normal and constricted sympathetic neurones. J Physiol 204: 583–592.LB GeffenA. Ostberg1969Distribution of granular vesicles in normal and constricted sympathetic neurones.J Physiol204583592
17. Gennerich A, Vale RD (2009) Walking the walk: how kinesin and dynein coordinate their steps. Curr Opin Cell Biol 21: 59–67.A. GennerichRD Vale2009Walking the walk: how kinesin and dynein coordinate their steps.Curr Opin Cell Biol215967
18. Snider J, Lin F, Zahedi N, Rodionov V, Yu CC, et al. (2004) Intracellular actin-based transport: how far you go depends on how often you switch. Proc Natl Acad Sci U S A 101: 13204–13209.J. SniderF. LinN. ZahediV. RodionovCC Yu2004Intracellular actin-based transport: how far you go depends on how often you switch. Proc Natl Acad Sci U S A1011320413209
19. Gross SP, Vershinin M, Shubeita GT (2007) Cargo transport: two motors are sometimes better than one. Curr Biol 17: R478–486.SP GrossM. VershininGT Shubeita2007Cargo transport: two motors are sometimes better than one.Curr Biol17R478486
20. Holleran EA, Ligon LA, Tokito M, Stankewich MC, Morrow JS, et al. (2001) beta III spectrin binds to the Arp1 subunit of dynactin. J Biol Chem 276: 36598–36605.EA HolleranLA LigonM. TokitoMC StankewichJS Morrow2001beta III spectrin binds to the Arp1 subunit of dynactin.J Biol Chem2763659836605

21. Aspengren S, Wallin M (2004) A role for spectrin in dynactin-dependent melanosome transport in Xenopus laevis melanophores. Pigment Cell Res 17: 295–301.S. AspengrenM. Wallin2004A role for spectrin in dynactin-dependent melanosome transport in Xenopus laevis melanophores.Pigment Cell Res17295301

22. Campas O, Leduc C, Bassereau P, Casademunt J, Joanny JF, et al. (2008) Coordination of Kinesin motors pulling on fluid membranes. Biophys J 94: 5009–5017.O. CampasC. LeducP. BassereauJ. CasademuntJF Joanny2008Coordination of Kinesin motors pulling on fluid membranes.Biophys J9450095017

CHAPTER

5

A STOCHASTIC MODEL FOR MICROTUBULE MOTORS DESCRIBES THE IN VIVO CYTOPLASMIC TRANSPORT OF HUMAN ADENOVIRUS

Mattia Gazzola[1], Christoph J. Burckhardt[2], Basil Bayati[1], Martin Engelke[2], Urs F. Greber[2], Petros Koumoutsakos[1]

[1]Chair of Computational Science, ETH Zurich, Zurich, Switzerland,

[2]Institute of Zoology, University of Zurich, Zurich, Switzerland

ABSTRACT

Cytoplasmic transport of organelles, nucleic acids and proteins on microtubules is usually bidirectional with dynein and kinesin motors mediating the delivery of cargoes in the cytoplasm. Here we combine live cell microscopy, single virus tracking and trajectory segmentation to systematically identify the parameters of a stochastic computational model of cargo transport by molecular motors on microtubules. The model parameters are identified using an evolutionary optimization algorithm to minimize the Kullback-Leibler divergence between the *in silico* and the *in vivo* run length and velocity distributions of the viruses on microtubules. The present stochastic model suggests that bidirectional transport of human adenoviruses can be explained without explicit motor coordination. The model enables the prediction of the number of motors active on the viral cargo during microtubule-dependent motions as well as the number of motor binding sites, with the protein hexon as the binding site for the motors.

AUTHOR SUMMARY

Molecular motors, due to their transportation function, are essential to the cell, but they are often hijacked by viruses to reach their replication site. Imaging of virus trajectories provides information about the patterns of virus transport in the cytoplasm, leading to improved understanding of the underlying mechanisms. In turn improved understanding may suggest actions that can be taken to interfere with the transport of pathogens in the cell. In this work we use *in vivo* imaging of virus trajectories to develop a computational model of virus transport in the cell. The model parameters are identified by an optimization procedure to minimize the discrepancy between *in vivo* and *in silico* trajectories. The model explains the *in vivo*trajectories as the result of a stochastic interaction between motors. Furthermore it enables predictions on the number of motors and binding sites on pathogens, quantities that are difficult to obtain experimentally. Beyond the understanding of mechanisms involved in pathogen transport, the present paper introduces a systematic parameter identification algorithm for stochastic models using *in vivo* imaging. The discrete and noisy characteristics of biological systems have led to increased attention in stochastic models and this work provides a methodology for their systematic development.

INTRODUCTION

The function of eukaryotic cells relies on the transport of macromolecules and organelles throughout the cytoplasm. Pathogenic viruses can exploit a cell's cytoplasmic transport mechanisms [1],[2] in order to reach their site of replication. Cytoplasmic transport involves three types of molecular motors. Kinesin and dynein motors use microtubule tracks to move cargo throughout the cytoplasm, while myosin motors interact with actin filaments to move their cargoes [3],[4]. Microtubule based transport is usually bidirectional and its mechanism can be explained by the exclusive binding of dynein and kinesin motors to the cargo, motor cooperation and regulation, or a stochastic tug-of-war [5]–[8]. Exclusive binding of motors has not been reported in cells, while in systems with cooperating motors, additional factors such as on/off switches or coordinators between motors have been postulated for bidirectional transport of large cargo, such as vesicles [7]. The mechanism of bidirectional motor transport by non-coordinated motors of opposite polarity has been the basis of tug-of-war models [7],[9].

In this work we propose a stochastic model for motor transport on microtubules and we systematically identify its parameters using virus trajectories obtained by *in vivo* imaging (Fig. 1). Trajectories are obtained by live cell microscopy of fluorescently labelled human adenovirus type 2 (Ad2) using confocal microscopy. Motility information extracted through single virus tracking [10], and trajectory segmentation [11] are implemented in order to study the properties of virus transport by employing a systems identification process [12] for a stochastic model of cargo transport on microtubules.

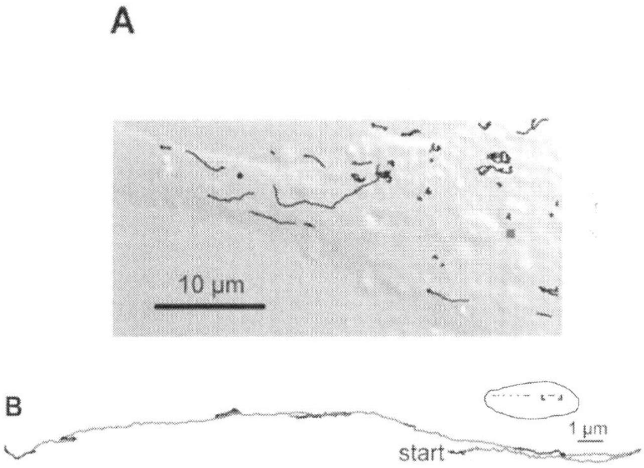

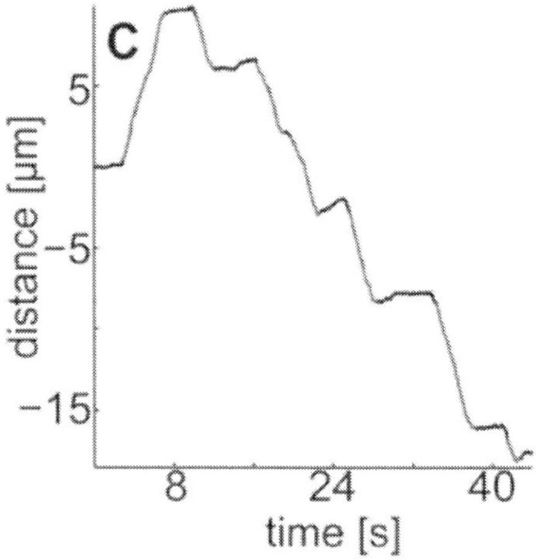

Figure 1. Imaging, tracking and trajectory segmentations of single adenoviruses. (A) HeLa cells were infected with fluorescent adenovirus type 2 for 30 min, and imaged by spinning disc confocal fluorescence microscopy [42]. Virus tracks (black lines) recorded by a single particle tracking algorithm [10] using the nucleus (red square) as a reference point are displayed over a phase contrast image of the infected cell. (B) Two-dimensional projection of a single virus trajectory with directed motion segments in red. (C) Distance travelled along the trajectory shown in Fig. 1B plotted as a function of time. Reduction to 1D is justified, since in cultured cells microtubules are largely flat and straight over distances of many micrometers [43]. Negative/positive values indicate displacements towards the cell centre/periphery.

A Stochastic Model for Cargo Transport

The small number of motor proteins involved in microtubule transport implies a system where the fluctuations in the behavior of motors and the randomness of molecular reactions are essential characteristics [13] suggesting a stochastic modeling of the governing processes. Here we propose a stochastic representation of the main events involved in motor transport, namely stepping along microtubules and binding and unbinding of molecular motors to the cargo.

The proposed model has six parameters, namely the binding, unbinding and stepping rates of plus-end and minus-end motors (herein presumed to be dynein and kinesin, respectively). The step sizes of the motors were set to −8/+8 nm for dynein/kinesin as suggested by the results of single molecule experiments [14],[15]. We note that we do not impose any geometrical information on the motors and their binding sites on the virus capsid. The motor protein binding sites on the adenovirus capsid are not known even though a recent cryo-EM image of the structure of the human adenovirus type 2 temperature sensitive mutant revealed the organization of the surface of the virus capsid [16].

The six model parameters are inferred through a system identification process using the velocity and displacement distributions of segmented trajectories as the cost function of our optimization. An evolutionary algorithm, capable of handling noisy cost functions, is used to obtain the rates that minimize the distance between the velocity and displacement distributions of the *in silico* and *in vivo* trajectories.

The velocity distribution in virus trajectories has led to several suggestions regarding the cooperation or lack thereof between molecular motors. High velocities, in the order of a few microns per second, were observed for intracellular viruses (Fig. 2E) [17]. Similar high speeds have been observed for vesicles moving along microtubules such as peroxisomes [18] and endosomes [19]. These velocities are above the maximum velocities measured for single motors without load (3 μm/s for dynein, [14]; 0.4 μm/s for kinesin-1, [20]; 3 μm/s for kinesin-1, [21]; 0.8 μm/s for kinesin-1, [22]; 0.8 μm/s kinesin-1 and 0.5 μm/s kinesin-2, [23] in *in vitro*experiments. It has also been reported for drosophila lipid droplets, that multiple processive motors do not move cargoes faster [24]. It is likely that yet unknown mechanisms account for the high velocities measured in *in vivo* biological systems. These mechanisms may involve motors which are able to increase their velocities up to few microns per second or motors are able to act additively to achieve higher speeds. Both assumptions have not been experimentally validated or discarded in *in vivo* experiments. Additive behaviour of motors is an underlying assumption in our model (Fig. 2A). The additive behaviour is inherent to the Stochastic Simulation Algorithm [25] used herein to simulate the model, since the time step to the next event depends on the total propensity (numbers and event rates). The proposed stochastic model does not impose any explicit coordination between motor proteins, e.g. a switching mechanism that selects a certain motor protein type to be active.

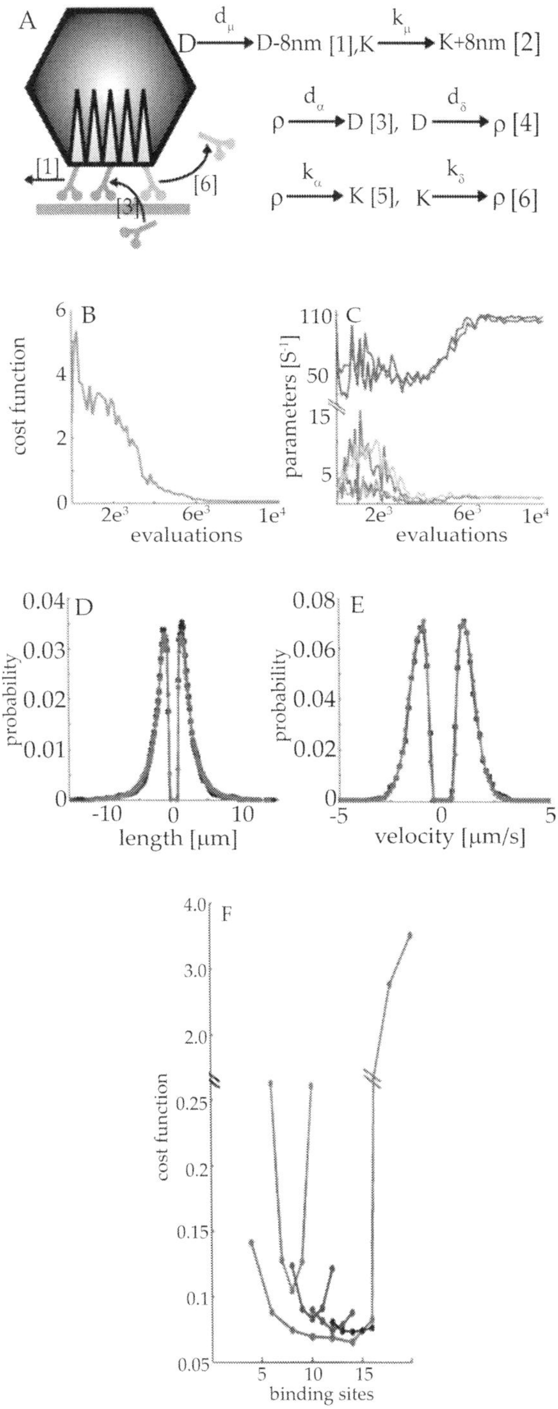
A
[1]
[3]
[6]
D → D-8nm [1], K → K+8nm [2]
ρ → D [3], D → ρ [4]
ρ → K [5], K → ρ [6]
B
cost function
evaluations
C
parameters [S-1]
evaluations
D
probability
length [μm]
E
probability
velocity [μm/s]
F
cost function
binding sites

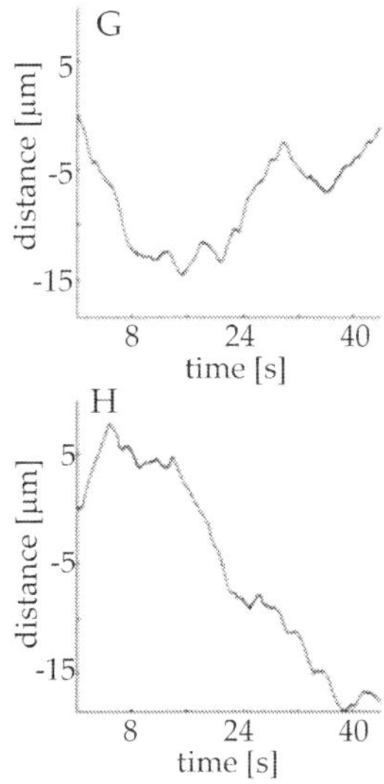

Figure 2. A stochastic model for microtubule-dependent movements of human adenovirus. The stochastic model reproduces directed motion length and velocity distributions of human adenovirus, and predicts the optimal number of either common or separate binding sites for dynein and kinesin motors on the capsid. (A) Dynein (D, blue) and kinesin (K, orange) bind to, and unbind from the capsid and transport it along a microtubule (green). Equations [1]–[6] describe the dynamics of the model: [3]/[4] dynein binding/unbinding and decrease/increase of the number (ρ) of the available motor binding sites on the virus capsid with d_α/d_δ binding/unbinding rates, [5]/[6] kinesin binding/unbinding and decrease/increase of the number(ρ) of the available motor binding sites on the virus capsid with $k_{\alpha/}k_\delta$ binding/unbinding rates, and [1]/[2]dynein/kinesin motor stepping with d_μ/k_μ stepping rates. Cost function (B) and parameter values (C) (blue=d_μ, green=k_μ, red=d_α, cyan=k_α, yellow=d_δ, magenta=k_δ) versus number of evaluations during the optimization of the 14 common binding sites model. Probability distribution of directed motion length (D) and velocity (E) for the *in vivo* and *in silico* (black/red) trajectories. (F) Plot of the cost function versus the number of motor binding sites for the common (red) and the separate binding sites model (blue, grey, dark blue, black colours). The separate binding sites have a total number of 8 (blue), 10 (grey), 12 (dark blue) and 14 (black) binding sites for dynein plus kinesin motors. The central dot in each curve represents 50% dynein and 50% kinesin

occupancy (e.g. black curve: 7+7). The remaining dots denote permutations with decreasing/increasing dynein binding sites (e.g. 6 dynein + 8 kinesin on the right and 8 dynein +6 kinesin on the left of the central dot of the black curve). (G, H) examples of segmented *in silico* 1D trajectories for the 14 common (G) or 7+7 separate (H) binding sites models. The distance (μm) traveled along the 1D microtubule is plotted against the time in seconds, and the directed motions are depicted in red.

We emphasize that our model does not aim at a mechanistic description at the motor level. Forces are known to affect motor properties, but it is not clear how they are distributed among multiple motors [26]. Furthermore while it is possible to obtain data relating forces for certain motors *in vitro*, there is no such data for *in vivo* experiments. In the present model the forces between molecular motors and cargo are implicitly taken into account through the binding/unbinding/stepping rates of the stochastic models.

RESULTS

The simulation of the stochastic model produces cargo trajectories that depend on the parameter settings. The model contains no a-priori assumptions on the existence of either a tug-of-war or coordination between molecular motors. In turn, the model parameters are systematically identified with a derandomized evolution strategy that minimizes the difference between the length and velocity distributions of directed motions (fast microtubule dependent runs [11]) produced by the model and those of fluorescently labelled human adenovirus type 2 (Ad2) as measured by confocal microscopy at 25 Hz temporal resolution. The two-dimensional virus trajectories are extracted by a single particle tracking algorithm [10] (Fig. 1A, B). Directed motions along microtubules are classified by trajectory segmentation [11] and the distance travelled along the microtubule is determined as a function of time (1D trajectory shown in Fig. 1C). The same analysis is applied to trajectories obtained by the simulation of our model using the Stochastic Simulation Algorithm (SSA) [27]. These trajectories are also subsequently segmented to classify directed motions [11]. In turn an optimization algorithm is employed to identify the parameters of the stochastic model [28].

Here the six model parameters (binding, unbinding and stepping for both kinesin and dynein, Fig. 2A) were identified by minimizing the

Symmetric Kullback-Leibler divergence between the *in silico* and *in vivo* length and velocity distributions using an Evolution Strategy with Covariance Matrix Adaptation (CMA-ES) [29] (Fig. 2B,C). The proposed de-randomized optimization algorithm is an essential aspect of our method. CMA-ES samples the six-dimensional multivariate normal distribution involving the parameters of this problem at each iteration and it encodes relations between the parameters of the model and the objective that is being optimized without requiring explicitly the gradients of the cost function [29]. The CMA-ES is a method capable of optimizing noisy cost functions (such as those from the present stochastic model) and its efficiency, reliability and robustness were demonstrated over a number of benchmark problems [30],[31]. The CMA-ES is particularly suitable to this optimization problem as it is know to perform best [29] in problems that are low dimensional (here six parameters), inherently noisy (here a stochastic model), multimodal and computationally expensive (for each parameter set thousands of trajectories are generated and segmented to collect reliable statistics). The algorithm identifies an optimal set of parameters and at the same time provides a sensitivity analysis of the model. The standard deviations of the six principal axes are shown to converge (Fig. 3), thus yielding a minimum.

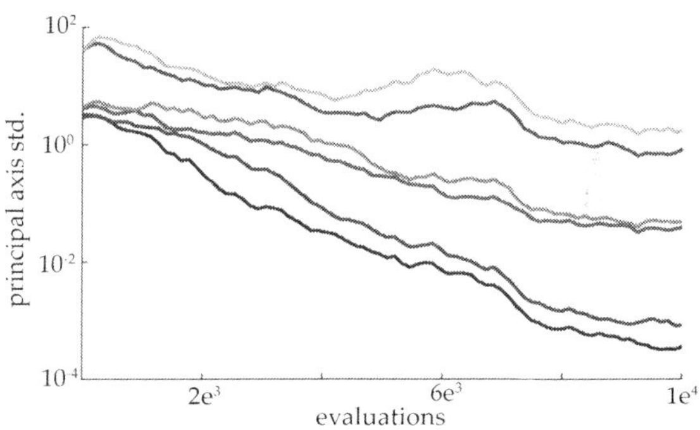

Figure 3. Convergence of the standard deviation of the principal axes. Convergence is shown for the six axes of the distribution which CMA-ES samples from. The evolution of the standard deviations during the optimization procedure is shown for the non-competing binding sites model with 14 receptors.

After the convergence of the optimization process (Fig. 3) we found that the directed motion length and velocity distributions of the *in silico* trajectories, under the optimal set of parameters, matched with high accuracy the experimental data (Fig. 2D, E).

The maximum number of motors attached to the viral cargo is limited by the number of binding sites on the virus. The present model enables predictions on the number of motor binding sites on the viral capsid, a quantity that is difficult to determine experimentally but important for understanding the mechanisms of transport. We first estimated the number (between 2 and 20) of motor binding sites on the virus by an optimization procedure (Fig. 2F). In models with 6–16 binding sites, the cost function values were almost constant around the minimum value obtained for 14 binding sites. For less than 6 motor binding sites, the optimization process did not converge to the experimentally observed directed motion length and velocity distributions. Above 16 binding sites, an unbalanced configuration of motors was feasible only at low binding and unbinding rates, and yielded largely unidirectional trajectories due to infrequent motor binding to the virus. We concluded that 14 common binding sites for dynein and kinesin correspond optimally to the experimental data.

Since it is not possible to differentiate between common and separate binding sites, we additionally investigated the possibility that the experimental data support separate binding sites for the different motors. We optimized a model where dynein and kinesin have distinct binding sites, namely 4+4, 5+5, 6+6, 7+7 binding sites, and various permutations thereof (Fig. 2F), and found that an equal number of motor binding sites was optimal in all cases. This is consistent with the observation that center and periphery directed length and velocity distributions were almost symmetric (Fig. 2D, E). We note that the optimal number of binding sites, i.e., 14, is the same for the models with common and separate binding sites (Fig. 2F, black curve).

Molecular motors carrying cargo on microtubules operate as individuals or as an ensemble. We found that, on average, during virus directed motions, 1.56±0.56 dynein or kinesin (for minus-end and plus-end directed motions, respectively) motors, and 0.15±0.22 motors of opposite polarity were bound to the virus (Fig. 4A). The probability of binding more than four motors to one virus particle was below 10^{-3}, and most often only one type of motor was bound (Fig. 4A, B). These data are in agreement with low number of motors estimated on vesicular cargo in squid axoplasm by cryo-EM [7]. For other organelles, the estimates for motor numbers range from a few to dozens based on immunological detections in chemically fixed cells.

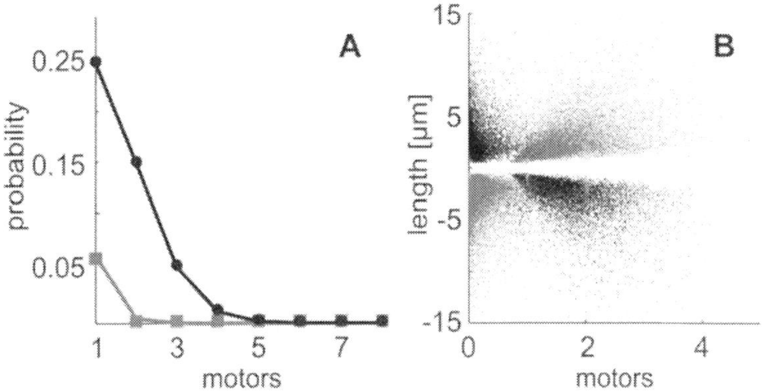

Figure 4. A low number of dynein and kinesin motors mediate directed motions of adenovirus. (A) *In silico* probabilities of the number of bound dynein (black) and kinesin (red) motors during periphery directed motions. Similar results were observed for center directed motions with dynein prevalence. Phase diagrams of the length (B) of directed motions versus the average number of bound dynein (black) and kinesin (red). Positive (negative) lengths correspond to periphery (center) directed motions. The results were obtained with the optimized model with 14 common binding sites.

In order to quantify the correlation between the number of bound motors and the directed motion length, the Sample Pearson Product Moment correlation coefficient (with a range of 0 to 1, where 1 is maximal correlation) between motor numbers and directed motion length was computed to be 0.51 for dynein and 0.49 for kinesin for minus-end and plus-end directed motions, respectively. This implies a weak correlation between the number of bound motors and the directed motion length, showing that long runs do not necessarily require many motors, as two or three already account for lengths in the order of micrometers (Fig. 4B). This result is consistent with the recent *in vitro* observation that two motors are sufficient to move a cargo over several micrometers [32].

The velocities, derived from optimized stepping rates, for single dynein and kinesin motors were 866 nm/s and 833 nm/s, respectively, consistent with values reported for dynein and conventional kinesin-1 or kinesin-2 [21]–[23],[33]. Although kinesins are currently not known to be involved in cytoplasmic transport of adenovirus [1], the model makes a clear prediction for a plus end directed motor in cytoplasmic transport of adenovirus.

Our findings indicate that microtubule-based motility of adenovirus requires a low number of bound motors compared to the number of binding sites on the capsid. This allows configurations where only one motor type is bound, and thereby produce directed motions. Low numbers of motors allow fast switches between directions and therefore, bidirectional motion. Importantly, the binding and unbinding rates were much smaller than the stepping rates, which is key for directed motion runs (Fig. 2C). Small perturbations of binding and unbinding rates greatly affect the model dynamics. The susceptibility of motor based cargo transport to these parameters has been reported in other theoretical studies [26] and hints to a possible mechanism to regulate the run length of the motors [32].

The present results enabled an assessment on the virus binding sites for motor proteins. The outer surface of adenovirus particles is composed of five polypeptides, three of which are still present on cytosolic viruses that have undergone stepwise disassembly [34]. Cytosolic particles contain the major protein hexon, a large fraction of the pentameric penton base at the icosahedral vertex, and protein IX (pIX), which stabilizes hexon. By considering the size (90 nm in diameter) and icosahedral geometry of the virus and the cylindrical microtubule (25 nm in diameter), we can postulate that the maximum number of microtubule motor-capsid interactions occurs along the edge of a capsid facet, in this case on hexon (Fig. 5A, B). This arrangement implies that 9 hexon trimers are aligned with the microtubule, giving a maximum of 27 hexon binding sites for the motors. This is above the value of 14 binding sites predicted from the simulations. If we assume, however, that the motor protein binding sites are located at the interface of two trimeric hexons, one microtubule filament could cover 1–15 sites (Fig. 5B, red lines), which is within the predicted range of 6–16. In addition to hexon, 6 to 8 binding sites were available for pIX, and less than 5 for penton base which detaches to a significant extent from the incoming virions before reaching the cytosol [34]. We analyzed trajectories of pIX-deficient adenoviruses to distinguish between hexon and pIX binding sites for motor proteins [35].

The directed motion length and velocity distributions of pIX-deleted adenovirus were similar to those from wild-type viruses without significant deviations or asymmetries, indicating that pIX may not provide a binding site for microtubule dependent motors during cytoplasmic transport (Fig. 5C, D). Therefore, we predict that hexon harbours the binding sites for dynein and kinesin motors.

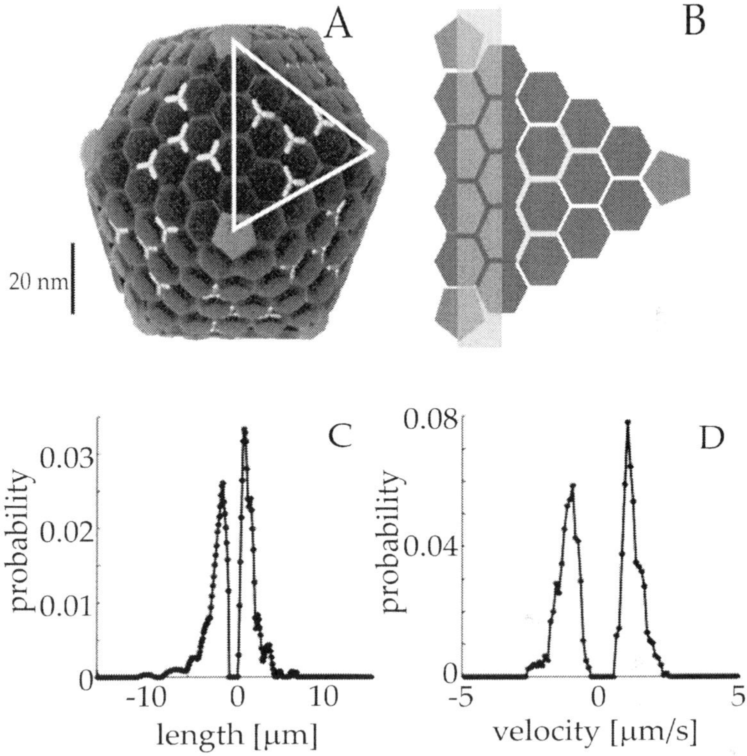

Figure 5. Hexon not protein IX is the predicted to be motor binding site on the adenovirus capsid. (A) Schematic model of the icosahedral adenovirus capsid with the major capsid protein hexon (blue hexagonal structures representing trimers), and pIX (yellow lines representing trimers). An icosahedral facette is enlarged in (B), where hexon-hexon trimer interfaces are depicted in red and in light green the microtubule orientation that maximizes the overlap with one facet. Note that the fourth trimer of pIX is covered by the red lines depicting hexon-hexon interfaces. (C, D) Directed motion length and velocity distributions for protein IX deficient adenovirus recorded in HeLa cells 30 to 90 min post infection.

DISCUSSION

In this study, we use *in vivo* imaging to identify a stochastic model of cargo transport by molecular motors on microtubules. The model parameters were systematically identified using live imaging data of virus trajectories and a de-randomized optimization algorithm to minimize the Kullback-Leibler divergence between the length and velocity distributions of adenovirus directed motions on microtubules with the *in silico* trajectories produced by the model. The model accounts for directed motions at μm/s speeds, processive stepping over hundreds of nanometres, and periods of stationary behaviour. The results show that the stochastic model can result in bidirectional support without an explicit coordination mechanism.

In our work kinetic rates of a stochastic model are determined via an evolutionary optimization approach using experimental data. The identified model enables a number of predictions. First, it shows that one to four motors are active on virus particles during microtubule-dependent motions, although the number of motor binding sites is estimated to be 6–16. The observation that the cost function value is constant within this range suggests that the virus may align with the microtubule in different orientations (Fig. 5B) and still preserve its motility. This range is consistent with the maximum of 15 hexon trimer-trimer interfaces along the edge of a capsid facet. The low number of motors involved in directed motions supports an emerging concept from wet lab experiments and *in silico* simulations, that key events of cell functions are in many cases executed by only a few polypeptides [36].

Second, if equal numbers of opposite motors are attached, the cargo oscillates and eventually stops, or remains confined to small areas. This may be an important mechanism for fine-tuning the subcellular velocity to achieve localized delivery of the cargo. We anticipate that viral transport is tuned by the binding and unbinding rates of motors to microtubules or the cargo, rather than by additional regulatory factors. Such tuning could be cell-type specific [17], and could control the number of engaged motors and motor configuration, and also provide specific segregation or guidance cues for traffic. In support of this, it has been suggested that the microtubule binding protein Tau can fine-tune the distance that the cargo travels by reducing microtubule binding of kinesin in distal parts of neuronal axons [7],[37]. In addition, motor properties can be tuned by post-translational modifications, such as phosphorylation of dynein or kinesin binding partners, which could affect their enzymatic functions and hence their stepping rates [7].

We close by noting that besides the results on motor transport on microtubules, the algorithm taken here is in line with reverse engineering and systems identification approaches [28], [38]–[40] which are gaining significance as discovery and model validation tools in systems biology. The CMA-ES algorithm is capable of handling noisy and multimodal cost functions that are inherently associated with stochastic models. The CMA-ES optimization algorithm along with image analysis of *in vivo* systems can be a robust process to help identify parameters of stochastic models employed in several areas of systems biology.

MATERIALS AND METHODS

HeLa cells were grown to 30% confluency on 18 mm glass cover slips (Menzel Glaser) and kept in Hanks buffered salt solution containing 0.5% BSA (Sigma) and 1 mg/ml ascorbic acid (Sigma). Adenovirus serotype 2 and protein IX deficient adenoviruses were grown, isolated, and labeled with atto565 (Atto-tec, Germany) as described by Nakano and Greber in [41] and Suomalainen et al. in [17].

HeLa cells were infected with Ad-atto565 and imaged between 30 and 90 minutes post infection at 25 Hz. Flat regions of the cell were chosen for imaging in order to minimize the cytoplasmic volume above the imaging plane. The center of the cell was detected by differential interference contrast imaging to assign directionality to the virus motions. Images were recorded using a spinning disc confocal microscope (Olympus IX81) fitted with an UplanApo100x objective of N.A. 1.35 on a back-illuminated monochrome Cascade 512 EM-CCD camera (Photometrics) containing a 512×512 pixel chip (with 16×16 micrometer large pixels).

ACKNOWLEDGMENTS

We thank Diego Rossinelli for his support during the code parallelization and Jort Vellinga for Figure 4A.

Author Contributions

Conceived and designed the experiments: MG CJB BB UFG PK. Performed the experiments: CJB ME. Analyzed the data: MG CJB BB. Contributed reagents/materials/analysis tools: PK. Wrote the paper: MG CJB BB UFG PK.

REFERENCES

1. Greber UF, Way M (2006) A superhighway to virus infection. Cell 124: 741–754.UF GreberM. Way2006A superhighway to virus infection.Cell124741754
2. Brandenburg B, Zhuang XW (2007) Virus trafficking-learning from single-virus tracking. Nature Reviews Microbiology 5: 97–208.B. BrandenburgXW Zhuang2007Virus trafficking-learning from single-virus tracking.Nature Reviews Microbiology597208
3. Schliwa M, Woehlke G (2003) Molecular motors. Nature 422: 759–765.M. SchliwaG. Woehlke2003Molecular motors. Nature422759765
4. Vale RD (2003) The molecular motor toolbox for intracellular transport. Cell 112: 467–480.RD Vale2003The molecular motor toolbox for intracellular transport.Cell112467480
5. Welte MA (2004) Bidirectional transport along microtubules. Current Biology 14: 525–537.MA Welte2004Bidirectional transport along microtubules.Current Biology14525537
6. Gross SP (2204) Hither and yon: a review of bi-directional microtubule-based transport. Physical Biology 1: 1–11.SP Gross2204Hither and yon: a review of bi-directional microtubule-based transport.Physical Biology1111
7. Gross SP, Vershinin M, Shubeita GT (2007) Cargo transport: Two motors are sometimes better than one. Current Biology 17: 478–486.SP GrossM. VershininGT Shubeita2007Cargo transport: Two motors are sometimes better than one.Current Biology17478486
8. Brown A, Wang L, Jung P (2005) Stochastic simulation of neurofilament transport in axons: The "stop-and-go" hypothesis. Molecular Biology of the Cell 16: 4243–4255.A. BrownL. WangP. Jung2005Stochastic simulation of neurofilament transport in

axons: The "stop-and-go" hypothesis.Molecular Biology of the Cell1642434255

9. Muller MJI, Klumpp S, Lipowsky R (2008) Tug-of-war as a cooperative mechanism for bidirectional cargo transport by molecular motors. Proceedings of the National Academy of Sciences of the United States of America 105: 4609–4614.MJI MullerS. KlumppR. Lipowsky2008Tug-of-war as a cooperative mechanism for bidirectional cargo transport by molecular motors. Proceedings of the National Academy of Sciences of the United States of America10546094614
10. Sbalzarini IF, Koumoutsakos P (2005) Feature point tracking and trajectory analysis for video imaging in cell biology. Journal of Structural Biology 151: 182–195.IF SbalzariniP. Koumoutsakos2005Feature point tracking and trajectory analysis for video imaging in cell biology.Journal of Structural Biology151182195
11. Helmuth JA, Burckhardt CJ, Koumoutsakos P, Greber UF, Sbalzarini IF (2007) A novel supervised trajectory segmentation algorithm identifies distinct types of human adenovirus motion in host cells. Journal of Structural Biology 159: 347–358.JA HelmuthCJ BurckhardtP. KoumoutsakosUF GreberIF Sbalzarini2007A novel supervised trajectory segmentation algorithm identifies distinct types of human adenovirus motion in host cells.Journal of Structural Biology159347358
12. Astrom KJ, Eykhoff P (1971) System Identification – Survey. Automatica 7: 123–162.KJ AstromP. Eykhoff1971System Identification – Survey.Automatica7123162
13. Fedoroff N, Fontana W (2002) Genetic networks: Small numbers of big molecules. Science 297: 1129–1131.N. FedoroffW. Fontana2002Genetic networks: Small numbers of big molecules. Science29711291131
14. Gennerich A, Carter AP, Reck-Peterson SL, Vale RD (2007) Force-induced bidirectional stepping of cytoplasmic dynein. Cell 131: 952–965.A. GennerichAP CarterSL Reck-PetersonRD Vale2007Force-induced bidirectional stepping of cytoplasmic dynein.Cell131952965

15. Svoboda K, Schmidt CF, Schnapp BJ, Block SM (1993) Direct observation of kinesin stepping by optical trapping interferometry. Nature 365: 721–727.K. SvobodaCF SchmidtBJ SchnappSM Block1993Direct observation of kinesin stepping by optical trapping interferometry.Nature365721727

16. Silvestry M, Lindert S, Smith JG, Maier O, Wiethoff CM, et al. (2009) Cryo-Electron Microscopy Structure of Adenovirus Type 2 Temperature-Sensitive Mutant 1 Reveals Insight into the Cell Entry Defect. Journal of Virology 83: 7375–7383.M. SilvestryS. LindertJG SmithO. MaierCM Wiethoff2009Cryo-Electron Microscopy Structure of Adenovirus Type 2 Temperature-Sensitive Mutant 1 Reveals Insight into the Cell Entry Defect.Journal of Virology8373757383

17. Suomalainen M, Nakano MY, Keller S, Boucke K, Stidwill RP, et al. (1999) Microtubule-dependent plus- and minus end-directed motilities are competing processes for nuclear targeting of adenovirus. Journal of Cell Biology 144: 657–672.M. SuomalainenMY NakanoS. KellerK. BouckeRP Stidwill1999Microtubule-dependent plus- and minus end-directed motilities are competing processes for nuclear targeting of adenovirus.Journal of Cell Biology144657672

18. Rapp S, Saffrich R, Anton M, Jakle U, Ansorge W, et al. (1996) Microtubule-based peroxisome movement. Journal of Cell Science 109: 837–849.S. RappR. SaffrichM. AntonU. JakleW. Ansorge1996Microtubule-based peroxisome movement.Journal of Cell Science109837849

19. Lakadamyali M, Rust MJ, Zhuang XW (2006) Ligands for clathrin-mediated endocytosis are differentially sorted into distinct populations of early endosomes. Cell 124: 997–1009.M. LakadamyaliMJ RustXW Zhuang2006Ligands for clathrin-mediated endocytosis are differentially sorted into distinct populations of early endosomes. Cell1249971009

20. Yildiz A, Tomishige M, Gennerich A, Vale RD (2008) Intramolecular strain coordinates kinesin stepping behavior along microtubules. Cell 134: 1030–1041.A. YildizM. TomishigeA. GennerichRD Vale2008Intramolecular strain coordinates kinesin stepping behavior along microtubules.Cell13410301041

21. Visscher K, Schnitzer MJ, Block SM (1999) Single kinesin molecules studied with a molecular force clamp. Nature 400: 184–189.K. VisscherMJ SchnitzerSM Block1999Single kinesin molecules studied with a molecular force clamp.Nature400184189
22. Carter NJ, Cross RA (2005) Mechanics of the kinesin step. Nature 435: 308–312.NJ CarterRA Cross2005Mechanics of the kinesin step.Nature435308312
23. Berezuk MA, Schroer TA (2007) Dynactin enhances the processivity of kinesin-2, Traffic 8: 124–129.MA BerezukTA Schroer2007Dynactin enhances the processivity of kinesin-2,Traffic8124129
24. Shubeita GT, Tran SL, Xu J, Vershinin M, Cermelli S, et al. (2008) Consequences of Motor Copy Number on the Intracellular Transport of Kinesin-1-Driven Lipid Droplets. Cell 135: 1098–1107.GT ShubeitaSL TranJ. XuM. VershininS. Cermelli2008Consequences of Motor Copy Number on the Intracellular Transport of Kinesin-1-Driven Lipid Droplets.Cell13510981107
25. Gillespie DT (1977) Exact stochastic simulation of coupled chemical reactions. Journal of Physical Chemistry 81: 2340–2361. DT Gillespie1977Exact stochastic simulation of coupled chemical reactions.Journal of Physical Chemistry8123402361
26. Kunwar A, Vershinin M, Xu J, Gross SP (2008) Stepping, strain gating, and an unexpected force-velocity curve for multiple-motor-based transport. Current Biology 18: 1173–1183.A. KunwarM. VershininJ. XuSP Gross2008Stepping, strain gating, and an unexpected force-velocity curve for multiple-motor-based transport. Current Biology1811731183
27. Gillespie DT (1976) General method for numerically simulating stochastic time evolution of coupled chemical reactions. Journal of Computational Physics 22: 403–434.DT Gillespie1976General method for numerically simulating stochastic time evolution of coupled chemical reactions.Journal of Computational Physics22403434
28. Banga JR (2008) Optimization in computational systems biology. BMC Systems Biology 2: 47.JR Banga2008Optimization in computational systems biology.BMC Systems Biology247

29. Hansen N, Muller SD, Koumoutsakos P (2003) Reducing the time complexity of the derandomized evolution strategy with covariance matrix adaptation (CMA-ES). Evolutionary Computation 11: 1–18.N. HansenSD MullerP. Koumoutsakos2003Reducing the time complexity of the derandomized evolution strategy with covariance matrix adaptation (CMA-ES).Evolutionary Computation11118
30. Kern S, Koumoutsakos P (2006) Simulations of optimized anguilliform swimming. Journal of Experimental Biology 209: 4841–4857.S. KernP. Koumoutsakos2006Simulations of optimized anguilliform swimming.Journal of Experimental Biology20948414857
31. Hansen N, Niederberger ASP, Guzzella L, Koumoutsakos P (2009) A Method for Handling Uncertainty in Evolutionary Optimization With an Application to Feedback Control of Combustion. IEEE Transactions on Evolutionary Computation 13: 180–197.N. HansenASP NiederbergerL. GuzzellaP. Koumoutsakos2009A Method for Handling Uncertainty in Evolutionary Optimization With an Application to Feedback Control of Combustion.IEEE Transactions on Evolutionary Computation13180197
32. Vershinin M, Carter BC, Razafsky DS, King SJ, Gross SP (2007) Multiple-motor based transport and its regulation by Tau. Proceedings of the National Academy of Sciences of the United States of America 104: 87–92.M. VershininBC CarterDS RazafskySJ KingSP Gross2007Multiple-motor based transport and its regulation by Tau.Proceedings of the National Academy of Sciences of the United States of America1048792
33. Mallik R, Petrov D, Lex SA, King SJ, Gross SP (2005) Building complexity: An in vitro study of cytoplasmic dynein with in vivo implications. Current Biology 15: 2075–2085.R. MallikD. PetrovSA LexSJ KingSP Gross2005Building complexity: An in vitro study of cytoplasmic dynein with in vivo implications.Current Biology1520752085
34. Greber UF, Willetts M, Webster P, Helenius A (1993) Stepwise dismantling of Adenovirus-2 during entry into cells. Cell 75: 477–486.UF GreberM. WillettsP. WebsterA. Helenius1993Stepwise dismantling of Adenovirus-2 during entry into cells.Cell75477486

35. Vellinga J, Van der Heijdt S, Hoeben RC (2005) The adenovirus capsid: major progress in minor proteins. Journal of General Virology 86: 1581–1588.J. VellingaS. Van der HeijdtRC Hoeben2005The adenovirus capsid: major progress in minor proteins.Journal of General Virology8615811588
36. Cai L, Friedman N, Xie XS (2006) Stochastic protein expression in individual cells at the single molecule level. Nature 440: 358–362.L. CaiN. FriedmanXS Xie2006Stochastic protein expression in individual cells at the single molecule level.Nature440358362
37. Dixit R, Ross JL, Goldman YE, Holzbaur ELF (2008) Differential regulation of dynein and kinesin motor proteins by tau. Science 319: 1086–1089.R. DixitJL RossYE GoldmanELF Holzbaur2008Differential regulation of dynein and kinesin motor proteins by tau.Science31910861089
38. Jaqaman K, Danuser G (2006) Linking data to models: data regression. Nature Reviews Molecular Cell Biology 7: 813–819.K. JaqamanG. Danuser2006Linking data to models: data regression. Nature Reviews Molecular Cell Biology7813819
39. Sontag ED (2008) Network reconstruction based on steady-state data. Essays in Biochemistry 45: 161–176.ED Sontag2008Network reconstruction based on steady-state data.Essays in Biochemistry45161176
40. Ingolia NT, Weissman JS (2008) Systems biology - Reverse engineering the cell. Nature 454: 1119–1122.NT IngoliaJS Weissman2008Systems biology - Reverse engineering the cell. Nature45411191122
41. Nakano MY, Greber UF (2000) Quantitative microscopy of fluorescent adenovirus entry. Journal of Structural Biology 129: 57–68.MY NakanoUF Greber2000Quantitative microscopy of fluorescent adenovirus entry.Journal of Structural Biology1295768
42. Gastaldelli M, Imelli N, Boucke K, Amstutz B, Meier O, et al. (2008) Infectious Adenovirus Type 2 Transport Through Early but not Late Endosomes. Traffic 9: 2265–2278.M. GastaldelliN. ImelliK. BouckeB. AmstutzO. Meier2008Infectious Adenovirus Type 2 Transport Through Early but not Late Endosomes.Traffic922652278

43. Scales SJ, Pepperkok R, Kreis TE (1997) Visualization of ER-to-Golgi transport in living cells reveals a sequential mode of action for COPII and COPI. Cell 90: 1137–1148.SJ ScalesR. PepperkokTE Kreis1997Visualization of ER-to-Golgi transport in living cells reveals a sequential mode of action for COPII and COPI. Cell9011371148

CHAPTER

6

THE ROLE OF MOLECULAR MICROTUBULE MOTORS AND THE MICROTUBULE CYTOSKELETON IN STRESS GRANULE DYNAMICS

Kristen M. Bartoli, Darryl L. Bishop, and William S. Saunders

Department of Biological Sciences, University of Pittsburgh, 4249 5th Avenue, Crawford Hall 274, Pittsburgh, PA 15213, USA

ABSTRACT

Stress granules (SGs) are cytoplasmic foci that appear in cells exposed to stress-induced translational inhibition. SGs function as a triage center, where mRNAs are sorted for storage, degradation, and translation reinitiation. The underlying mechanisms of SGs dynamics are still being characterized, although many key players have been identified. The main components of SGs are stalled 48S preinitiation complexes. To date, many other proteins have also been found to localize in SGs and are hypothesized to function in SG dynamics. Most recently, the microtubule cytoskeleton and associated motor proteins have been demonstrated to function in SG dynamics. In this paper, we will discuss current literature examining the function of microtubules and the molecular microtubule motors in SG assembly, coalescence, movement, composition, organization, and disassembly.

INTRODUCTION

Stress granules (SGs) are nonmembranous cytoplasmic foci that rapidly appear in cells exposed to various types of stress including oxidative stress, heat shock, viral infection, and UV irradiation, all of which impair translation initiation (as reviewed in [1–3]). In response to such cellular insults, the cell activates mechanisms to selectively repress the translation of house-keeping gene transcripts to conserve energy for repair of stress-induced damage, while upregulating translation of proteins required for the repair process such as DNA-repair proteins, chaperone proteins, and transcription factors (as reviewed in [3–6]).

In this paper, we will discuss the importance of the microtubule network and the microtubule motors involved in SG dynamics including the assembly, coalescence, and disassembly processes. There are at least three different elements that are known to affect SG assembly and dynamics: posttranslational modifications, protein-protein interactions, and the microtubule network. The first two factors have been previously reviewed in [5, 7], and in this paper we will focus our attention on the role of the microtubule network and associated molecular motors in SG dynamics.

Although the components of SGs often vary in different experimental systems or even under different types of stress, the formation of SGs is highly conserved [5, 7, 8]. SG formation has been observed in various systems including yeast [9–11], trypanosomatid [12], and mammalian [4, 13, 14]. Furthermore, SGs are not limited to in vitro cell culture models, as SGs have been observed in the tissues of stressed animals. Animals subjected to gentamicin-induced toxicity [15], ionizing radiation (IR) [16], and ischemic reperfusion injuries exhibit several hallmarks of SG formation [17, 18]. These results indicate that SG formation is part of a cell's native physiological response, rather than an in vitro artifact of cell culture systems.

At the onset of stress, SG formation typically begins through a mechanism involving phosphorylation of eIF2αmediated by one of four eIF2α kinases (PKR, PERK, HRI, or GCN2) [1, 5, 8, 14]. Each of these kinases are activated in response to specific types of stress, for example PKR is activated by double-strand RNAs in response to heat shock, UV, or viral infection [19, 20], PERK is activated in response to unfolded protein accumulation in the endoplasmic reticulum lumen [21, 22], HRI in response to oxidative stress [23], and GCN2 responds to amino acid deprivation and low nutrient levels [24]. eIF2α is a component of the ternary complex required for translation initiation. Phosphorylation of eIF2α causes a reduction in the availability of , thus preventing assembly of the 48S preinitiation complex required for normal translation initiation [19]. The lack of available preinitiation complexes does not affect those transcripts actively undergoing elongation, allowing the translating ribosomes to terminate and run off, disassembling the polysomes [1, 2, 25]. The eIF2/eIF5 deficient "stalled" 48S preinitiation complexes consisting of bound polyadenylated mRNAs, the small ribosomal subunits, as well as various translation initiation factors including eIF3, eIF4E, and eIF4G, are organized into the newly forming SGs [26, 27]. These defective 48S preinitiation complexes are critical substrates for the assembly process of SGs to occur [26]. This formation is promoted through a variety of RNA-binding proteins (TIA-1/R [14], Fragile X Mental Retardation protein FMRP/FXR1 [28], G3BP [29], TTP [30], BRF1 [31], CPEB [32], and SMN [33]), some of which bind to RNA and oligomerize, leading to the initiation of SG assembly [3, 5].

Phosphorylation of eIF2α is not exclusively required for SG formation, as treatment with the translation initiation inhibitors pateamine A [34] or hippuristanol cause SG formation independent of eIF2αphosphorylation [35]. Both translation initiation inhibitors target the eIF4A helicase, thus preventing ribosome scanning and ultimately leading to inhibition of translation initiation and polysome disassembly [36, 37]. Polysomes and SGs are thought to be in equilibrium, as polysome stabilization with elongation inhibitors leads to disassembly of SGs even in the presence of eIF2α phosphorylation [5, 38]. Additionally, overexpression alone of specific RNA-binding proteins whose translation is typically repressed [5, 31, 32, 39, 40] or even the inhibition of certain initiation factors results in spontaneous SG formation [5, 41].

SG DYNAMICS

Within minutes of stress, the buildup of defective 48S pre-initiation complexes overwhelms the cells' ability to deal with the stalled translation, and cells begin to store these complexes temporarily [25]. This initiates the first stages of SG formation, where SGs begin to develop as small foci, which then coalesce or fuse to form larger foci that persist while the damage is repaired, and ultimately disassemble within a few hours of the stress recovery [4, 5]. As SGs disassemble, the mRNAs and associated proteins are able to resume normal translation.

Table 1: Cytoskeletal-associated proteins that participate in SG dynamics. All proteins in this table localize to SGs. Assembly factors are shown in (a), disassembly factors are shown in (b), and proteins associated with SGs but with additional functions in SG dynamics are shown in (c). Cytoskeletal-associated proteins that functioned in SG coalescence were not identified, but publications demonstrating the importance of microtubule function in SG coalescence are suggestive of molecular motors aiding in this process.

(a) Assembly factors/phenotypes

Protein	Function	Cytoskeletal-association
CPEB	Translational regulator; OX induces SG formation; binding partners: RCK, eIF4E, FXR1P [32]	MT-associated [42]
DIC1/DHC1	Required for SG formation; transport [43–45]	MT-associated molecular motor [46]
DIS1	May function as a translational regulator; OX promotes assembly of SGs; binding partner eIF3h [47]	MT-associated [48]
eIF4A	An RNA helicase required for ribosome recruitment; inhibition induces SG formation [35]	Cytoskeletal-associated [49]
FMRP	Functions in mRNA transport or translation; OX induces SG formation; binding partners Ago2, RISC [28]	Cytoskeletal-associated [50, 51]
Grb7	Translational regulator; K/D inhibits SG formation; stimulates SG formation by stabilizing TIA-1 aggregates and enhancing SG or RNP integrity [52]	Cytoskeletal-associated [53]
Pumilio 2	Translational inhibitor; OX induces SGs; K/D interferes with SG formation [54]	MT-associated [54]
Smaug1	Translational repressor; OX induces SG formation [55]	Cortical cytoskeletal-associated [55]
Staufen	mRNA binding, transport, and decay; OX impairs SG formation; depletion facilitates SG assembly [56]	MT-associated [56, 57]
SMN	Assembly of small mRNP complexes; OX induces SG [33]	Role in actin dynamics [58]
TDP-43	OX induces SG; K/D has no effect on formation [59]	Role in microtubule organization [60]

(b) Disassembly factors/phenotypes

Protein	Function	Cytoskeletal-association
FAK	K/D impairs SG disassembly; FAK activity causes Grb7 phosphorylation and SG disassembly in recovering cells [52]	Role in actin assembly/microtubule organization [61, 62]
KHC/KLC	Required for SG dissolution; K/D inhibits SG disassembly; transport [44]	MT-associated molecular motor [63]
Dynein	Inhibition affects dissolution; transport [43–45]	MT-associated molecular motor [46]

(c) Factors with additional roles in SG dynamics

Protein	Function	Cytoskeletal-association
eIF2B	Guanine nucleotide exchange factor for eIF2 [27]	Cytokeletal-associated [64]
eIF3	Recruited to SGs during disassembly [5, 27]	MT-associated [49]
eIF4E	mRNA 5′ cap binding protein [65]	Cytoskeletal-associated [66]
eIF4G	Early translation initiation factor [67]	Roles in F-actin localization/microtubule organization [68]
FXR1P	FRMP-associated protein [28]	MT-associated [50]
FXR2P	FMRP-associated protein [28]	MT-associated [50]
Hsp27	Granzyme-associated molecular chaperone [69]	Cytoskeletal-associated [70, 71]
HuD	ELAV/Hu family of RNA-binding proteins, mRNA localization [72]	MT-associated [73]
NAT1/p97	Related to eIF4G, translational repressor activated in apoptosis [74]	MT-associated [75]
PABP-1	Initiation factor binds poly(A) tail [38]	Binds mRNA to MT [76]
PMR1	mRNA decay endonuclease forms a complex with TIA-1 [77]	Cytoskeletal-associated through binding partners [78]
RACK1	Allows survival of stressed cells [1, 79]	Required for astral MT length [80]
Sam68	Nuclear RNA-binding protein [81]	Roles in cytoskeleton organization [82]
TRAF2	Allows survival of stressed cells [1, 83]	Regulates actin cytoskeleton [84]
Xrn1	5′-3′ exoribonuclease I [31, 85]	MT-associated [86]
YB1	SG stability [87, 88]	Binds mRNA to microtubules [76]
ZBP1	RNA-binding protein important for localized translation [89]	MT-associated [90]

The process of SG disassembly does not appear to simply be the process of assembly in reverse. Although certain markers of SG assembly/disassembly appear to correlate (i.e., phosphorylation status of eIF2α), the most compelling argument for distinct mechanisms of these two processes is that the series of events that occurs during the formation of SGs (small foci that coalesce into larger foci) is not reciprocated during the disassembly process. Rather, disassembly is observed as a dissolution of SGs rather than a breaking apart into smaller distinct foci [4, 5, 38]. These phenotypic differences are suggestive of different mechanisms underlying SG assembly and disassembly. Table 1 lists proteins that associate or function with the cytoskeleton under nonstressed conditions and participate in SG dynamics. These proteins are characterized based on their function in SG dynamics as assembly factors, disassembly factors, or those proteins with additional functions in SG dynamics.

Despite the fact that SGs can persist for several hours, SGs are highly dynamic. Fluorescence recovery after photobleaching (FRAP) experiments has demonstrated that certain components of SGs continuously exchange with cytosolic pools [7, 31, 38]. Some proteins stay associated with SGs until disassembly, however other proteins, exhibiting half-lives of 2–60 seconds, rapidly shuttle between the cytoplasm and SGs [4, 31, 38, 91–94]. Some of these rapidly shuttling proteins include RNA-binding proteins known to control mRNA stability, structure, and function (TIA1, TIAR-1, G3BP, PABP, HuR, and TTP) and those known to regulate mRNA translation and decay. This shuttling of proteins and mRNA in and out of SGs implies that they function as mRNA triage centers that sort sequestered mRNAs for storage, degradation, or translation reinitiation [5, 6, 95]. An RNAi screen completed by Ohn et al. identified over a hundred proteins which function in SG assembly [96]. Additionally, in a recent survey of the literature Buchan and Parker identified over 80 proteins which have been confirmed to localize to SGs [7]. Some of these proteins were not only found to be involved in mRNA metabolism, but some of which have functions in other cellular processes. Those proteins not involved in RNA metabolism span diverse functions as transcription factors, RNA-binding proteins, helicases, nucleases, molecular motors, and other signaling proteins [2, 3, 5, 7]. Because of this diversity, one cannot, with knowledge of protein function alone, predict if a protein will be found in SGs or not. This is exemplified by the fact that different experimental conditions produce SGs that contain different components.

SG DYNAMICS REQUIRE THE MICROTUBULE NETWORK

Investigation of microtubule function in SG dynamics is made possible through the use of chemical inhibitors that allow disruption or stabilization of microtubules. Multiple investigations have concluded that the microtubule network is required for SG dynamics including the assembly, the coalescence, and the disassembly process as will be discussed below. Furthermore, recent data is emerging that is beginning to unravel which specific molecular microtubule motors participate in SG dynamics.

SG ASSEMBLY

Ivanov et al. were the first to investigate the effects of microtubule disruption on SG assembly [97]. Using the mammalian CV-1 cell line and two different inhibitors that depolymerize microtubules, nocodazole or vinblastine, a striking inhibition of SG formation in the presence of aresnite was observed. In agreement, Kwon et al. showed a similar phenomenon in HeLa cells following nocodazole treatment [43]. However, in contrast with these results, Kolobova et al., and Fujimura et al., amongst others, reported that microtubule disruption not only delayed SG formation, but when SGs did form, they were of significantly smaller size, greater in number and variable in distribution, rather than observing an abolishment of SG formation [44, 98, 99]. Some of these differences may be attributed to the concentration of the SG-inducing agent (arsenite) used, cell line-specific effects, or the markers used to measure SG formation.

SGs typically range in size from 0.2 to 5 μm [1, 98, 100], however after microtubule disruption by nocodazole or colchicine, SGs were found to remain small (<0.6 μm) [98], with a complete loss of large SGs [44, 98, 99]. Additionally, Kolobova et al. demonstrated that the number of individual SGs per cell was found to significantly increase, depending on the SG marker used [98]. Similar effects were observed by different groups in a variety of cell lines including CHO, HeLa, and COS-7 [44, 99, 100]. A decrease in SG size and an increase in SG number suggest a loss of SG coalescence in the absence of microtubules. These experiments demonstrate that microtubules are not required for the initial steps of SG formation, but are required for secondary steps in SG coalescence/aggregation [98].

Fujimura et al. confirmed the loss of SG coalescence after microtubule disruption by immunofluorescence while Kolobova et al. confirmed this coalescence defect by live cell imaging of cells transfected with mCherry-G3BP, a SG component, and RNA-binding protein [98, 99]. Depolymerization of microtubules led to decreased SG movement and an obvious loss in SG fusion/coalescence [98, 100]. Closer examination of fixed cells with various antibodies demonstrated changes in the composition of SGs [98]. A majority of SGs consistently contained many of the typical markers of SGs such as TIA1, G3BP, TIAR1, and other RNA-binding proteins [98, 99], but were deficient in a variety of proteins including CCAR-1, AKAP350A, eIF2α, HuR, and CUGBP, suggesting that loss of microtubules causes asymmetrical assembly of SGs, not just slowed assembly [8, 98, 99]. Therefore, coalescence is a microtubule-based phenomenon and is required to incorporate specific markers of SGs as the complete and proper assembly and maturation of SGs is affected after loss of microtubules.

Typically, after SG formation and during the process of coalescence, the fully formed SGs are transported and organized around the perinuclear region [99, 100]. However after microtubule depolymerization, SGs were observed in fixed cells to be diffusely distributed throughout the cell in no discernable order, suggestive of microtubules functioning in transport and the reorganization process of SGs [44, 98–100].

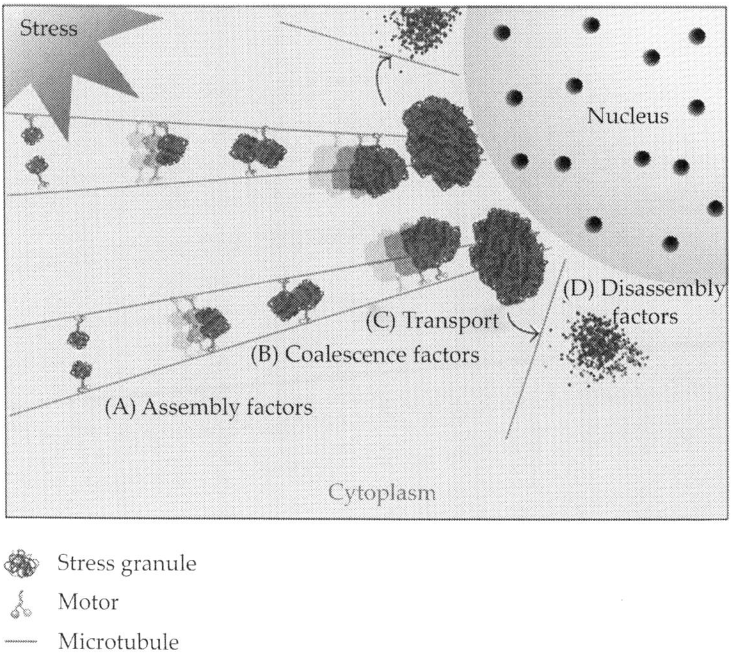

Figure 1: Model of microtubules and molecular motors in stress granule dynamics. When cells are exposed to stress-induced translational inhibition, SGs begin to form as small foci (A), assisted by assembly factors, and are transported along microtubules by molecular motors, enabling coalescence of SGs into larger foci (B). This process of coalescence continues while SGs are transported along microtubules to the perinuclear region where larger granules reside (C). Once the stress is removed SGs disassemble (D) aided by molecular motors as well as other disassembly factors. See Table 1 for the list of cytoskeletal proteins that function in these processes.

Investigation of SG transport by live cell imaging of CHO cells, selected for their distinct radial microtubule network, provided compelling evidence that SGs are indeed transported along microtubules [100]. Following double transfection of mCherry-tagged PABP and GFP-tagged tubulin revealed that SGs traveled along microtubules relocating to the perinuclear region [100]. This movement of SGs along microtubules tracked by time-lapse imaging was noncontinuous, but directional. SGs were found to move in one direction, stop, and sometimes travel again but not always in the same direction, consistent with the movement of microtubule motors. However, through its movement, the SGs continuously remained associated with microtubules, as confirmed in fixed cells [100]. When microtubules were disrupted, the dynamic movement of SGs was severely inhibited and ultimately SGs were not transported to the perinuclear region, confirming that microtubules are involved in the spatial intracellular placement of SGs. The above studies suggest that SGs are transported in a microtubule-dependent manner as small foci coalesce with other foci to form large SGs while on their way to their final destination of the perinuclear region (Figure 1). Loss of SG coalescence and variability in SG composition after microtubule depolymerization supports a microtubule-mediated transport of SGs.

SG assembly in the presence of the microtubule stabilizing drug paclitaxel was also investigated by three groups with different conclusions. Nadezhdina et al. showed an increase in SG movement when microtubules were stabilized and Ivanov et al. demonstrated that microtubule stabilization allowed for enhanced SG formation, favoring large SGs consistent with enhanced coalescence [97, 100]. However, Kwon et al. described microtubule stabilization as having no impact on SG assembly [43]. The reason for the discrepancy in these results is not immediately clear.

DISASSEMBLY

Microtubule function in SG disassembly has also been proposed [100]. SG disassembly has been shown to depend not only on the activity of HSP70 molecular chaperone that accumulates during stress induction, but also on the activity of certain phospho-specific proteins present in SGs [3, 8, 29, 52, 101]. Nadezhdina et al. took advantage of cycloheximides' ability to facilitate SG disassembly; cells treated with arsenite and cycloheximide disassembled SGs gradually starting 30 min after its addition. However, analysis of SG disassembly in cells treated with nocodazole prior to arsenite and cycloheximide treatment illustrated persistent SGs that did not disassemble or change in size [100]. This demonstrates a complete inhibition of SG disassembly when microtubules were disrupted and confirms the importance of the microtubule network in SG dynamics.

MOLECULAR MICROTUBULE MOTORS IN SG DYNAMICS

Molecular microtubule motors are a class of proteins that transport intracellular cargoes along microtubules using ATP hydrolysis; there are approximately 50 different microtubule motors in humans. As discussed above, microtubules were shown to function in SG coalescence and disassembly processes; most likely through the association of microtubule molecular motors. Microtubule disruption resulted in a number of SG abnormalities including SG movement, coalescence, composition, relocation, and disassembly. Each of which can be explained through a molecular motor-based transport of SGs along microtubules towards the perinuclear region using SG coalescence to retain proper SG composition.

Recently, several publications have emerged demonstrating the importance of specific molecular motors in SG dynamics [43–45]. Loschi et al. identified the subunits of two molecular microtubule motors, cytoplasmic dynein and conventional kinesin, that localize to SGs and function in both SG assembly and disassembly [44].

DYNEIN

Dynein is a minus-end directed macromolecular motor complex functioning in numerous processes including mRNA movement [102], mitosis [103], and most recently SG assembly [44, 45]. Components of dyneins' intermediate chain (DIC) and heavy chain (DHC) were identified to localize to SGs by immunofluorescence analysis [44]. Upon closer examination of SG formation in the absence of dynein activity (by p50 overexpression [44], siRNA directed against DHC or DLC2 [44], or pharmacological-inhibition [43]) a ~3-fold impairment in SG formation was observed to occur after treatment with arsenite [43, 44]. The cells that contained SGs in the absence of dynein activity were found to be small in size and in number, suggesting that dynein is required for the initial stages of SG assembly [43, 44].

Studying SG dynamics in primary neurons by immunohistochemistry, Tsai et al. showed that dynein also functions in SG disassembly [45]. Additionally, assays measuring SG integrity by protease sensitivity of TIA-1 complexes [39] demonstrated that dynein enhances the integrity of SGs by assisting SG formation [45]. After transient expression of dynein, the TIA-1 complexes in stressed cells became more resistant to protease treatment than control-stressed cells, whereas in the absence of dynein, these TIA-1 complexes became more sensitive to proteases. These data demonstrate the importance of dynein function in SG dynamics in primary neurons not only in the assembly/disassembly process but also in the integrity of TIA-1 complexes [45].

KINESIN

Conventional kinesin subunits, KIF5B (kinesin heavy chain) and KLC1 (kinesin light chain 1), were also identified by Loschi et al. to localize to SGs and to function in SG dynamics [44]. Investigation of KIF5B in SG assembly after depletion revealed that this kinesin was not involved in SG formation; rather, after removal of arsenite treatment in cells depleted of KIF5B or KLC1, SGs were found to persist relative to control cells [44]. This suggests that conventional kinesin subunits, KIF5B and KLC1, participate in SG disassembly.

Motor proteins are known to work in concert, at times working together and at other times actively opposing each other to achieve the proper balance of transport. This coordinated relationship extends to SG dynamics as well. In the background of dynein knockdown, Loschi et al. was able to demonstrate that kinesin actually functions to inhibit SG formation. As mentioned above, the inhibition of dynein results in decreased SGs [43–45]. However, when cells were subjected to simultaneous inhibition of dynein (DHC1) and kinesin (KIF5B), SG formation resembled that of control cells [44]. Furthermore, the SG defects observed following KIF5B depletion were also partially reverted back to control cell levels following double knockdown, suggesting a possible redundancy of motors in the disassembly of SGs [44]. These results suggest that SG dynamics are in part orchestrated by a finely tuned balance of forces exerted by various molecular motors acting simultaneously.

CONCLUDING REMARKS

Not all of the defects in SG dynamics after loss of microtubules are mimicked by inhibition of either dynein or kinesin, individually or together. After impairment of dynein activity, a complete abolishment in SG formation was not observed, rather at least a third of the population of cells was still able to form SGs suggesting additional motors may participate in the assembly process [43–45]. Furthermore, neither depletion of dynein or kinesin was found to cause a loss in SG coalescence; this is in contrast to the disruption of microtubules which exhibited defects in SG coalescence. Although Tsai et al. speculated that the motor dynein plays a role in SG coalescence, the data is not as compelling or complete as its roles in assembly or disassembly [45]. Given that disruption of microtubules acts as a surrogate for pan-molecular motor inhibition, it is likely that other yet to be discovered molecular motors contribute to SG formation and/or participate in coalescence. Another motor, KIF18A, was identified in an RNAi-based screen to possibly function in SG assembly [96], but has not thoroughly been tested to date. Thus a continued search for molecular motors which participate in SG dynamics is needed to fully understand the underlying SG biology.

REFERENCES

1. P. Anderson and N. Kedersha, "Stress granules," Current Biology, vol. 19, no. 10, pp. R397–R398, 2009.
2. N. Kedersha and P. Anderson, "Mammalian stress granules and processing bodies," Methods in Enzymology, vol. 431, pp. 61–81, 2007.
3. P. Anderson and N. Kedersha, "RNA granules," Journal of Cell Biology, vol. 172, no. 6, pp. 803–808, 2006.
4. P. Anderson and N. Kedersha, "Stressful initiations," Journal of Cell Science, vol. 115, no. 16, pp. 3227–3234, 2002.
5. P. Anderson and N. Kedersha, "Stress granules: the Tao of RNA triage," Trends in Biochemical Sciences, vol. 33, no. 3, pp. 141–150, 2008.
6. N. Kedersha and P. Anderson, "Stress granules: sites of mRNA triage that regulate mRNA stability and translatability," Biochemical Society Transactions, vol. 30, no. 6, pp. 963–969, 2002.
7. J. R. Buchan and R. Parker, "Eukaryotic stress granules: the ins and outs of translation," Molecular Cell, vol. 36, no. 6, pp. 932–941, 2009.

8. M. G. Thomas, M. Loschi, M. A. Desbats, and G. L. Boccaccio, "RNA granules: the good, the bad and the ugly," Cell Signaling, vol. 23, pp. 324–334, 2011.
9. M. Brengues and R. Parker, "Accumulation of polyadenylated mRNA, Pab1p, eIF4E, and eIF4G with P-bodies in Saccharomyces cerevisiae," Molecular Biology of the Cell, vol. 18, no. 7, pp. 2592–2602, 2007.
10. J. R. Buchan, T. Nissan, and R. Parker, "Analyzing P-bodies and stress granules in Saccharomyces cerevisiae," Methods in Enzymology, vol. 470, pp. 619–640, 2010.
11. T. Grouši, P. Ivanov, I. Frýdlová et al., "Robust heat shock induces eIF2α-phosphorylationin-dependent assembly of stress granules containing eIF3 and 40S ribosomal subunits in budding yeast, Saccharomyces cerevisiae," Journal of Cell Science, vol. 122, no. 12, pp. 2078–2088, 2009.
12. S. Kramer, R. Queiroz, L. Ellis et al., "Heat shock causes a decrease in polysomes and the appearance of stress granules in trypanosomes independently of eIF2α phosphorylation at Thr169," Journal of Cell Science, vol. 121, no. 18, pp. 3002–3014, 2008.
13. A. P. Arrigo, J. P. Suhan, and W. J. Welch, "Dynamic changes in the structure and intracellular locale of the mammalian low-molecular-weight heat shock protein," Molecular and Cellular Biology, vol. 8, no. 12, pp. 5059–5071, 1988.
14. N. L. Kedersha, M. Gupta, W. Li, I. Miller, and P. Anderson, "RNA-binding proteins TIA-1 and TIAR link the phosphorylation of eIF-2α to the assembly of mammalian stress granules," Journal of Cell Biology, vol. 147, no. 7, pp. 1431–1441, 1999.
15. D. A. Mangiardi, K. McLaughlin-Williamson, K. E. May, E. P. Messana, D. C. Mountain, and D. A. Cotanche, "Progression of hair cell ejection and molecular markers of apoptosis in the avian cochlea following gentamicin treatment," Journal of Comparative Neurology, vol. 475, no. 1, pp. 1–18, 2004.
16. B. J. Moeller, Y. Cao, C. Y. Li, and M. W. Dewhirst, "Radiation activates HIF-1 to regulate vascular radiosensitivity in tumors: role of reoxygenation, free radicals, and stress granules," Cancer Cell, vol. 5, no. 5, pp. 429–441, 2004.

17. F. Kayali, H. L. Montie, J. A. Rafols, and D. J. DeGracia, "Prolonged translation arrest in reperfused hippocampal cornu Ammonis 1 is mediated by stress granules," Neuroscience, vol. 134, no. 4, pp. 1223–1245, 2005.
18. D. J. DeGracia, J. A. Rafols, S. J. Morley, and F. Kayali, "Immunohistochemical mapping of total and phosphorylated eukaryotic initiation factor 4G in rat hippocampus following global brain ischemia and reperfusion," Neuroscience, vol. 139, no. 4, pp. 1235–1248, 2006.
19. S. P. Srivastava, K. U. Kumar, and R. J. Kaufman, "Phosphorylation of eukaryotic translation initiation factor 2 mediates apoptosis in response to activation of the double-stranded RNA-dependent protein kinase," Journal of Biological Chemistry, vol. 273, no. 4, pp. 2416–2423, 1998.
20. B. R. Williams, "Signal integration via PKR," Sciences STKE, vol. 2001, no. 89, article re2, 2001.
21. H. P. Harding, Y. Zhang, A. Bertolotti, H. Zeng, and D. Ron, "Perk is essential for translational regulation and cell survival during the unfolded protein response," Molecular Cell, vol. 5, no. 5, pp. 897–904, 2000.
22. H. P. Harding, I. Novoa, Y. Zhang et al., "Regulated translation initiation controls stress-induced gene expression in mammalian cells," Molecular Cell, vol. 6, no. 5, pp. 1099–1108, 2000.
23. E. McEwen, N. Kedersha, B. Song et al., "Heme-regulated inhibitor kinase-mediated phosphorylation of eukaryotic translation initiation factor 2 inhibits translation, induces stress granule formation, and mediates survival upon arsenite exposure," Journal of Biological Chemistry, vol. 280, no. 17, pp. 16925–16933, 2005.
24. S. A. Wek, S. Zhu, and R. C. Wek, "The histidyl-tRNA synthetase-related sequence in the eIF-2α protein kinase GCN2 interacts with tRNA and is required for activation in response to starvation for different amino acids," Molecular and Cellular Biology, vol. 15, no. 8, pp. 4497–4506, 1995.
25. N. Kedersha, S. Tisdale, T. Hickman, and P. Anderson, "Real-time and quantitative imaging of mammalian stress granules and

processing bodies," Methods in Enzymology, vol. 448, pp. 521–552, 2008.

26. N. Kedersha, S. Chen, N. Gilks et al., "Evidence that ternary complex (eIF2-GTP-tRNA)-Deficient preinitiation complexes are core constituents of mammalian stress granules," Molecular Biology of the Cell, vol. 13, no. 1, pp. 195–210, 2002.
27. S. R. Kimball, R. L. Horetsky, D. Ron, L. S. Jefferson, and H. P. Harding, "Mammalian stress granules represent sites of accumulation of stalled translation initiation complexes," American Journal of Physiology, vol. 284, no. 2, pp. C273–C284, 2003.
28. R. Mazroui, M. E. Hout, S. Tremblay, C. Fillion, Y. Labelle, and E. W. Khandjian, "Trapping of messenger RNA by Fragile X Mental Retardation protein into cytoplasmic granules induces translation repression," Human Molecular Genetics, vol. 11, no. 24, pp. 3007–3017, 2002.
29. H. Tourrière, K. Chebli, L. Zekri et al., "The RasGAP-associated endoribonuclease G3BP assembles stress granules," Journal of Cell Biology, vol. 160, no. 6, pp. 823–831, 2003.
30. G. Stoecklin, T. Stubbs, N. Kedersha et al., "MK2-induced tristetraprolin:14-3-3 Complexes prevent stress granule association and ARE-mRNA decay," EMBO Journal, vol. 23, no. 6, pp. 1313–1324, 2004.
31. N. Kedersha, G. Stoecklin, M. Ayodele et al., "Stress granules and processing bodies are dynamically linked sites of mRNP remodeling," Journal of Cell Biology, vol. 169, no. 6, pp. 871–884, 2005.
32. A. Wilczynska, C. Aigueperse, M. Kress, F. Dautry, and D. Weil, "The translational regulator CPEB1 provides a link between dcp1 bodies and stress granules," Journal of Cell Science, vol. 118, no. 5, pp. 981–992, 2005.
33. Y. Hua and J. Zhou, "Survival motor neuron protein facilitates assembly of stress granules," FEBS Letters, vol. 572, no. 1-3, pp. 69–74, 2004.
34. Y. Dang, N. Kedersha, W. K. Low et al., "Eukaryotic initiation factor 2α-independent pathway of stress granule induction by the

natural product pateamine A," Journal of Biological Chemistry, vol. 281, no. 43, pp. 32870–32878, 2006.

35. R. Mazroui, R. Sukarieh, M. E. Bordeleau et al., "Inhibition of ribosome recruitment induces stress granule formation independently of eukaryotic initiation factor 2α phosphorylation," Molecular Biology of the Cell, vol. 17, no. 10, pp. 4212–4219, 2006.
36. M. E. Bordeleau, A. Mori, M. Oberer et al., "Functional characterization of IRESes by an inhibitor of the RNA helicase eIF4A," Nature Chemical Biology, vol. 2, no. 4, pp. 213–220, 2006.
37. M. E. Bordeleau, J. Matthews, J. M. Wojnar et al., "Stimulation of mammalian translation initiation factor eIF4A activity by a small molecule inhibitor of eukaryotic translation," Proceedings of the National Academy of Sciences of the United States of America, vol. 102, no. 30, pp. 10460–10465, 2005.
38. N. Kedersha, M. R. Cho, W. Li et al., "Dynamic shuttling of TIA-1 accompanies the recruitment of mRNA to mammalian stress granules," Journal of Cell Biology, vol. 151, no. 6, pp. 1257–1268, 2000.
39. N. Gilks, N. Kedersha, M. Ayodele et al., "Stress granule assembly is mediated by prion-like aggregation of TIA-1," Molecular Biology of the Cell, vol. 15, no. 12, pp. 5383–5398, 2004.
40. F. De Leeuw, T. Zhang, C. Wauquier, G. Huez, V. Kruys, and C. Gueydan, "The cold-inducible RNA-binding protein migrates from the nucleus to cytoplasmic stress granules by a methylation-dependent mechanism and acts as a translational repressor," Experimental Cell Research, vol. 313, no. 20, pp. 4130–4144, 2007.
41. S. Mokas, J. R. Mills, C. Garreau et al., "Uncoupling stress granule assembly and translation initiation inhibition," Molecular Biology of the Cell, vol. 20, no. 11, pp. 2673–2683, 2009.
42. I. Groisman, YI. S. Huang, R. Mendez, Q. Cao, W. Theurkauf, and J. D. Richter, "CPEB, maskin, and cyclin B1 mRNA at the mitotic apparatus: implications for local translational of cell division," Cell, vol. 103, no. 3, pp. 435–447, 2000.
43. S. Kwon, YU. Zhang, and P. Matthias, "The deacetylase HDAC6 is a novel critical component of stress granules involved in the stress

response," Genes and Development, vol. 21, no. 24, pp. 3381–3394, 2007.

44. M. Loschi, C. C. Leishman, N. Berardone, and G. L. Boccacio, "Dynein and kinesin regulate stress-granule and P-body dynamics," Journal of Cell Science, vol. 122, no. 21, pp. 3973–3982, 2009.
45. N. P. Tsai, Y. C. Tsui, and L. N. Wei, "Dynein motor contributes to stress granule dynamics in primary neurons," Neuroscience, vol. 159, no. 2, pp. 647–656, 2009.
46. S. Ma and R. L. Chisholm, "Cytoplasmic dynein-associated structures move bidirectionally in vivo," Journal of Cell Science, vol. 115, no. 7, pp. 1453–1460, 2002.
47. F. Ogawa, M. Kasai, and T. Akiyama, "A functional link between disrupted-in-Schizophrenia 1 and the eukaryotic translation initiation factor 3," Biochemical and Biophysical Research Communications, vol. 338, no. 2, pp. 771–776, 2005.
48. H. Ohkura, M. A. Garcia, and T. Toda, "Dis1/TOG universal microtubule adaptors—one MAP for all?" Journal of Cell Science, vol. 114, no. 21, pp. 3805–3812, 2001.
49. J. G. Howe and J. W. B. Hershey, "Translational initiation factor and ribosome association with the cytoskeletal framework fraction from HeLa cells," Cell, vol. 37, no. 1, pp. 85–93, 1984.
50. M. E. Huot, R. Mazroui, P. Leclerc, and E. W. Khandjian, "Developmental expression of the fragile X-related 1 proteins in mouse testis: association with microtubule elements," Human Molecular Genetics, vol. 10, no. 24, pp. 2803–2811, 2001.
51. L. N. Antar, J. B. Dictenberg, M. Plociniak, R. Afroz, and G. J. Bassell, "Localization of FMRP-associated mRNA granules and requirement of microtubules for activity-dependent trafficking in hippocampal neurons," Genes, Brain and Behavior, vol. 4, no. 6, pp. 350–359, 2005.
52. N. P. Tsai, P. C. Ho, and L. N. Wei, "Regulation of stress granule dynamics by Grb7 and FAK signalling pathway," EMBO Journal, vol. 27, no. 5, pp. 715–726, 2008.
53. S. Siamakpour-Reihani, T. A. Peterson, A. M. Bradford et al., "Grb7 binds to Hax-1 and undergoes an intramolecular domain association

that offers a model for Grb7 regulation," Journal of Molecular Recognition, vol. 24, pp. 314–321, 2011.

54. J. P. Vessey, A. Vaccani, Y. Xie et al., "Dendritic localization of the translational repressor Pumilio 2 and its contribution to dendritic stress granules," Journal of Neuroscience, vol. 26, no. 24, pp. 6496–6508, 2006.
55. M. V. Baez and G. L. Boccaccio, "Mammalian smaug is a translational repressor that forms cytoplasmic foci similar to stress granules," Journal of Biological Chemistry, vol. 280, no. 52, pp. 43131–43140, 2005.
56. M. Köhrmann, M. Luo, C. Kaether, L. DesGroseillers, C. G. Dotti, and M. A. Kiebler, "Microtubule-dependent recruitment of Staufen-green fluorescent protein into large RNA-containing granules and subsequent dendritic transport in living hippocampal neurons," Molecular Biology of the Cell, vol. 10, no. 9, pp. 2945–2953, 1999.
57. L. Wickham, T. Duchaîne, M. Luo, I. R. Nabi, and L. DesGroseillers, "Mammalian staufen is a double-stranded-RNA- and tubulin-binding protein which localizes to the rough endoplasmic reticulum," Molecular and Cellular Biology, vol. 19, no. 3, pp. 2220–2230, 1999.
58. J. Van Bergeijk, K. Rydel-Könecke, C. Grothe, and P. Claus, "The spinal muscular atrophy gene product regulates neurite outgrowth: importance of the C terminus," FASEB Journal, vol. 21, no. 7, pp. 1492–1502, 2007.
59. L. Liu-Yesucevitz, A. Bilgutay, Y. J. Zhang et al., "Tar DNA binding protein-43 (TDP-43) associates with stress granules: analysis of cultured cells and pathological brain tissue," PLoS One, vol. 5, article e13250, 2010.
60. V. K. Godena, G. Romano, M. Romano et al., "TDP-43 regulates drosophila neuromuscular junctions growth by modulating futsch/MAP1B levels and synaptic microtubules organization," PLoS One, vol. 6, article e17808, 2011.
61. B. Serrels, A. Serrels, V. G. Brunton et al., "Focal adhesion kinase controls actin assembly via a FERM-mediated interaction with the

Arp2/3 complex," Nature Cell Biology, vol. 9, no. 9, pp. 1046–1056, 2007.

62. Z. Xie, K. Sanada, B. A. Samuels, H. Shih, and L. H. Tsai, "Serine 732 phosphorylation of FAK by Cdk5 is important for microtubule organization, nuclear movement, and neuronal migration," Cell, vol. 114, no. 4, pp. 469–482, 2003.
63. M. A. Seeger and S. E. Rice, "Microtubule-associated protein-like binding of the kinesin-1 tail to microtubules," Journal of Biological Chemistry, vol. 285, no. 11, pp. 8155–8162, 2010.
64. S. E. Hardt, H. Tomita, H. A. Katus, and J. Sadoshima, "Phosphorylation of eukaryotic translation initiation factor 2Bε by glycogen synthase kinase-3β regulates β-adrenergic cardiac myocyte hypertrophy," Circulation Research, vol. 94, no. 7, pp. 926–935, 2004.
65. Y. Suzuki, M. Minami, M. Suzuki et al., "The Hsp90 inhibitor geldanamycin abrogates colocalization of eIF4E and eIF4E-transporter into stress granules and association of eIF4E with eIF4G," Journal of Biological Chemistry, vol. 284, no. 51, pp. 35597–35604, 2009.
66. S. Lindemann, N. D. Tolley, J. R. Eyre, L. W. Kraiss, T. M. Mahoney, and A. S. Weyrich, "Integrins regulate the intracellular distribution of eukaryotic initiation factor 4E in platelets. A checkpoint for translational control," Journal of Biological Chemistry, vol. 276, no. 36, pp. 33947–33951, 2001.
67. N. P. Hoyle, L. M. Castelli, S. G. Campbell, L. E. A. Holmes, and M. P. Ashe, "Stress-dependent relocalization of translationally primed mRNPs to cytoplasmic granules that are kinetically and spatially distinct from P-bodies," Journal of Cell Biology, vol. 179, no. 1, pp. 65–74, 2007.
68. L. Hashemzadeh-Bonehi, P. S. Curtis, S. J. Morley, J. R. Thorpe, and V. M. Pain, "Overproduction of a conserved domain of fission yeast and mammalian translation initiation factor eIF4G causes aberrant cell morphology and results in disruption of the localization of F-actin and the organization of microtubules," Genes to Cells, vol. 8, no. 2, pp. 163–178, 2003.
69. P. J. Beresford, M. Jaju, R. S. Friedman, M. J. Yoon, and J.

Lieberman, "A role for heat shock protein 27 in CTL-mediated cell death," Journal of Immunology, vol. 161, no. 1, pp. 161–167, 1998.

70. M. Hino, K. Kurogi, M. A. Okubo, M. Murata-Hori, and H. Hosoya, "Small heat shock protein 27 (HSP27) associates with tubulin/microtubules in HeLa cells," Biochemical and Biophysical Research Communications, vol. 271, no. 1, pp. 164–169, 2000.
71. S. K. Van Why, A. S. Mann, T. Ardito et al., "Hsp27 associates with actin and limits injury in energy depleted renal epithelia," Journal of the American Society of Nephrology, vol. 14, no. 1, pp. 98–106, 2003.
72. R. W. Burry and C. L. Smith, "HuD distribution changes in response to heat shock but not neurotrophic stimulation," Journal of Histochemistry and Cytochemistry, vol. 54, no. 10, pp. 1129–1138, 2006.
73. Y. Fujiwara, K. Kasashima, K. Saito et al., "Microtubule association of a neuronal RNA-binding protein HuD through its binding to the light chain of MAP1B," Biochimie, vol. 93, pp. 817–822, 2011.
74. M. Nousch, V. Reed, R. J. Bryson-Richardson, P. D. Currie, and T. Preiss, "The eIF4G-homolog p97 can activate translation independent of caspase cleavage," RNA, vol. 13, no. 3, pp. 374–384, 2007.
75. N. Ohkawa, S. Sugisaki, E. Tokunaga et al., "N-acetyltransferase ARD1-NAT1 regulates neuronal dendritic development," Genes to Cells, vol. 13, no. 11, pp. 1171–1183, 2008.
76. K. G. Chernov, P. A. Curmi, L. Hamon, A. Mechulam, L. P. Ovchinnikov, and D. Pastré, "Atomic force microscopy reveals binding of mRNA to microtubules mediated by two major mRNP proteins YB-1 and PABP," FEBS Letters, vol. 582, no. 19, pp. 2875–2881, 2008.
77. F. Yang, Y. Peng, E. L. Murray, Y. Otsuka, N. Kedersha, and D. R. Schoenberg, "Polysome-bound endonuclease PMR1 is targeted to stress granules via stress-specific binding to TIA-1," Molecular and Cellular Biology, vol. 26, no. 23, pp. 8803–8813, 2006.
78. Y. Peng, E. L. Murray, M. Sarkar, L. I. U. Xiaoqiang, and D. R. Schoenberg, "The cytoskeleton-associated Ena/VASP proteins are

unanticipated partners of the PMR1 mRNA endonuclease," RNA, vol. 15, no. 4, pp. 576–587, 2009.

79. K. Arimoto, H. Fukuda, S. Imajoh-Ohmi, H. Saito, and M. Takekawa, "Formation of stress granules inhibits apoptosis by suppressing stress-responsive MAPK pathways," Nature Cell Biology, vol. 10, no. 11, pp. 1324–1332, 2008.
80. E. Ai, D. S. Poole, and A. R. Skop, "RACK-1 directs dynactin-dependent RAB-11 endosomal recycling during mitosis in Caenorhabditis elegans," Molecular Biology of the Cell, vol. 20, no. 6, pp. 1629–1638, 2009.
81. J. Henao-Mejia and J. J. He, "Sam68 relocalization into stress granules in response to oxidative stress through complexing with TIA-1," Experimental Cell Research, vol. 315, no. 19, pp. 3381–3395, 2009.
82. G. Lazer, L. Pe'er, V. Schapira, S. Richard, and S. Katzav, "The association of Sam68 with Vav1 contributes to tumorigenesis," Cellular Signalling, vol. 19, no. 12, pp. 2479–2486, 2007.
83. J. K. Woo, H. B. Sung, V. Kim, I. Ryu, and K. J. Sung, "Sequestration of TRAF2 into stress granules interrupts tumor necrosis factor signaling under stress conditions," Molecular and Cellular Biology, vol. 25, no. 6, pp. 2450–2462, 2005.
84. K. Taira, M. Umikawa, K. Takei et al., "The traf2- and nck-interacting kinase as a putative effector of Rap2 to regulate actin cytoskeleton," Journal of Biological Chemistry, vol. 279, no. 47, pp. 49488–49496, 2004.
85. K. D. Swisher and R. Parker, "Localization to, and effects of Pbp1, Pbp4, Lsm12, Dhh1, and Pab1 on stress granules in Saccharomyces cerevisiae," PLoS One, vol. 5, article e10006, 2010.
86. J. A. Solinger, D. Pascolini, and W. D. Heyer, "Active-site mutations in the Xrn1p exoribonuclease of Saccharomyces cerevisiae reveal a specific role in meiosis," Molecular and Cellular Biology, vol. 19, no. 9, pp. 5930–5942, 1999.
87. H. Onishi, Y. Kino, T. Morita, E. Futai, N. Sasagawa, and S.

Ishiura, "MBNL1 associates with YB-1 in cytoplasmic stress granules," Journal of Neuroscience Research, vol. 86, no. 9, pp. 1994–2002, 2008.

88. W. H. Yang and D. B. Bloch, "Probing the mRNA processing body using protein macroarrays and "autoantigenomics"," RNA, vol. 13, no. 5, pp. 704–712, 2007.
89. N. Stöhr, M. Lederer, C. Reinke et al., "ZBP1 regulates mRNA stability during cellular stress," Journal of Cell Biology, vol. 175, no. 4, pp. 527–534, 2006.
90. D. Scholz, C. F. Baicu, W. J. Tuxworth et al., "Microtubule-dependent distribution of mRNA in adult cardiocytes," American Journal of Physiology, vol. 294, no. 3, pp. H1135–H1144, 2008. ·
91. S. Mollet, N. Cougot, A. Wilczynska et al., "Translationally repressed mRNA transiently cycles through stress granules during stress," Molecular Biology of the Cell, vol. 19, no. 10, pp. 4469–4479, 2008.
92. T. S. Eisinger-Mathason, J. Andrade, A. L. Groehler et al., "Codependent functions of RSK2 and the apoptosis-promoting factor TIA-1 in stress granule assembly and cell survival," Molecular cell, vol. 31, no. 5, pp. 722–736, 2008.
93. K. Chalupníková, S. Lattmann, N. Selak, F. Iwamoto, Y. Fujiki, and Y. Nagamine, "Recruitment of the RNA helicase RHAU to stress granules via a unique RNA-binding domain," Journal of Biological Chemistry, vol. 283, no. 50, pp. 35186–35198, 2008.
94. K. Fujimura, F. Kano, and M. Murata, "Identification of PCBP2, a facilitator of IRES-mediated translation, as a novel constituent of stress granules and processing bodies," RNA, vol. 14, no. 3, pp. 425–431, 2008.
95. P. Anderson and N. Kedersha, "Visibly stressed: the role of eIF2, TIA-1, and stress granules in protein translation," Cell Stress and Chaperones, vol. 7, no. 2, pp. 213–221, 2002.
96. T. Ohn, N. Kedersha, T. Hickman, S. Tisdale, and P. Anderson, "A functional RNAi screen links O-GlcNAc modification of ribosomal proteins to stress granule and processing body assembly," Nature Cell Biology, vol. 10, no. 10, pp. 1224–1231, 2008.

97. P. A. Ivanov, E. M. Chudinova, and E. S. Nadezhdina, "Disruption of microtubules inhibits cytoplasmic ribonucleoprotein stress granule formation," Experimental Cell Research, vol. 290, no. 2, pp. 227–233, 2003.

98. E. Kolobova, A. Efimov, I. Kaverina et al., "Microtubule-dependent association of AKAP350A and CCAR1 with RNA stress granules," Experimental Cell Research, vol. 315, no. 3, pp. 542–555, 2009.

99. K. Fujimura, J. Katahira, F. Kano, Y. Yoneda, and M. Murata, "Microscopic dissection of the process of stress granule assembly," Biochimica et Biophysica Acta, vol. 1793, no. 11, pp. 1728–1737, 2009.

100. E. S. Nadezhdina, A. J. Lomakin, A. A. Shpilman, E. M. Chudinova, and P. A. Ivanov, "Microtubules govern stress granule mobility and dynamics," Biochimica et Biophysica Acta, vol. 1803, no. 3, pp. 361–371, 2010.

101. R. Mazroui, S. Di Marco, R. J. Kaufman, and I. E. Gallouzi, "Inhibition of the ubiquitin-proteasome system induces stress granule formation," Molecular Biology of the Cell, vol. 18, no. 7, pp. 2603–2618, 2007.

102. J. R. Hughes, S. L. Bullock, and D. Ish-Horowicz, "Inscuteable mRNA localization is dynein-dependent and regulates apicobasal polarity and spindle length in Drosophila neuroblasts," Current Biology, vol. 14, no. 21, pp. 1950–1956, 2004.

103. E. A. Vaisberg, M. P. Koonce, and J. R. McIntosh, "Cytoplasmic dynein plays a role in mammalian mitotic spindle formation," Journal of Cell Biology, vol. 123, no. 4, pp. 849–858, 1993.

CHAPTER

7

STRUCTURE AND FUNCTION OF THE BI-DIRECTIONAL BACTERIAL FLAGELLAR MOTOR

Yusuke V. Morimoto [1,2] and Tohru Minamino [2]

[1]Quantitative Biology Center, RIKEN, 6-2-3 Furuedai, Suita, Osaka 565-0874, Japan

[2]Graduate School of Frontier Biosciences, Osaka University 1-3 Yamadaoka, Suita, Osaka 565-0871, Japan

ABSTRACT

The bacterial flagellum is a locomotive organelle that propels the bacterial cell body in liquid environments. The flagellum is a supramolecular complex composed of about 30 different proteins and consists of at least three parts: a rotary motor, a universal joint, and a helical filament. The flagellar motor of Escherichia coli and Salmonella enterica is powered by an inward-directed electrochemical potential difference of protons across the cytoplasmic membrane. The flagellar motor consists of a rotor made of FliF, FliG, FliM and FliN and a dozen stators consisting of MotA and

MotB. FliG, FliM and FliN also act as a molecular switch, enabling the motor to spin in both counterclockwise and clockwise directions. Each stator is anchored to the peptidoglycan layer through the C-terminal periplasmic domain of MotB and acts as a proton channel to couple the proton flow through the channel with torque generation. Highly conserved charged residues at the rotor–stator interface are required not only for torque generation but also for stator assembly around the rotor. In this review, we will summarize our current understanding of the structure and function of the proton-driven bacterial flagellar motor.

Keywords: Bacterial flagellum; rotary motor; motility; stator; rotor; torque generation; mechanosensor

INTRODUCTION

Many bacteria swim in liquid media by rotating bacterial flagella. The bacterial flagellum is a supramolecular complex made of about 30 different proteins with copy numbers ranging from a few to a few tens of thousands (Figure 1). The flagellum consists of at least three parts: the basal body, the hook and the filament. The basal body is embedded within the cell membranes and acts as a rotary motor. The hook and filament extend outwards in the cell exterior. The filament works as a helical screw to produce the thrust. The hook exists between the basal body and filament and functions as a universal joint to smoothly transmit torque produced by the motor to the helical filament [1,2,3].

The flagellar motor of Escherichia coli and Salmonella enterica is fueled by the electrochemical potential of protons (H^+) across the cytoplasmic membrane, i.e., the proton motive force [1,3,4,5]. The flagellar motor can operate in either a counter-clockwise (CCW, viewed from filament to motor), or clockwise (CW) direction. When most of the motors rotate in the CCW direction, the filaments form a bundle and propel the cell smoothly. When one or more motors spin in the CW direction, the bundle is disrupted and hence the cell tumbles and changes the swimming direction [6,7]. E. coli and Salmonella can move towards more favorable conditions and escape from undesirable ones for their survival by sensing temporal variations of environmental stimuli, such as chemical attractants and repellents, temperature and pH via methyl-accepting chemotaxis proteins (MCP). MCPs are transmembrane proteins with a large cytoplasmic domain involved in interactions with a histidine kinase CheA, and an adaptor protein CheW. The MCPs control CheA autophosphorylation. Phosphorylated CheA transfers its phosphate group to a response regulator CheY. Phosphorylated CheY (CheY-P) binds to the cytoplasmic face of the flagellar motor, allowing the motor to spin in the CW direction [8].

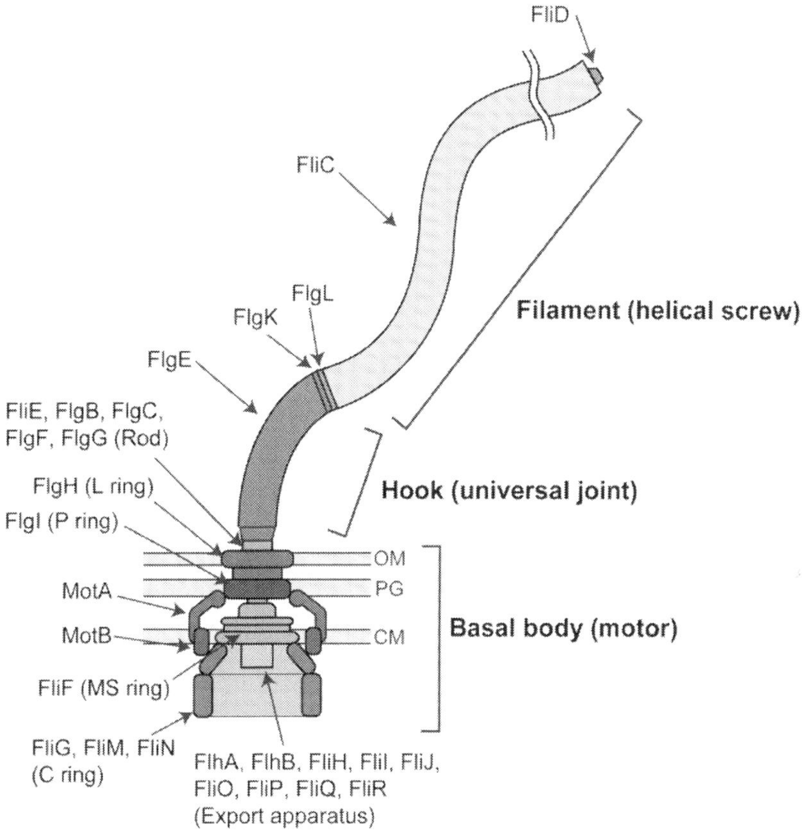

Figure 1. Schematic diagram of the bacterial flagellum. The flagellum is a locomotive organelle for bacterial propulsion. The flagellum consists of the basal body, which acts as a reversible rotary motor, the hook, which functions as a universal joint and the filament, which works as a helical screw. OM, outer membrane; PG, peptidoglycan layer; CM, cytoplasmic membrane.

The flagellar motor consists of a rotor and a dozen stators (Figure 2). The rotor consists of four flagellar proteins, FliF, FliG, FliM and FliN. FliF forms the MS ring within the cytoplasmic membrane [9,10]. FliG, FliM and FliN form the C ring on the cytoplasmic face of the MS ring [11,12]. FliG, FliM and FliN are also responsible for switching the direction of motor rotation, enabling the motor to rotate in either CCW or CW [13]. Each stator consists of two integral membrane proteins, MotA and MotB [14,15,16] and is postulated to be anchored to the peptidoglycan (PG) layer through a well-conserved peptidoglycan binding (PGB) motif within the C-terminal periplasmic domain of MotB ($MotB_C$) [17]. MotA and MotB form a proton channel complex to couple proton flow through the channel with torque generation [18,19,20,21]. Torque is generated by interactions between MotA and FliG [22]. This article covers our current discoveries of the proton-driven flagellar motor of E. coli and Salmonella with the emphasis being on the structure, function and dynamic properties of the motor.

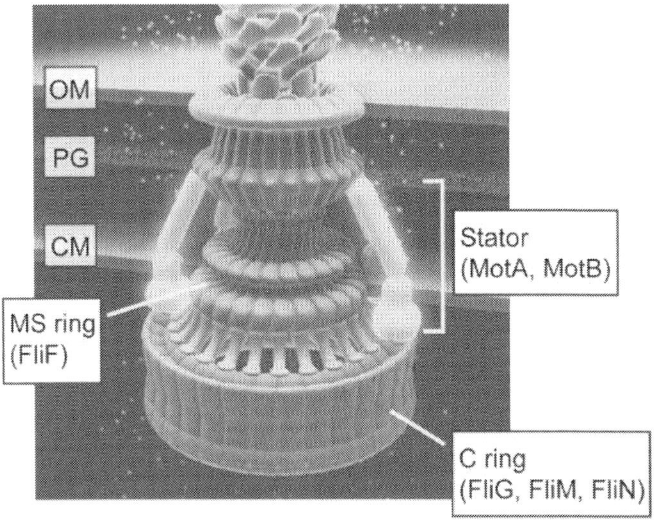

Figure 2. Schematic diagram of the proton-driven bacterial flagellar motor. The flagellar motor consists of a reversible rotor made of FliF, FliG, FliM and FliN and a dozen stators, each of which consists of MotA and MotB. FliF forms the MS ring within the cytoplasmic membrane. FliG, FliM and FliN form the C ring on the cytoplasmic face of the MS ring. OM, outer membrane; PG, peptidoglycan layer; CM, cytoplasmic membrane.

ROTOR

The rotor consists of two ring structures, the MS and C rings (Figure 2). The MS ring is formed by 26 copies of FliF and acts as an assembly platform [10]. Using the MS ring as a template, FliG, FliM and FliN assemble into the C ring onto the cytoplasmic face of the MS ring [12]. FliG directly associates with the cytoplasmic face of the MS ring with a 1 FliG : 1 FliF stoichiometry [23,24], suggesting that the FliG ring has the 26-fold symmetry. FliG consists of three domains, $FliG_N$, $FliG_M$ and $FliG_C$ [25]. $FliG_N$ directly binds to the MS ring [26,27]. $FliG_C$ is directly involved in the interaction with a stator protein MotA [22]. This suggests that $FliG_C$ is located at the upper part of the C ring. $FliG_M$ provides a binding site for FliM [28,29,30,31]. FliM and FliN form a stable complex with a 1 FliM: 4 FliN stoichiometry and occupy most part of the C ring [32]. The C ring has rotational symmetry that varies from 32-fold to 36-fold [33], and the diameter actually shows a range of variability (45 ~ 49.0 nm) [34]. Crystal structures of FliG [25,35,36,37], FliM [38], FliN [32,39] and the FliG-FliM complex [29,30,31] have been solved, and possible models for their organization have been proposed. Please see a recent review article focusing on the current models of the rotor ring structure [40].

Certain mutations in FliG, FliM and FliN affect the switching between CW and CCW rotation, indicating that these three proteins are responsible for switching the direction of motor rotation [13]. CheY-P binds to FliM and FliN, inducing cooperative conformational changes of the FliG ring to allow the motor to spin in the CW direction [41,42]. The switching between the CW and CCW states is highly cooperative [43]. The cooperative switching mechanism can be explained by a conformational spread model, in which a switching event is mediated by conformational changes in a ring of subunits that spread from subunit to subunit via nearest-neighbor interactions [44,45]. When motor speed is increased up to 100–150 Hz, both the CW-to-CCW and CCW-to-CW switching rates increase linearly and then decrease with speed to an equivalent degree, without changing the motor bias [46]. This suggests that the motor switch can sense the load on the stator–rotor interaction as well as the cytoplasmic level of CheY-P [46]. A mathematical model to explain this load-switching relationship of the motor has been proposed [47]. Recently, it has been reported that the nucleotide second messenger cyclic di-GMP (c-di-GMP), which promotes transition of single motile cells to surface-attached multicellular biofilm, regulates flagella-driven motility. A c-di-GMP binding protein YcgR binds to MotA, FliG and FliM in the presence of c-di-GMP, resulting not only in a significant reduction in motor speed but also in the CCW bias in the motor rotation [48,49,50].

Mutations located in and around helix$_{MC}$, which connects FliG$_{M}$ and FliG$_{C}$, generate a diversity of phenotype, including motors that are strongly CW biased, infrequent switchers, rapid switchers, and transiently or permanently paused. This suggests that helix$_{MC}$ is involved in switching of the direction of flagellar motor rotation [51]. The Salmonella FliG(ΔPAA) mutant motor, which has three-amino-acid deletion at positions 169 to 171, are extremely CW biased [52]. This mutant motor remains in the CW rotation even in the absence of CheY-P, indicating that the motor is locked in the CW state [37]. This suggests that the PAA deletion induces a conformational change of FliG at the rotor–stator interface in a way similar to the binding of CheY-P to FliM and FliN. Limited proteolysis has shown that the PAA deletion results in a conformational change in a hinge between FliG$_{M}$ and helix$_{MC}$. This result is in agreement with the crystal structure of Thermotoga maritima FliG$_{MC}$(ΔPEV) showing that the orientation of helix$_{MC}$ relative to FliG$_{M}$ is distinct from that seen in the wild-type T. maritima FliG$_{MC}$ structure [36,37]. However, it remains unclear how such a change in the orientation of helix$_{MC}$ relative to FliG$_{M}$ affects the switching process.

The flagellar motor containing a single stator unit rotates with 26 steps per revolution, which is in agreement with the stoichiometry of the FliG ring [53,54]. The elementary process of torque generation by stator–rotor interactions is symmetric in CCW and CW rotation [54], although the torque-speed curves are distinct between them [55]. To achieve the symmetric elementary process of torque generation in both CCW and CW rotation, FliG$_{C}$ may need to go through a 180° rotation to reorient its charged residues in the two opposite directions. However, it remains unknown how it occurs.

FliM-GFP (or YPet) and GFP-FliN show turnovers between the motor and their cytoplasmic pools [56,57], indicating that the C ring is a highly dynamic structure rather than a static one. The turnover of FliM occurs only in the presence of CheY-P, suggesting that this turnover is directly involved in the switching process. The number of FliM subunits in the C ring increases in response to a reduction in the concentration of CheY-P, suggesting that the flagellar motor adapts to changes in the steady-state level of CheY-P by adjusting the number of FliM molecules to which CheY-P binds [58,59].

STATOR

Proton Channel of the MotA/B Complex

The energy for flagellar motor rotation is supplied by the electrochemical potential of specific ions across the cytoplasmic membrane [60,61,62,63]. Based on the coupling ions, the flagellar motors are divided into at least two types (Figure 3): one is the H^+-driven type found in E. coli and Salmonella[18,21,52,65,66] and the other is the Na^+-driven type in marine Vibrio and extremely alkalophilic Bacillus[67,68,69]. Recently, it has been reported that extremely alkaliphilic Bacillus alcalophilus Vedder 1934 utilizes both Na^+ and K^+ as the coupling ions for flagellar rotation [64]. The stator of the H^+-driven motor consists of four copies of MotA and two copies of MotB [70] and acts as a proton channel to couple proton flow through the channel with torque generation (Figure 4) [18]. Terahara et al. [71] have reported that the MotA/B complex of alkaliphilic Bacillus clausii can conduct bothH^+ and Na^+ at different pH ranges. The Na^+-coupled stator complex of Vibrio alginolyticus consists of four proteins, PomA, PomB, MotX and MotY [72]. PomA and PomB form a Na^+-conducting pathway within the cytoplasmic membrane in a way similar to the MotA/B proton channel [73]. MotX and MotY form the T ring located beneath the P ring for efficient assembly and stable anchoring of the PomA/B complex to its binding site on the motor [74]. Interestingly, both H^+-coupled and Na^+-coupled stators are incorporated into a single flagellar motor in Bacillus subtilis and Shewanella oneidensis MR-1 [75,76]. Please see a review article summarizing the current understanding of the Na^+-driven flagellar motor in Vibrio alginolyticus [72].

MotA consists of four transmembrane spans (TMs), two short periplasmic loops and two extensive cytoplasmic regions. MotB consists of an N-terminal cytoplasmic region, one TM and $MotB_C$ containing a well-conserved PGB motif. Asp-33 of Salmonella MotB, which corresponds to Asp-32 of E. coli MotB, is a highly conserved aspartic residue [66,77]. This aspartic acid residue is located at the proximal end of MotB-TM and is critical for proton translocation through the channel [52,65,66]. MotB exists as a homo-dimer in the stator and these aspartic acid residues are positioned on the surface of the MotB-TM dimer facing MotA-TMs [78]. These observations suggest that the MotA/B complex has two proton-conducting pathways (Figure 4). MotA-TM3 and MotA-TM4 are close to MotB-TM while MotA-TM1 and MotA-TM2 are in the outer positions of the MotA/B complex [79]. This suggests that MotB-TM forms a proton channel along with the TM-3 and TM-4 helices of MotA (Figure 4). The V35F mutation, which lies in MotA-TM2, has been isolated as an extragenic suppressor of the motB(D33E) mutant [80]. It is impossible for the side chain of Val-35 in MotA-TM2 to directly access the proton-conducting pathway formed at the interface between MotA and MotB because it is likely to be located within the hydrophobic core of the four transmembrane α-helix bundle of MotA. This suggests that TM2 is presumably close to TM3 and TM4 to maintain the proton pathway. This is supported by site-directed disulfide cross-linking experiments [81].

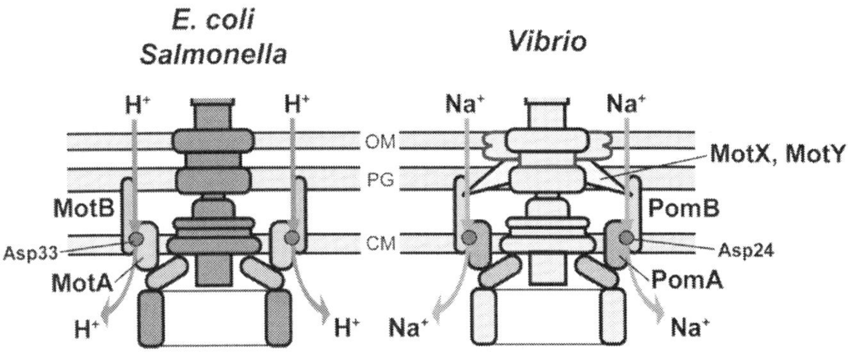

Figure 3. Schematic diagrams of the bacterial H^+-driven and Na^+-driven flagellar motors. E. coli and Salmonella have the H^+-driven flagellar motor (left) whereas Vibrio alginolyticus has the Na^+-driven polar flagellar motor (right). The stator of the flagellar motor consists of MotA and MotB in the H^+-driven motor and of PomA and PomB in the Na^+-driven motor. MotX and MotY are required for efficient assembly of the PomA/B complex around the motor in Vibrio.

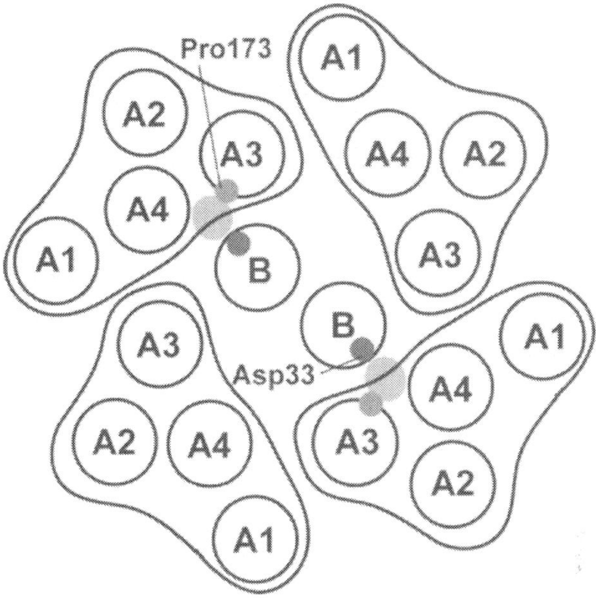

Figure 4. Arrangement of transmembrane segments of the MotA/B complex, which consists of four copies of MotA and two copies of MotB. The view is from the periplasmic side of the membrane. The complex has two proton conducting pathway shown by orange ellipsoids [79,80,81].

Pro-173 of MotA, which is highly conserved among MotA orthologues, is located at the proximal end of MotA-TM3 (Figure 4) [82,83]. Site-directed disulfide cross-linking experiments have shown that Pro-173 of E. coli MotA is in relative close proximity to Asp-32 of E. coli MotB [81]. Co-overproduction of MotA with the MotB-TetA chimera results in proton leakage, arresting cell growth [18]. The E. coliMotB(D32N) mutation suppresses the proton leakage through the MotA/B-TetA proton channel. Interestingly, introduction of the MotA(P173D) mutation into the MotA/B(D32N)-TetA complex causes massive proton flow again, largely impairing the cell growth. This indicates that this Asp-173 residue of MotA allows protons to go through the channel in the absence of Asp-32 of MotB, probably by presenting a carboxyl residue as an alternative proton-binding site [83]. These observations suggest that Pro-173 of MotA forms a part of the proton-conducting pathway along with Asp-32 of MotB. The Salmonella MotA(P173A) mutant motor produces torque at the wild-type level when the motor operates under a very high load condition [84]. However, the MotA(P173A) mutation causes ca. 90% reduction in the maximum rotation rate, suggesting that the P173A mutation significantly reduces the rate of the mechanochemical reaction cycle of the motor [84]. Simulation of the torque-speed curve of the MotA(P173A) mutant motor by a simple kinetic model indicates that the P173A mutation slows the rates of conformational changes of the MotA/B complex by 25 times [84]. Because a prolyl residue induces bending in TM to facilitate ion transport [85], Pro-173 of MotA is proposed to facilitate the conformational changes of the MotA/B complex that support the rapid mechanocmemical cycle when the motor rotates at high speed [81,84].

Over-expression of the MotA/B complex from a plasmid does not affect cell growth at all. In contrast, in-frame deletion of a plug segment formed by residues 53 to 66 in MotB causes massive proton flow through a proton channel, thereby arresting cell growth [21,86]. This suggests that the plug segment interferes with proton channel formation, thereby suppressing undesirable proton flow through the channel when the MotA/B complex is not installed into the motor. Interestingly, however, the cell growth is not significantly impaired by in-frame deletion of residues 51 to 100 in Salmonella MotB, which contains the plug segment although MotB(Δ51–100) is partially functional [87,88]. This indicates that proton channel formation of the MotA/B(Δ51–100) complex does not occur prior to its assembly around the motor although the plug is missing. The L119P and L119E substitutions allow unassembled MotA/B(Δ51–100) complex to conduct protons to a significant degree, impairing the cell growth [21,88]. This suggests that not only the plug segment but also some other region within $MotB_C$ regulate proper proton channel formation.

Role of $MotB_C$ *in Stator Assembly around a Rotor*

$MotB_C$ forms a homo-dimer and its dimerization is required for MotB function [88,89]. $MotB_C$ is required not only for proper anchoring of the stator to the PG layer [90] but also for proper alignment of the stator relative to the rotor [91]. The crystal structure of the core domain of $MotB_C$, which consists of residues 99 to 276 in MotB, shows considerable structural similarities with the PGB domain of Pal (Figure 5a) [88]. Interestingly, a chimeric MotB-Pal protein, in which the PGB domain of E. coli MotB is replaced by that of E. coli Pal, exerts MotB function to a considerable degree, suggesting that $MotB_C$ binds to the PG layer through its PGB motif in a way similar to Pal [92]. Site-directed disulfide crosslinking experiments have shown that $MotB_C$ is in relatively close proximity to the P ring of the basal body [93].

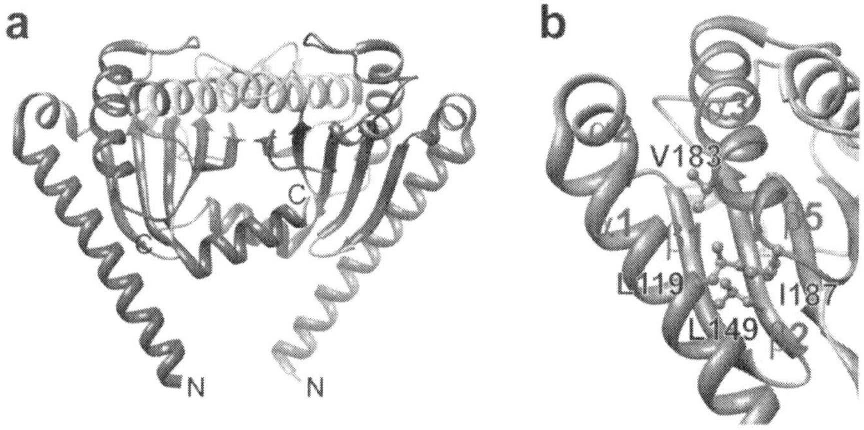

Figure 5. The crystal structure of $MotB_C$ from Salmonella. (**a**) Cα ribbon representation of Salmonella $MotB_C$ dimer in the crystal (PDB ID: 2ZVY). The peptidoglycan binding (PGB) motif is shown by orange. (**b**) The Mot-B_C structure in the vicinity of Leu119. Leu119 in helix α1 shows hydrophobic interactions with Leu149 in strand β2 and Val183 and Ile187 after helix α3.

Upon induction of the MotA/B complex from a plasmid in a motAB null mutant, a stepwise increment in speed is observed at high load, estimating that there are as many as 11 stators around the rotor [96]. Interestingly, abrupt, stepwise drops and restorations of the rotation speed of the motor are also observed even in steadily rotating motors, indicating that the stators are replaced frequently [14,15,53]. Consistently, GFP-fused MotB shows rapid exchanges between the motor and the cytoplasmic membrane pool, indicating that an interaction between $MotB_C$ and the PG layer is highly dynamic [97]. These observations suggest that the MotA/B complexes are no permanently fixed in place around the motor during motor rotation.

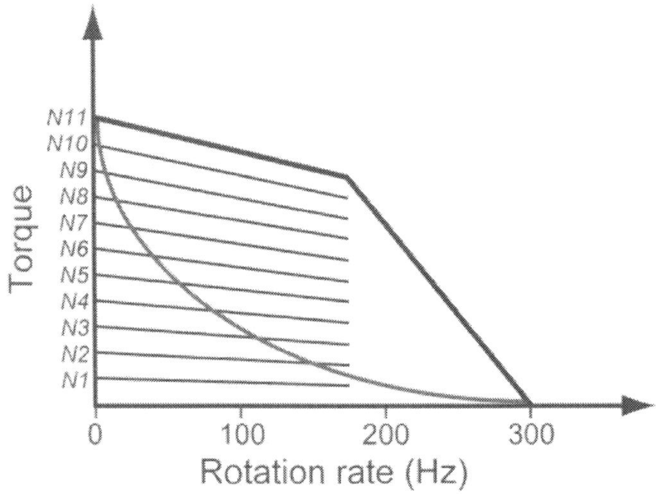

Figure 6. Torque-speed curves of the wild-type (black line) and MotB(Δ72–100) mutant motors (red line). Torque-speed curves of the wild-type motor with different number of stator units are indicated (N_1–N_{11}).

The torque-speed curve of the flagellar motor consists of two regimes: a high-load, low-speed regime and a low-load, high-speed regime (Figure 6). In the high-load, low-speed regime, the speed of the motor is proportional to the number of stators, whereas in the low-load, high-speed regime, one stator unit can spin the motor at the maximum speed [94,95]. Recently, it has been reported that a few stators spin the motor at low loads and that 6–11 stators drive the same motor at high loads. This suggests that the flagellar motor controls the number of functional stators in the motor in response to changes in external load [98,99]. In contrast to the torque-speed curve of the wild-type motor with a gradual decrease in the high-load regime and a rapid drop in the low-load regime, in-frame deletion of 72 to 100 in MotB causes an unusual torque-speed curve of the motor with a rapid drop in the high-load regime and a slow decrease in the low-load regime (Figure 6) [100]. At a low speed near stall, the MotB(Δ72–100) mutant motor produces torque at the wild-type level, indicating that the deletion does not affect an actual torque-generation step by stator-rotor interactions coupled with proton translocation through the proton channel. In contrast, as motor speed increases, torque produced by the MotB(Δ72–100) mutant motor drops off at much higher load than that of the wild-type motor due to a rapid decrease in the number of stators driving the motor at much higher load than that the wild-type motor responds to dissociate its stator from the motor [100]. As torque generation by the motor applies an equal and opposite force on the PG layer through $MotB_C$, $MotB_C$ is proposed to act as a load-sensitive structural switch to regulate the assembly and disassembly cycle of the stators in response to the load changes and that an appropriate length of a linker connecting the PGB domain to MotB-TM may stabilize this structural switch during the torque generation cycle [100].

Load-Sensitive Coupling between Proton Translocation and Torque Generation

Both protonation and deprotonation of MotB-Asp33 induce conformational changes of the cytoplasmic loop between TM2 and TM3 of MotA ($MotA_C$), which is directly involved in the interaction with FliG. This suggests that proton translocation through the proton channel is coupled with cyclic conformational changes of the MotA/B complex for torque generation [101]. The MotB(D33E) mutation in MotB causes a slow motile phenotype [52]. This mutation results in about 50% reduction in near-stall torque. Interestingly, the suppressor mutations, which exist in MotA-TM2, MotA-TM3, MotB-TM and $MotB_C$, restore stall torque to the wild-type level [80]. The D33E mutation does not decrease the number of stators around the rotor, indicating that the decrease in stall torque is not due to the reduction in the number of stators driving the motor [102]. Since the rotation rate of the flagellar motor is not limited by the rate of the mechanochemical cycle of the motor at low speed near stall [103], it is suggested that the D33E mutation misaligns $MotA_C$ relative to FliG at the rotor–stator interface, causing the 50% reduction in the energy coupling efficiency of the motor and that the suppressor mutations readjusts the alignment of $MotA_C$ relative to FliG, thereby restoring the coupling efficiency to the wild-type level [102].

The MotB(D33E) mutation causes a 90% reduction in the maximum rotation speed of the flagellar motor at a very low load, indicating that this mutation results in a considerable reduction in the maximum rate of the mechanochemical cycle of the motor. The suppressor mutations do not support the maximal rotation rate at low load [80]. Because the proton conductivity of the unplugged proton channels of the MotA/B(D33E) and its suppressor MotA(V35F)/B(D33E) complexes are measured to be ca. 50% and ca. 60% of that through unplugged wild-type channel, the decreased proton conductivity through the channel itself is not the main cause of the marked reduction in the maximal rotation speed at low load [102]. The rate of proton flow through the MotA/B proton channel is limited by the rate of conformational change of the MotA/B complex, suggesting that the 90% reduction in the maximal rotation speed of these mutant motors is due to a considerable decrease in rates of the conformational changes of the MotA/B complex coupled with proton translocation through the channel [102].

The stator has a high duty ratio and hence remains attached to the rotor for most of the mechanochemical reaction cycle [94,95]. The rotation speed of the wild-type motor is very stable over a wide range of rotation rates [95]. However, the rotation speeds of the MotA/B(D33E) and MotA(V35F)/B(D33E) motors show large fluctuations at low load although they are stable at high load. Interestingly, frequent pauses are also observed for these two mutant motors at low load [102]. These observations suggest that the wild-type MotA/B complex can sense even a small load to continue the mechanochemical reaction cycle for torque generation by efficiently coupling proton translocation with its conformational changes. Because the stators are dynamic mechanosensors and change their structure to control the number of stators driving the motor in response to changes in external load [98,99], it is proposed that the MotA/B complex is a load-sensitive proton channel that efficiently couples proton translocation with torque generation in response to changes in the load and that MotB-Asp33 is important for this coordinated proton translocation [102].

ROLE OF STATOR-ROTOR INTERACTIONS IN EFFICIENT ASSEMBLY OF THE STATORS INTO THE MOTOR

MotA-Arg90 and MotA-Glu98, which are located in $MotA_C$, are highly conserved residues among MotA orthologues and are critical for motor function [104]. The extragenic FliG(D289A) and FliG(D289K) suppressor mutations partially rescue the motility of the motA(R90E) mutant whereas the extragenic FliG(R281V) and FliG(R281W) suppressor mutations restore the motility of the motA(E98K)mutant, indicating genetic interactions between MotA-Arg90 and FliG-Asp289 and between MotA-Glu98 and FliG-Arg281 (Figure 7) [22]. This suggests that these interactions at the rotor–stator interface are directly involved in flagellar rotation. The MotA(R90E), MotA(E98K), FliG(D289K) and FliG(R281V) mutations considerably affect the assembly of GFP-labeled MotA/B complexes around the rotor, suggesting that these interactions are also critical for stator assembly around the motor [105,106]. The FliG(D289K) mutation significantly recovers the localization of GFP-MotB to the motor in the motA(R90E) mutant whereas the FliG(R281V) mutation does not recover the GFP-MotB localization in the motA(E98K) mutant. These suggest that the MotA-Arg90−FliG-Asp289 interaction is critical for the proper positioning of the stators around the rotor, whereas the MotA-Glu98−FliG-Arg281 interaction is more important for torque generation [106].

Salmonella MotB missing residues 51 to 100 retains the ability to form the functional stator with MotA to a considerable degree [87]. The distance between the surface of the hydrophobic core layer of the cytoplasmic membrane and that of the PG layer is about 100 Å. The crystal structure of a SalmonellaMotB$_C$ fragment corresponding to residues 99 to 276 of MotB is only about 50 Å tall [88] and hence it cannot reach the PG layer if connected directly to MotB-TM by the deletion of residues 51–100. Therefore, a large conformational change must be required for the PGB sites on the top surface of MotB$_C$to reach the PG layer. Leu119 in helix α1 shows hydrophobic interactions with Leu149 in strand β2 and Val183 and Ile187 after helix α3 (Figure 5b) [88]. The L119P and L119E mutations, which not only cause massive proton flow through the proton channel of the MotA/B(Δ51–100) complex but also increase the assembly efficiency of the MotA/B(Δ51–100) complex into the motor, would destabilize these hydrophobic interactions [21,88]. This raises a plausible hypothesis that these hydrophobic interactions act as a structural switch to regulate the assembly and disassembly of the stators [88,100]. Over-expression of MotA inhibits wild-type motility due to a reduction in the number of the functional stator around the rotor [105]. Consistently, MotA-mCherry co-localizes with GFP-FliG even in the absence of MotB [105]. These observations indicate that MotA alone can be installed into the motor. Because conserved charged residues at the rotor–stator interface are directly involved in stator assembly around the motor [105,106], interactions between MotA and FliG may trigger a large conformational change in $MotB_C$ that open the proton channel and allow the stator to be anchored to the PG layer.

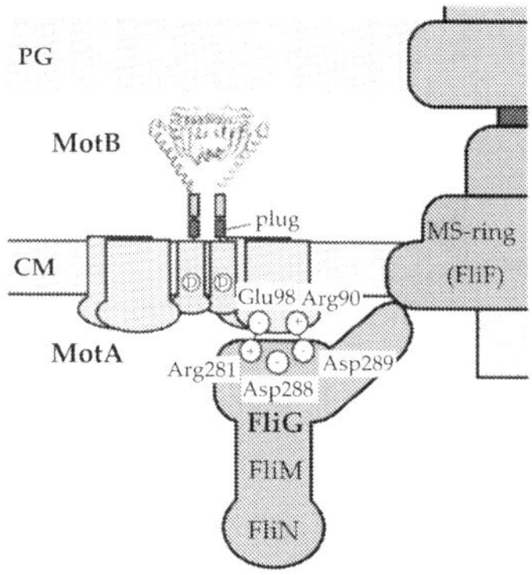

Figure 7. Schematic diagram of interactions at the rotor–stator interface. FliG is postulated to occupy the upper part of the C ring. The interactions between MotA-Arg90 and FliG-Asp289 and between MotA-Glu98 and FliG-Arg281 are directly involved not only in torque generation but also in stator assembly around the rotor.

CONCLUSIONS

Both the rotor and stator of the flagellar motor are highly dynamic structures rather than static ones. [56,57,58,97,98,99] Two rotor proteins, FliM and FliN, show highly dynamic turnovers between the motor and their cytoplasmic pools. The FliM turnover plays an important role in switching the direction of flagellar motor rotation [56,58,59]. However, little is known about the switching mechanism of the rotor ring complex. The stators are machanosensors to regulate the number of the stators driving the motor in response to changes in external load [98,99,100]. Each stator is also a load-sensitive proton channel that efficiently couples proton translocation with torque generation in response to the load changes [102]. However, it remains unknown how the stators sense the amount of external load to control the number of functional stators around the rotor as well as the proton flow through the channel. Interactions at the rotor–stator interface are important, not only for torque generation [22], but also for efficient stator assembly around the rotor [105,106]. However, it remains unknown how the rotor–stator interactions allow the stator to open the proton channel and to be anchored to the PG layer. We need to investigate the switching mechanism of the flagellar motor in much more detail by biophysical and structural techniques combined with genetic and biochemical approaches. Super-resolution fluorescence microscopic techniques would be required for understanding of the mechanosensing mechanism of the stator. Furthermore, not only in vitro reconstitution experiments but also high-resolution structural analysis by X-ray crystallography and electron cryomicroscopy would be essential to advance our mechanistic understanding of the energy coupling mechanism of the flagellar motor.

ACKNOWLEDGMENTS

We acknowledge Keiichi Namba and Masahiro Ueda for continuous support and encouragement. Y.V.M. has been supported by the Special Postdoctoral Researchers program of RIKEN. Our research is supported in part by the Japanese Ministry of Education, Culture, Sports, Science and Technology.

CONFLICT OF INTEREST

The authors declare no conflict of interest.

REFERENCES

1. Berg, H.C. The rotary motor of bacterial flagella. Ann. Rev. Biochem. 2003, 72, 19–54.
2. Macnab, R.M. How bacteria assemble flagella. Annu. Rev. Microbiol. 2003, 57, 77–100.
3. Minamino, T.; Imada, K.; Namba, K. Molecular motors of the bacterial flagella. Curr. Opin. Struct. Biol. 2008, 18, 693–701.
4. Blair, D.F. Flagellar movement driven by proton translocation. FEBS Lett. 2003, 545, 86–95.
5. Sowa, Y.; Berry, R.M. Bacterial flagellar motor. Q. Rev. Biophys. 2008, 41, 103–132.

6. Macnab, R.M.; Ornston, M.K. Normal-to-curly flagellar transitions and their role in bacterial tumbling. Stabilization of an alternative quaternary structure by mechanical force. J. Mol. Biol.1977, 112, 1–30.
7. Turner, L.; Ryu, W.S.; Berg, H.C. Real-time imaging of fluorescent flagellar filaments. J. Bacteriol.2000, 182, 2793–2801.
8. Sourjik, V.; Armitage, J.P. Spatial organization in bacterial chemotaxis. EMBO J. 2010, 29, 2724–2733.
9. Ueno, T.; Oosawa, K.; Aizawa, S. M ring, S ring and proximal rod of the flagellar basal body of Salmonella typhimurium are composed of subunits of a single protein, FliF. J. Mol. Biol. 1992, 227, 672–677.
10. Suzuki, H.; Yonekura, K.; Namba, K. Structure of the rotor of the bacterial flagellar motor revealed by electron cryomicroscopy and single-particle image analysis. J. Mol. Biol. 2004, 337, 105–113.
11. Khan, I.H.; Reese, T.S.; Khan, S. The cytoplasmic component of the bacterial flagellar motor. Proc. Natl. Acad. Sci. USA 1992, 89, 5956–5960.
12. Francis, N.R.; Sosinsky, G.E.; Thomas, D.; DeRosier, D.J. Isolation, characterization, and structure of bacterial flagellar motors containing the switch complex. J. Mol. Biol. 1994, 235, 1261–1270.
13. Yamaguchi, S.; Aizawa, S.; Kihara, M.; Isomura, M.; Jones, C.J.; Macnab, R.M. Genetic evidence for a switching and energytransducing complex in the flagellar motor of Salmonella typhimurium. J. Bacteriol. 1986, 168, 1172–1179.
14. Block, S.M.; Berg, H.C. Successive incorporation of force generating units in the bacterial rotary motor. Nature 1984, 309, 470–472.
15. Blair, D.F.; Berg, H.C. Restoration of torque in defective flagellar motors. Science 1988, 242, 1678–1681.
16. Wilson, M.L.; Macnab, R.M. Co-overproduction and localization of the Escherichia coli motility proteins MotA and MotB. J. Bacteriol. 1990, 172, 3932–3939.
17. De Mot, R.; Vanderleyden, J. The C-terminal sequence conservation between OmpA-related outer membrane proteins and MotB suggests a common function in both gram-positive and gram-negative bacteria, possibly in the interaction of these domains with peptidoglycan. Mol. Microbiol.1994, 12, 333–334.

18. Blair, D.F.; Berg, H.C. The MotA protein of E. coli is a proton-conducting component of the flagellar motor. Cell 1990, 60, 439–449.
19. Stolz, B.; Berg, H.C. Evidence for interactions between MotA and MotB, torque-generating elements of the flagellar motor of Escherichia coli. J. Bacteriol. 1991, 173, 7033–7037.
20. Tang, H.; Braun, T.F.; Blair, D.F. Motility protein complexes in the bacterial flagellar motor. J. Mol. Biol. 1996, 261, 209–221.
21. Morimoto, Y.V.; Che, Y.-S.; Minamino, T.; Namba, K. Proton-conductivity assay of plugged and unplugged MotA/B proton channel by cytoplasmic pHluorin expressed in Salmonella. FEBS lett.2010, 584, 1268–1272.
22. Zhou, J.; Lloyd, S.A.; Blair, D.F. Electrostatic interactions between rotor and stator in the bacterial flagellar motor. Proc. Natl. Acad. Sci. USA 1998, 95, 6436–6441.
23. Francis, N.R.; Irikura, V.M.; Yamaguchi, S.; DeRosier, D.J.; Macnab, R.M. Localization of the Salmonella typhimurium flagellar switch protein FliG to the cytoplasmic M-ring face of the basal body. Proc. Natl. Acad. Sci. USA 1992, 89, 6304–6308.
24. Oosawa, K.; Ueno, T.; Aizawa, S. Overproduction of the bacterial flagellar switch proteins and their interactions with the MS ring complex in vitro. J. Bacteriol. 1994, 176, 3683–3691.
25. Lee, L.K.; Ginsburg, M.A.; Crovace, C.; Donohoe, M.; Stock, D. Structure of the torque ring of the flagellar motor and the molecular basis for rotational switching. Nature 2010, 466, 996–1000.
26. Kihara, M.; Miller, G.U.; Macnab, R.M. Deletion analysis of the flagellar switch protein FliG of Salmonella. J. Bacteriol. 2000, 182, 3022–3028.
27. Levenson, R.; Zhou, H.; Dahlquist, F.W. Structural insights into the interaction between the bacterial flagellar motor proteins FliF and FliG. Biochemistry 2012, 51, 5052–5060.
28. Brown, P.N.; Terrazas, M.; Paul, K.; Blair, D.F. Mutational analysis of the flagellar protein FliG: sites of interaction with FliM and implications for organization of the switch complex. J. Bacteriol.2007, 189, 305–312.

29. Paul, K.; Gonzalez-Bonet, G.; Bilwes, A.M.; Crane, B.R.; Blair, D. Architecture of the flagellar rotor. EMBO J. 2011, 30, 2962–2971.
30. Vartanian, A.S.; Paz, A.; Fortgang, E.A.; Abramson, J.; Dahlquist, F.W. Structure of flagellar motor proteins in complex allows for insights into motor structure and switching. J. Biol. Chem. 2012, 287, 35779–35783.
31. Lam, K.H.; Lam, W.W.; Wong, J.Y.; Chan, L.C.; Kotaka, M.; Ling, T.K.; Jin, D.Y.; Ottemann, K.M.; Au, S.W. Structural basis of FliG-FliM interaction in Helicobacter pylori. Mol Microbiol. 2013, 88, 798–812.
32. Brown, P.N.; Mathews, M.A.; Joss, L.A.; Hill, C.P.; Blair, D.F. Crystal structure of the flagellar rotor protein FliN from Thermotoga maritima. J. Bacteriol. 2005, 187, 2890–2902.
33. Thomas, D.R.; Morgan, D.G.; DeRosier, D.J. Rotational symmetry of the C ring and a mechanism for the flagellar rotary motor. Proc. Natl. Acad. Sci. USA 1999, 96, 10134–10139.
34. Thomas, D.R.; Francis, N.R.; Xu, C.; DeRosier, D.J. The three-dimensional structure of the flagellar rotor from a clockwise-locked mutant of Salmonella enterica serovar Typhimurium. J. Bacteriol.2006, 188, 7039–7048.
35. Lloyd, S.A.; Whitby, F.G.; Blair, D.F.; Hill, C.P. Structure of the C-terminal domain of FliG, a component of the rotor in the bacterial flagellar motor. Nature 1999, 400, 472–475.
36. Brown, P.N.; Hill, C.P.; Blair, D.F. Crystal structure of the middle and C-terminal domains of the flagellar rotor protein FliG. EMBO J. 2002, 21, 3225–3234.
37. Minamino, T.; Imada, K.; Kinoshita, M.; Nakamura, S.; Morimoto, Y.V.; Namba, K. Structural insight into the rotational switching mechanism of the bacterial flagellar motor. PLoS Biol. 2011, 9, e1000616.
38. Park, S.Y.; Lowder, B.; Bilwes, A.M.; Blair, D.F.; Crane, B.R. Structure of FliM provides insight into assembly of the switch complex in the bacterial flagella motor. Proc. Natl. Acad. Sci. USA 2006, 103, 11886–11891.
39. Sircar, R.; Greenswag, A.R.; Bilwes, A.M.; Gonzalez-Bonet, G.;

Crane, B.R. Structure and activity of the flagellar rotor protein FliY: a member of the CheC phosphatase family. J. Biol. Chem. 2013, 288, 13493–13502.

40. Stock, D.; Namba, K.; Lee, L.K. Nanorotors and self-assembling macromolecular machines: the torque ring of the bacterial flagellar motor. Curr. Opin. Biotechnol. 2012, 23, 545–554.
41. Dyer, C.M.; Vartanian, A.S.; Zhou, H.; Dahlquist, F.W. A molecular mechanism of bacterial flagellar motor switching. J. Mol. Biol. 2009, 388, 71–84.
42. Sarkar, M.K.; Paul, K.; Blair, D. Chemotaxis signaling protein CheY binds to the rotor protein FliN to control the direction of flagellar rotation in Escherichia coli. Proc. Natl. Acad. Sci. USA 2010, 107, 9370–9375.
43. Cluzel, P.; Surette, M.; Leibler, S. An ultrasensitive bacterial motor revealed by monitoring signaling proteins in single cells. Science 2000, 287, 1652–1655.
44. Duke, T.A.; Le Novere, N.; Bray, D. Conformational spread in a ring of proteins: a stochastic approach to allostery. J. Mol. Biol. 2001, 308, 541–553.
45. Bai, F.; Branch, R.W.; Nicolau, D.V., Jr.; Pilizota, T.; Steel, B.C.; Maini, P.K.; Berry, R.M. Conformational spread as a mechanism for cooperativity in the bacterial flagellar switch. Science2010, 327, 685–689.
46. Yuan, J.; Fahrner, K.A.; Berg, H.C. Switching of the bacterial flagellar motor near zero load. J. Mol. Biol. 2009, 390, 394–400.
47. Bai, F.; Minamino, T.; Wu, Z.; Namba, K.; Xing, J. Coupling between switching regulation and torque generation in bacterial flagellar motor. Phys. Rev. Lett. 2012, 108, 178105.
48. Boehm, A.; Kaiser, M.; Li, H.; Spangler, C.; Kasper, C.A.; Ackermann, M.; Kaever, V.; Sourjik, V.; Roth, V.; Jenal, U. Second messenger-mediated adjustment of bacterial swimming velocity. Cell2010, 141, 1–10.
49. Paul, K.; Nieto, V.; Carlquist, W.C.; Blair, D.F.; Harshey, R.M. (2010) The c-di-GMP binding protein YcgR controls flagellar motor

direction and speed to affect chemotaxis by a "backstop brake" mechanism. Mol. Cell. 2010, 38, 128–139.

50. Fang, X.; Gomelsky, M. A post-translational, c-di-GMP-dependent mechanism regulating flagellar motility. Mol. Microbiol. 2010, 76, 1295–1305.
51. Van Way, S.M.; Millas, S.G.; Lee, A.H.; Manson, M.D. Rusty, jammed, and well-oiled hinges: Mutations affecting the interdomain region of FliG, a rotor element of the Escherichia coli flagellar motor. J. Bacteriol. 2004, 186, 3173–3181.
52. Togashi, F.; Yamaguchi, S.; Kihara, M.; Aizawa, S.-I.; Macnab, R.M. An extreme clockwise switch bias mutation in fliG of Salmonella typhimurium and its suppression by slow-motile mutations in motA and motB. J. Bacteriol. 1997, 179, 2994–3003.
53. Sowa, Y.; Rowe, A.D.; Leake, M.C.; Yakushi, T.; Homma, M.; Ishijima, A.; Berry, R.M. Direct observation of steps in rotation of the bacterial flagellar motor. Nature 2005, 437, 916–919.
54. Nakamura, S.; Kami-ike, N.; Yokota, J.P.; Minamino, T.; Namba, K. Evidence for symmetry in the elementary process of bidirectional torque generation by the bacterial flagellar motor. Proc. Natl. Acad. Sci. USA 2010, 107, 17616–17620.
55. Yuan, J.; Fahrner, K.A.; Turner, L.; Berg, H.C. Asymmetry in the clockwise and counterclockwise rotation of the bacterial flagellar motor. Proc. Natl. Acad. Sci. USA 2010, 107, 12846–12849.
56. Delalez, N.J.; Wadhams, G.H.; Rosser, G.; Xue, Q.; Brown, M.T.; Dobbie, I.M.; Berry, R.M.; Leake, M.C.; Armitage, J.P. Signal-dependent turnover of the bacterial flagellar switch protein FliM. Proc. Natl. Acad. Sci. USA 2010, 107, 11347–11351.
57. Fukuoka, H.; Inoue, Y.; Terasawa, S.; Takahashi, H.; Ishijima, A. Exchange of rotor components in functioning bacterial flagellar motor. Biochem. Biophys. Res. Commun. 2010, 394, 130–135.
58. Yuan, J.; Branch, R.W.; Hosu, B.G.; Berg, H.C. Adaptation at the output of the chemotaxis signalling pathway. Nature 2012, 484, 233–236.
59. Yuan, J.; Berg, H.C. Ultrasensitivity of an adaptive bacterial motor. J. Mol. Biol. 2013, 425, 1760–1764.

60. Larsen, S.H.; Adler, J.; Gargus, J.J.; Hogg, R.W. Chemomechanical coupling without ATP: the source of energy for motility and chemotaxis in bacteria. Proc. Natl. Acad. Sci. USA 1974, 71, 1239–1243.
61. Manson, M.D.; Tedesco, P.; Berg, H.C.; Harold, F.M.; Van der Drift, C. A protonmotive force drives bacterial flagella. Proc. Natl. Acad. Sci. USA 1977, 4, 3060–3064.
62. Matsuura, S.; Shioi, J.; Imae, Y. Motility in Bacillus subtilis driven by an artificial protonmotive force. FEBS Lett. 1977, 82, 187–190.
63. Ravid, S.; Eisenbach, M. Minimal requirements for rotation of bacterial flagella. J. Bacteriol. 1984, 158, 1208–1210.
64. Terahara, N.; Sano, M.; Ito, M. A Bacillus flagellar motor that can use both Na^+ and K^+ as a coupling ion is converted by a single mutation to use only Na^+. PLoS One 2012, 7, e46248.
65. Sharp, L.L.; Zhou, J.; Blair, D.F. Tryptophan scanning mutagenesis of MotB, an integral membrane protein essential for flagellar rotation in Escherichia coli. Biochemistry 1995, 34, 9166–9171.
66. Zhou, J.; Sharp, L.L.; Tang, H.L.; Lloyd, S.A.; Billings, S.; Braun, T.F.; Blair, D.F. Function of protonatable residues in the flagellar motor of Escherichia coli: A critical role for Asp 32 of MotB. J. Bacteriol. 1998, 180, 2729–2735.
67. Hirota, N.; Kitada, M.; Imae, Y. Flagellar motors of alkalophilic Bacillus are powered by an electrochemical potential gradient of Na^+. FEBS Lett. 1981, 132, 278–280.
68. Hirota, N.; Imae, Y. Na^+-driven flagellar motors of an alkalophilic Bacillus strain Yn-1. J. Biol. Chem. 1983, 258, 10577–10581.
69. Chernyak, B.V.; Dibrov, P.A.; Glagolev, A.N.; Sherman, M.Y.; Skulachev, V.P. A novel type of energetics in a marine alkali-tolerant bacterium: $\Delta\mu$-Na-driven motility and sodium cycle. FEBS Lett. 1983, 164, 38–42.
70. Kojima, S.; Blair, D.F. Solubilization and purification of the MotA/MotB complex of Escherichia coli. Biochemistry 2004, 43, 26–34.
71. Terahara, N.; Krulwich, T.A.; Ito, M. Mutations alter the sodium versus proton use of a Bacillus clausii flagellar motor and confer

dual ion use on Bacillus subtilis motors. Proc. Natl. Acad. Sci. USA2008, 105, 14359–14364.

72. Li, N.; Kojima, S.; Homma, M. Sodium-driven motor of the polar flagellum in marine bacteria Vibrio. Genes Cells. 2011, 16, 985–999.
73. Sato, K.; Homma, M. Functional reconstitution of the Na^+-driven polar flagellar motor component of Vibrio alginolyticus. J. Biol. Chem. 2000, 275, 5718–5722.
74. Terashima, H.; Fukuoka, H.; Yakushi, T.; Kojima, S.; Homma, M. The Vibrio motor proteins, MotX and MotY, are associated with the basal body of Na^+-driven flagella and required for stator formation. Mol. Microbiol. 2006, 62, 1170–1180.
75. Ito, M.; Hicks, D.B.; Henkin, T.M.; Guffanti, A.A.; Powers, B.D.; Zvi, L.; Uematsu, K.; Krulwich, T.A. MotPS is the stator-force generator for motility of alkaliphilic Bacillus and its homologue is a second functional Mot in Bacillus subtilis. Mol. Microbiol. 2004, 53, 1035–1049.
76. Paulick, A.; Koerdt, A.; Lassak, J.; Huntley, S.; Wilms, I.; Narberhaus, F.; Thormann, K.M. Two different stator systems drive a single polar flagellum in Shewanella oneidensis MR-1. Mol. Microbiol. 2009, 71, 836–850.
77. Hosking, E.R.; Manson, M.D. Clusters of charged residues at the C terminus of MotA and N terminus of MotB are important for function of the Escherichia coli flagellar motor. J. Bacteriol.2008, 190, 5517–5521.
78. Braun, T.; Blair, D.F. Targeted disulfide cross-linking of the MotB protein of Escherichia coli: evidence for two H^+ channels in the stator complex. Biochemistry 2001, 40, 13051–13059.
79. Braun, T.F.; Al-Mawasawi, L.Q.; Kojima, S.; Blair, D.F. Arrangement of core membrane segments in the MotA/MotB protein-channel complex of Escherichia coli. Biochemistry 2004, 43, 35–45.
80. Che, Y.-S.; Nakamura, S.; Kojima, S.; Kami-ike, N.; Namba, K.; Minamino, T. Suppressor analysis of the MotB(D33E) mutation to probe the bacterial flagellar motor dynamics coupled with proton translocation. J. Bacteriol. 2008, 190, 6660–6667.
81. Kim, E.A.; Price-Carter, M.; Carlquist, W.C.; Blair, D.F. Membrane segment organization in the stator complex of the flagellar motor:

implications for proton flow and proton-induced conformational change. Biochemistry 2008, 47, 11332–11339.

82. Blair, D.F.; Berg, H.C. Mutations in the MotA protein of Escherichia coli reveal domains critical for proton conduction. J. Mol. Biol. 1991, 221, 1433–1442.

83. Braun, T.; Poulson, S.; Gully, J.B.; Empey, J.C.; Van Way, S.; Putnam, A.; Blair, D.F. Function of proline residues of MotA in torque generation by the flagellar motor of Escherichia coli. J. Bacteriol.1999, 181, 3542–3551.

84. Nakamura, S.; Morimoto, Y.V.; Kami-ike, N.; Minamino, T.; Namba, K. Role of a conserved prolyl residue (Pro-173) of MotA in the mechanochemical reaction cycle of the proton-driven flagellar motor of Salmonella. J. Mol. Biol. 2009, 393, 300–307.

85. Brandl, C.J.; Deber, C.M. Hypothesis about the function of membrane-buried proline residues in transport proteins. Proc. Natl. Acad. Sci. USA 1986, 83, 917–921.

86. Hosking, E.R.; Vogt, C.; Bakker, E.P.; Manson, M.D. The Escherichia coli MotAB proton channel unplugged. J. Mol. Biol. 2006, 364, 921–937.

87. Muramoto, K.; Macnab, R.M. Deletion analysis of MotA and MotB, components of the force-generating unit in the flagellar motor of Salmonella. Mol. Microbiol. 1998, 29, 1191–202.

88. Kojima, S.; Imada, K.; Sakuma, M.; Sudo, Y.; Kojima, C.; Minamino, T.; Homma, M.; Namba, K. Stator assembly and activation mechanism of the flagellar motor by the periplasmic region of MotB. Mol. Microbiol. 2009, 73, 710–718.

89. Kojima, S.; Furukawa, Y.; Matsunami, H.; Minamino, T.; Namba, K. Characterization of the periplasmic domain of MotB and implications for its role in the stator assembly of the bacterial flagellar motor. J. Bacteriol. 2008, 190, 2259–2266.

90. Blair, D.F.; Kim, D.Y.; Berg, H.C. Mutant MotB proteins in Escherichia coli. J. Bacteriol. 1991, 173, 4049–4055.

91. Garza, A.G.; Biran, R.; Wohlschlegel, J.; Manson, M.D. Mutations in motB suppressible by changes in stator or rotor components of the bacterial flagellar motor. J. Mol. Biol. 1996, 25, 270–285.

92. Hizukuri, Y.; Morton, J.F.; Yakushi, T.; Kojima, S.; Homma, M. The peptidoglycan-binding (PGB) domain of the Escherichia coli pal protein can also function as the PGB domain in E. coli flagellar motor protein MotB. J. Biochem. 2009, 146, 219–229.

93. Hizukuri, Y.; Kojima, S.; Homma, M. Disulfide cross-linking between the stator and the bearing components in the bacterial flagellar motor. J. Biochem. 2010, 148, 309–318.

94. Ryu, W.S.; Berry, R.M.; Berg, H.C. Torque-generating units of the flagellar motor of Escherichia colihave a high duty ratio. Nature 2000, 403, 444–447.

95. Yuan, J.; Berg, H.C. Resurrection of the flagellar rotary motor near zero load. Proc. Natl Acad. Sci. USA 2008, 105, 1182–1185.

96. Reid, S.W.; Leake, M.C.; Chandler, J.H.; Lo, C.J.; Armitage, J.P.; Berry, R.M. The maximum number of torque-generating units in the flagellar motor of Escherichia coli is at least 11. Proc. Natl. Acad. Sci. USA 2006, 103, 8066–8071.

97. Leake, M.C.; Chandler, J.H.; Wadhams, G.H.; Bai, F.; Berry, R.M.; Armitage, J.P. Stoichiometry and turnover in single, functioning membrane protein complexes. Nature 2006, 443, 355–358.

98. Lele, P.P.; Hosu, B.G.; Berg, H.C. Dynamics of mechanosensing in the bacterial flagellar motor. Proc. Natl. Acad. Sci. USA 2013, 110, 11839–11844.

99. Tipping, M.J.; Delalez, N.J.; Lim, R.; Berry, R.M.; Armitage, J.P. Load-dependent assembly of the bacterial flagellar motor. MBio 2013, 4, e00551-13.

100. Castillo, D.J.; Nakamura, S.; Morimoto, Y.V.; Che, Y.-S.; Kamiike, N.; Kudo, S.; Minamino, T.; Namba, K. The C-terminal periplasmic domain of MotB is responsible for load-dependent control of the number of stators of the bacterial flagellar motor. Biophysics 2013, 9, 173–181.

101. Kojima, S.; Blair, D.F. Conformational change in the stator of the bacterial flagellar motor. Biochemistry 2001, 40, 13041–13050.

102. Che, Y.-S.; Nakamura, S.; Morimoto, Y.V.; Kami-ike, N.; Namba, K.; Minamino, T. Load-sensitive coupling of proton translocation and torque generation in the bacterial flagellar motor. Mol. Microbiol. 2014, 91, 175–184.

103. Chen, X.; Berg, H.C. Solvent-isotope and pH effects on flagellar rotation in Escherichia coli. Biophys. J. 2000, 78, 2280–2284.
104. Zhou, J.; Blair, D.F. Residues of the cytoplasmic domain of MotA essential for torque generation in the bacterial flagellar motor. J. Mol. Biol. 1997, 273, 428–439.
105. Morimoto, Y.V.; Nakamura, S.; Kami-ike, N.; Namba, K.; Minamino, T. Charged residues in the cytoplasmic loop of MotA are required for stator assembly into the bacterial flagellar motor. Mol. Microbiol. 2010, 78, 1117–1129.
106. Morimoto, Y.V.; Nakamura, S.; Hiraoka, K.D.; Namba, K.; Minamino, T. Distinct roles of highly conserved charged residues at the MotA-FliG interface in bacterial flagellar motor rotation. J. Bacteriol. 2013, 195, 474–481.

CHAPTER

8

THE ROLE OF THE BACTERIAL FLAGELLUM IN ADHESION AND VIRULENCE

Johanna Haiko and Benita Westerlund-Wikström

Division of General Microbiology, Department of Biosciences, University of Helsinki, Finland

ABSTRACT

The bacterial flagellum is a complex apparatus assembled of more than 20 different proteins. The flagellar basal body traverses the cell wall, whereas the curved hook connects the basal body to the whip-like flagellar filament that protrudes several μm from the bacterial cell. The flagellum has traditionally been regarded only as a motility organelle, but more recently it has become evident that flagella have a number of other biological functions. The major subunit, flagellin or FliC, of the flagellum plays a well-documented role in innate immunity and as a dominant antigen of the adaptive immune response. Importantly, flagella have also been reported to function as adhesins. Whole flagella have been indicated as significant in bacterial adhesion to and invasion into host cells. In various pathogens, e.g., Escherichia coli, Pseudomonas aeruginosa and Clostridium difficile, flagellin and/or the distally located flagellar cap protein have been reported to function as adhesins. Recently, FliC of Shiga-toxigenic E. coli was shown to be involved in cellular invasion via lipid rafts. Here, we examine the latest or most important findings regarding flagellar adhesive and invasive properties, especially focusing on the flagellum as a potential virulence factor.

Keywords: bacterial flagella; flagellin; FliD; adhesion; invasion

INTRODUCTION

Many Gram-positive and Gram-negative bacterial species and also Archaea as well as some eukaryotic cells have a flagellum ('whip' in Latin). Flagellum is primarily a motility organelle that enables movement and chemotaxis. Bacteria can have one flagellum or several, and they can be either polar (one or several flagella at one spot) or peritrichous (several flagella all over the bacterium). In addition to motility, flagella possess several other functions that differ between bacteria and during the bacterial life cycle: a flagellum can, for example, participate in biofilm formation, protein export, and adhesion. This review mainly focuses on the adhesive functions of flagella, and studies on the direct interaction of flagella with the host targets will be emphasized, though flagellar motility is often a prerequisite for adhesion and/or invasion. Duan et al. [1] have recently summarized in more detail the various roles of bacterial flagella in pathogenicity. For reviews on the role of flagella in biofilms, see Prüss et al. [2] and Conrad [3].

The length of the typical flagellum of Escherichia coli is about 10 μm and the diameter is 20 nm. Over 60 structural and regulatory proteins are required for flagellum assembly and function. Flagellum consists of a cytoplasmic export apparatus, a basal body embedded in the cell membrane (CM), a hook that connects the basal body to the filament, and a filament that functions as a propeller (Figure 1A). Flagellar assembly starts with the CM-associated components of the basal body and the secretion apparatus, through which the other flagellar proteins are then secreted, first the remaining basal body components, then the hook and the hook-filament junction proteins. Filament assembly starts after the hook (120 FlgE molecules) has been completed and the filament capping proteins have been positioned. The filament is composed of about 20,000 flagellin (FliC) proteins that are incorporated below the distal pentameric FliD cap, which functions as a plug and is required for assembly of nascent monomeric flagellin. A two-component signaling cascade involving chemotaxis-related proteins affects flagellar rotation, which is facilitated by the engine consisting of the basal body-associated stator proteins and the basal body, which functions as a rotor. Some proteins relevant to this review and their functions are listed in Table 1. For a detailed description of flagellar assembly, see Chevance and Hughes [4]. For a review on chemotaxis, the reader is referred to Sourjik and Wingreen [5].

Table 1. Overview of flagella proteins relevant for the review [(a)].

Role in/Part of flagellum	Protein name/Number of proteins	Function
Regulation of flagellar biosynthesis	FlhC, FlhD, FlbA	Regulators
	FliI	Export-related ATPase
	FliA	Flagellar sigma factor
Motility/chemotaxis	CheA	Smooth swimming
	CheB	Tumbling
	CheW, CheV	Chemotaxis
	MotA, MotB	Rotation of flagellum
Export machinery	9 different proteins	Protein export
	FlhE	Chaperone or plug
Basal body	20 different proteins	Cell-wall anchor, rotor, holds export apparatus
	FlgJ	Muramidase
Hook	FlgE	Hook
Hook-filament junction	FlgK, FlgL	Connect filament to hook
Filament	FliC (alternative names FljB, FlaA, FlaB)	Main structural subunit
	FliD	Filament cap

[(a)] Information combined from [2,4±7]. The designations used are mainly the names in enteric bacteria; for homologs in other species, see [8].

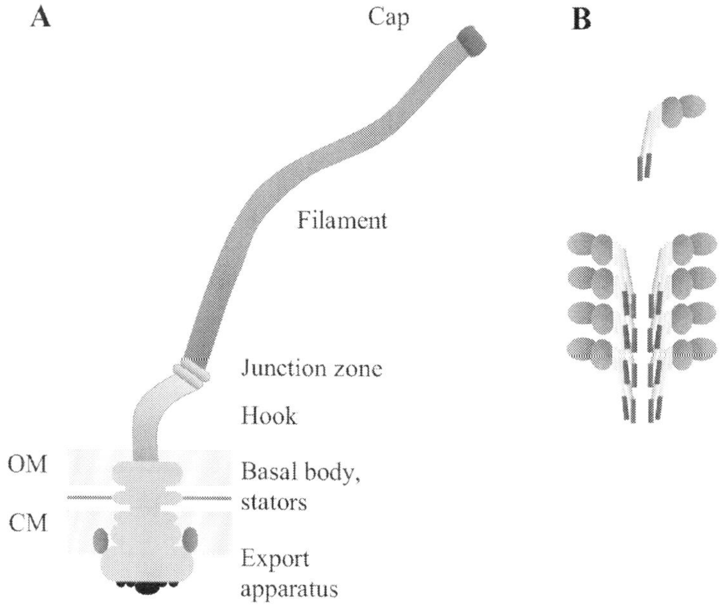

Figure 1. (**A**) Schematic presentation of bacterial flagellum structure. The pentameric FliD cap at the distal end of the filament, the hollow filament composed of about 20,000 identical flagellin subunits, the junction zone between filament and hook, and the hook connecting the filament to the basal body, represent extracellular parts (green shades) of the flagellum. The basal body (grey) in the cell wall consists of a centrally located hollow rod that connects different rings embedded in the outer membrane (OM), the peptidoglycan layer, and the cytoplasmic membrane (CM). Stator complexes (dark grey), composed of membrane proteins MotA and MotB, are associated with the CM-bound ring and the cytoplasmic ring below the CM, and provide motility-required energy. The cytoplasmic export machinery (black) that secretes the extracellular subunits is located within the cytoplasmic ring. Note that OM-associated parts of the basal body are absent in the flagella of Gram-positive bacteria. (**B**) Schematic presentation of flagellin monomer (upper panel) and flagellin polymerization (lower panel). The variable, exposed, globular domains of flagellin are shown in green. The conserved N- and C-terminal regions involved in flagellum polymerization are indicated (blue, light blue, yellow, red) as well as the regions binding to TLR5 (light blue, yellow) and those involved in inflammasome formation (blue, red).

Bacterial adhesion is a critical initiation step in bacterial colonization and persistence, both for pathogens and commensals. Bacteria express various adhesive surface structures such as capsule, fimbriae or pili, and several surface proteins (for examples the reader is referred to Klemm et al. [9]). Typically the adhesive structures are not expressed at the same time as the flagellum, so that movement and attachment occur one at a time. Thus, bacteria switch from motile to sessile lifestyle and vice versa, and these changes are triggered by different environmental conditions—such as temperature, osmolarity, and pH—which regulate the expression of the flhDC flagellar master operon [2]. The regulation of flagellar expression occurs temporally at both the level of transcription and assembly [2,4]. The flagellum has also been shown to function as an export apparatus that mediates extracellular secretion of non-flagellar virulence-associated effector proteins and biotechnologically important heterologous polypeptides [1,10,11].

From the mammalian host perspective, the flagellum is relevant for immune defense: The immune system recognizes flagellin, which triggers adaptive and innate immune response. The conserved N- and C-termini of monomeric flagellin (Figure 1B) involved in flagella assembly interact with cytoplasmic NOD-like receptors in eukaryotic cells and induce the formation of inflammasome, which leads to pyroptosis [12,13,14]. Flagellin belongs to molecules containing a pathogen-associated molecular pattern (PAMP) that is recognized by toll-like receptor 5 (TLR5) [12]. TLR5 is mostly expressed at the basolateral surface of intestinal epithelial cells and by monocytes and fibroblasts, and binds the conserved termini of flagellin, which leads to the activation of cytokine secretion in host cells [12,15]. The central region of FliC is variable in sequence, and this region is exposed in native flagellin [16,17]. Sequence variation explains the observed differences in e.g., antigenic variation and the adhesive functions of flagellins from different bacterial strains or species (see text and Table 2 below).

Table 2. Direct and indirect roles of flagella in bacterial adhesion.

Role of flagella/ Bacterial species	Flagellum/ Protein/Gene	Effect/Role in virulence	Target/Administration route	Receptor	Reference
Adhesion					
Bordetella bronchiseptica, *Bordetella pertussis*	Flagellum	Adhesion	HeLa cells	ND	[18]
Burkholderia pseudomallei	Flagellum	Adhesion; invasion	*Acanthamoeba astronyxis*	ND	[19]
Campylobacter jejuni	Flagellum	Adhesion	Intestine-407	ND	[20]
Clostridium difficile	FliC, FliD; flagellum	Binding; no effect	Mouse cecal mucus, hamster	ND	[21,22]
Escherichia coli	Flagellum, FliC	Adhesion; Microcolony formation	HeLa cells	ND	[23]
	Flagellum, FliC	Binding	Mucins [(a)], bovine intestinal mucus, laminin, collagen	ND	[24]
	Flagellum	Adhesion; invasion	Polarized Caco-2BBe, T-84; BREC; Caco-2; T-84; IPEC-J2; IPEC-1	ND	[25–30]
	Flagellum	Association	HBMEC	ND	[31]
	Flagellum, FliC	Adhesion	Caco-2, mice intestine	EtpA	[32]
	FliC	Adhesion	Human intestinal cryosections, LS174T	gluconate	[33]
Pseudomonas aeruginosa	FliC	Adhesion; virulence	1HAEo; mice	GM1, GD_{1a}, asialo-GM1	[34]
	FliC; FliD	Adhesion	mucin, MUC1	ND	[35–37]
	FliD	Adhesion	human respiratory mucin	Lewis x glycotype	[38]
	Flagellum	Adhesion	Calu-3 basolateral surface	HSPGs [(b)]	[39]
Salmonella enterica serovar Dublin	Flagellum	Adhesion; virulence	Intestine-407; orally in mice	ND	[40]
Salmonella enterica serovar Typhimurium	Flagellum	Adhesion	Intestine-407	ND	[40]
Stenotrophomonas maltophilia	Flagellin	Adhesion	Mouse tracheal mucus	ND	[41]
Motility					
Cronobacter sakazakii	Flagellum	Adhesion; biofilm formation	Caco-2	ND	[42]
Escherichia coli	Flagellum	Mucus penetration; persistence; invasion	Hep-2, HT2916E, HT2919A; orally in chicks; HCT-8	asialo-GM1, lipid rafts	[43–48]
	FliC	Virulence	Orally in Sm-treated mice	ND	[49]
Helicobacter pylori	Flagellum	Colonization	Gastric epithelium, orally in mice	ND	[49]
Proteus mirabilis	Flagellum	Invasion; colonization	Human renal proximal tubular epithelial cells, EJ/28	ND	[50,51]
Salmonella enterica serovar Dublin	Flagellum	Increased invasion	Intestine-407	ND	[40]
Salmonella enterica serovar Enteritidis	Flagellum	Increased invasion	Caco-2, Hep-2, Div-1	ND	[52–54]
Salmonella enterica serovar Typhimurium	Flagellum	Increased adhesion; invasion	Intestine-407	ND	[40]
Regulation					
Escherichia coli	*flhDC*	Adhesion, invasion	Intestine-407	ND	[55]
	flhC	Colonization	Orally in cattle	ND	[56]
	FliA	Adhesion, invasion	Intestine-407	ND	[57]
Helicobacter pylori	*flhA*	Adhesion	Kato III, primary human gastric epithelial cells	ND	[58]

[(a)] bovine submaxillary gland type I, porcine stomach type II; [(b)] heparan sulfate proteoglycans; ND, not determined.

The bacterial flagellum thus affects bacterial virulence in various ways, i.e., by providing motility towards host targets, promoting early biofilm formation and thus bacterial survival, secreting virulence factors, triggering the adaptive and innate immune defense, and by promoting adherence and invasion.

MOTILITY AND VIRULENCE

In many bacterial species, the flagellum is an acknowledged virulence factor, and non-flagellated strains have in several cases been observed to be less virulent. The flagellum can act directly as an adhesin, as detailed in Section 3 and Section 4, but can also affect virulence by other means. Motility towards a host cell is a prerequisite for adhesion and invasion, and flagella can play an essential role in colonization by facilitating bacterial motility even if the flagella do not directly participate in the adhesion or invasion. Flagella can also contribute to virulence by regulating the expression of other virulence factors [2] and, as discussed below, the flagellum in some cases affects virulence in more than one manner.

Flagellum Affects Virulence Mainly by Facilitating Motility

Early studies have shown that the single polar flagellum of Vibrio cholerae, the causative agent of cholera, is crucial for its virulence: non-motile V. cholerae had reduced virulence in mice, and their adsorption to cross sections of mice intestine was decreased compared to motile V. cholerae [59].

In Salmonella enterica serovar Enteritidis, a common cause of food-borne diarrhea, studies with non-flagellated ΔfliC and flagellated but non-motile motA mutants have shown that functional flagella enhance the invasive capacity of the bacterium into Intestine-407 and Caco-2 cells (see Table 3 for cell lines mentioned in the text) by enabling efficient motility, but the flagella filament per se is not required for adhesion and invasion [52,53]. However, non-flagellated mutants of S. Enteritidis have been shown to cause significantly less of the typical invasion-associated membrane ruffling than the wild-type strain in cultured human Hep-2 and avian Div-1 epithelial cells, indicating that flagella are involved in the early events of S. Enteritidis invasion in a still uncharacterized manner [54].

Table 3. Cell lines mentioned in the review [(a)].

Abbreviation	Cell line	Origin
1HAEo	airway epithelial cells	human
A549	lung epithelial cells	human
BREC	primary rectal epithelial cells	bovine
Caco-2	colorectal adenocarcinoma epithelial cells	human
Caco-2BBe	colorectal adenocarcinoma epithelial cells expressing enterocyte-like brush border	human
Calu-3	lung adenocarcinoma epithelial cells	human
CHO Lec-2	relatively ganglioside deficient CHO derivative	Chinese hamster
CHO-Muc1	mucin 1-expressing ovary cells	Chinese hamster
Div-1	gastrointestinal mucosal cells	avian
EJ/28	urothelial cells	human
16HBE	polarized airway epithelial cells	human
HBMEC	brain microvascular endothelial cells	human
HCT-8	ileocecal colorectal adenocarcinoma cells	human
HeLa	cervical tumor cells	human
Hep-2	epidermoid cancer cell line contaminated by HeLa cells	human
HT29	colorectal adenocarcinoma epithelial cells	human
HT2916E	mucus-secreting subclone of HT29	human
HT2919A	non-mucus-secreting subclone of HT29	human
Int(estine)-407	intestinal embryonic jejunum and ileum epithelial cells contaminated by HeLa cells	human
IPEC-1	neonatal ileal and jejunal epithelial cells	piglet
IPEC-J2	neonatal jejunal columnar epithelial cells	piglet
Kato III	gastric adenocarcinoma cells	human
LS174T	mucin-secreting colorectal adenocarcinoma epithelial cells	human
RAW264.7	macrophage cells	murine
T24	bladder epithelial cells	human
T-84	colorectal adenocarcinoma cells	human

Recently, Olsen and colleagues [40] revealed serovar-specific differences in the involvement of flagella and chemotaxis genes in the adhesion and invasion of Salmonella. In the cattle-adapted Salmonella entericaserovar Dublin, the flagella filament was required for adhesion and invasion of Intestine-407 cells, whereas adhesion was unaffected in the constitutively tumbling cheB mutant and the constitutively smoothly swimming cheA mutant indicating that targeted motility is not essential for the adhesive capacity. In the broad-host-range Salmonella enterica serovar Typhimurium, adhesion to Intestine-407 was decreased by the deletion of flagellin and chemotaxis-related genes. Co-infection of mice with wild-type and mutant strains of both serovars revealed that the chemotaxis-related genes were dispensable in vivoin both serovars. The virulence of the S. Dublin fliC mutant was decreased with oral but not with intraperitoneal administration indicating that flagellin is important at the early colonizations stages of S. Dublin infection. In S. Typhimurium, deletion of the flagellin genes fliC and fljB rendered the mutant more virulent when administered intraperitoneally, whereas no effect was observed with oral infection indicating that flagellin is not required at the systemic phase of S. Typhimurium infection. Taken together, the role of flagella in Salmonella pathogenesis appears to be mainly related to motility, but flagellum affects adhesion and invasion in a serovar- and/or host-target dependent manner that reflects the lifestyles of the different serovars.

Cronobacter spp. are opportunistic food-borne pathogens that can cause serious diseases like bacteremia and meningitis in neonates. A study of a random transposon mutant library of the clinical Cronobacter sakazakii strain ES5 showed that peritrichous flagella are involved in biofilm formation on polystyrene and adhesion to Caco-2 cells [42]. In the study, five mutants carried transposons in flagellum-associated genes fliD, flgJ, and flhE. Phenotypically the fliD mutant had shorter flagella than the wild-type, the flgJ mutant lacked flagella, and the flhE mutant had flagella morphologically similar to the wild-type. Motility could not be examined, but the biofilm formation was drastically reduced in fliDand flhE mutants. Dramatic decrease in adhesion to Caco-2 cells was observed with flgJ mutant, and flhEmutant adhered as well as the wild-type, whereas the adherence of the fliD mutant was not measured. The results apparently indicate that motility is relevant in C. sakazakii adhesion to Caco-2 cells, but direct interaction of flagella and Caco-2 cannot be excluded.

P. mirabilis, a cause of urinary tract infections, requires peritrichous flagella for invasion into renal epithelial cells and urothelial cells, thus enabling colonization of the urinary tract of mice [50,51]. Contradictory findings of the importance of flagella in P. mirabilis pathogenesis have also been reported, indicating that a flagellum is not necessary in all infection models or for all P. mirabilis strains causing urinary tract infections [60,61], however nowadays flagella are regarded as a crucial virulence factor for P. mirabilis [62].

Flagellum Affects Virulence by Regulating Other Virulence Factors

Polar flagella are important virulence factors for Helicobacter pylori, a cause of gastric ulcers, because motility enables the bacteria to reach the gastric epithelium, adhere to it with several adhesins, and colonize the epithelium (reviewed by Sheu et al. [63]). Studies with a non-motile fliD mutant showed that FliD and thus a functional flagellum is required for the colonization of mice [64]. Clyne et al. [58] have studied whether H. pylori flagella are directly involved in adhesion by constructing flagellin (flaA and/or flaB) mutants and a flagellar regulator (flbA) mutant, and it appeared that all mutants adhered to gastric cells, indicating that flagella do not play a direct role in adhesion of H. pylori and suggesting that in addition to regulating flagella, FlbA may regulate some H. pylori adhesins.

FLAGELLUM-MEDIATED ADHESION

In several bacterial species, such as E. coli, Pseudomonas aeruginosa, and Clostridium difficile, the flagellum is multifunctional and plays a role in motility and is an adhesive organelle. A comprehensive review of studies on the direct role of the flagellum as an adhesin is presented below.

Escherichia coli

E. coli Pathovars and Role of Flagella in Virulence

The Gram-negative enterobacterial species E. coli comprises a heterogeneous group of strains belonging on the one hand to commensals in the gut microbiota and on the other to pathogens. Two major groups of pathogenic E. coli exist, namely extraintestinal pathogenic E. coli (ExPEC) and intestinal pathogenic E. coli (IPEC). Human pathogenic E. coli strains are divided into pathovars on the basis of the symptoms and the disease they cause as well as some specific virulence determinants (Table 4). IPEC strains, which originate in domestic animals and infect human beings mainly via contaminated food or water, are divided into six main pathovars: enterohaemorrhagic E. coli (EHEC) causing outbreaks of severe gastroenteritis that can lead to haemorrhagic colitis or hemolytic-uremic syndrome; enteropathogenic E. coli (EPEC) which can cause diarrhea in young infants mainly in developing countries and also includes atypical enteropathogenic E. coli (aEPEC) that is an emerging enteropathogen; enterotoxigenic E. coli (ETEC) that causes diarrhea in infants and traveler's diarrhea; enteroinvasive E. coli (EIEC) that causes dysentery; enteroaggregative E. coli (EAEC) that, depending on expressed virulence factors, can represent non-pathogenic strains or even cause hemolytic uremic syndrome; and diffusely adherent E. coli (DAEC), which causes diarrhea in young children [65]. The term Shiga-toxin producing E. coli (STEC) can include all variants of a particular pathovar, e.g., all EHEC strains produce the Shiga-toxin and are hence STECs, or only some types of a pathovar, e.g., Shiga-toxin producing EAEC strains are referred to as STEAEC. The Crohn's disease associated pathovar is called adherent-invasive E. coli (AIEC), the avian pathogenic E. coli (APEC) resembles human ExPEC and causes colibacillosis in poultry, and fimbriated pathogenic E. coli is responsible for serious infections in piglets are pathovars not as well characterized as the IPECs and ExPECs [65,66,67,68,69].

E. coli express peritrichous flagella, which are classified into H-serotypes on the basis of sero-reactivity of the variable antigenic domain of FliC [70] (see Figure 1B). One pathovar may contain strains of different H serotypes and similar H serotypes can be found in more than one pathovar [69]. Early reports regarding the putative role of the flagella of APEC and Shiga-toxin producing EHEC in bacterial virulence indicated that E. coli flagella were not directly involved in bacterial adhesion but played a role in invasion. Experiments performed by inactivation of the flagellin gene and assessment of adhesive properties of the mutant in comparison to the wild-type strain indicated that the flagella of an APEC isolate did not significantly affect adhesion to Hep-2 cells and non-mucus-secreting HT2919A cells. The flagella mainly contributed to APEC penetration through the mucus in mucus-secreting HT2916E cells, and to persistence of APEC in the intestine of newly hatched chicks as well as of Shiga-toxin-negative EHEC O157:H7 in specific-pathogen-free chicks [43,44,45]. In Shiga-toxin producing EHEC O113:H21, a ΔfliC mutant colonized the intestine of streptomycin-treated mice as efficiently as the wild-type strain, but showed a reduced capacity to form the typical intimate association with colonic epithelium and was less virulent than the parental strain [49]. The flagellin of STEC O113:H21 was reported as essential for bacterial invasion into HCT-8 cells in a TLR5-, MyD88-, and IRAK-independent manner, although STEC adhesion was not mediated by FliC [47]. The results indicated that STEC interacted with lipid rafts, as pretreatment of HCT-8 monolayers with anti-asialoganglioside antibodies, cholesterol-depleting methyl-β-cyclodextrin, or tyrosine kinase-inhibiting genistein prior to invasion reduced STEC invasion. Invading STEC were also localized in the regions of lipid raft markers on HCT-8 cells. However, the pretreatment of HCT-8 cells reduced invasion of the STEC ΔfliC mutant, and the authors concluded that bacterial interaction with lipid rafts was not dependent solely on flagellin [47]. Conclusively, these aforementioned findings regarding APEC and especially Shiga-toxin producing EHEC O113:H21 indicate that flagellum-mediated motility is crucial in invasion, but that flagellin is not required for adhesion.

Table 4. E. coli pathovars relevant for the review.

Pathovar	Abbreviation	Disease/symptoms	Major site of colonization	Reference
IPEC				
Enterohaemorrhagic *E. coli*	EHEC	Gastroenteritis; hemorrhagic colitis; HUS [(a)]	Large bowel	[65,69]
Enteropathogenic *E. coli*	EPEC	Infant diarrhea in developing countries	Small bowel	[65,69]
Atypical enteropathogenic *E. coli*	aEPEC	Persistent diarrhea in children	Small bowel	[69]
Enterotoxigenic *E. coli*	ETEC	Infant diarrhea; travelers' diarrhea	Small bowel	[65,69]
Enteroinvasive *E. coli*	EIEC	Dysentery	Large bowel	[65,69]
Enteroaggregative *E. coli*	EAEC	Variable (from non-symptomatic to HUS)	Large bowel, small bowel	[65,69]
Diffusely adherent *E. coli*	DAEC	Diarrhea in young children; nosocomial adult diarrhea	Small bowel	[65,69]
Shiga-toxin producing *E. coli*	STEC	Dependent on Shiga-toxin expressing pathovar	Dependent on pathovar	[69]
ExPEC				
Newborn-meningitis causing *E. coli*	NMEC	Newborn meningitis; septicaemia	Brain, blood	[65,68,71]
Uropathogenic *E. coli*	UPEC	Urinary tract infections; septicaemia	Bladder, kidney, blood	[65,68]
Other				
Adherent-invasive *E. coli*	AIEC	Crohn's disease	Ileum of small bowel	[65,67–69]
Avian pathogenic *E. coli*	APEC	Avian colibacillosis	Systemic infection	[67,68]
Porcine pathogenic *E. coli*	Porcine STEC	Post-weaning diarrhea; edema disease	Intestine;vasculature	[28,29,48]

Flagellum-mediated IPEC Adhesion

Several reports, however, show that flagella of some E. coli pathovars do mediate direct bacterial adhesion, and the majority of these reports focus on IPEC strains. Girón and colleagues [23] have shown that wild-type EPEC expressing H2 or H6 flagella adhere to HeLa cells and form microcolonies on them. Insertional inactivation of the $fliC_{H6}$ gene in EPEC significantly decrease the adhesion, whereas the flagellated but non-motile isogenic motB mutant adheres to HeLa as efficiently as the wild-type strain thus demonstrating that the flagellar filament indeed is involved in adhesion, not only in the movement towards a target. Complementation of the fliC mutation with plasmid-cloned $fliC_{H6}$ partially restored the adhesiveness of the mutant, and expression of $FliC_{H6}$ in a non-adhesive laboratory strain of E. colirendered it adhesive to HeLa cells. The interaction with HeLa cells was also demonstrated with purified H6 flagella, whereas purified H7 flagella did not bind to the cells. It was also shown that flagella expression in all motile EPEC strains tested (expressing e.g., H2, H6 and H34 flagella) was enhanced when the bacteria were grown on eukaryotic cells and that the flagella extended within and between the bacterial microcolonies. In a non-motile EPEC strain, low-level flagella expression was observed as well, whereas EHEC carrying the genes encoding H7 flagella and ETEC having the $fliC_{H9}$ gene did not express flagella when the bacteria were grown on HeLa cells. Flagella of aEPEC O51:H40 were also observed to be adhesive: the wild-type aEPEC strain adhered to the colorectal cell line T-84, and invaded T-84 and Caco-2 cells, whereas an isogenic fliC mutant showed a significantly reduced adhesive and invasive capacity [27]. Taken together, the reports show that EPEC and aEPEC, but not EHEC (O157:H7) and ETEC (O8:H9), adhere to HeLa, T-84 and Caco-2 cells and that EPEC express flagella efficiently when cultured on HeLa cells. The authors speculate that a soluble molecule of eukaryotic origin regulates flagella production and thereby flagellum-mediated adhesion in EPEC.

Using immunodot analysis, Erdem and colleagues [24] showed that purified flagella and denatured monomeric flagellin of both EPEC O127:H6 and EHEC O157:H7 bound to immobilized commercially available bovine and porcine mucins, as well as to mucus prepared from bovine colon in a dose-dependent manner. Preincubation of purified flagella with commercial mucin followed by molecular exclusion chromatography demonstrated simultaneous elution of flagella and mucin, thus verifying the binding of flagella to mucin. The adhesive capacity of FliC-deficient mutants of EPEC O127:H6 and EHEC O157:H7 to bovine intestinal tissue was significantly lower than that of corresponding isogenic wild-type strains, thus indicating that $FliC_{H6}$ but also $FliC_{H7}$ mediate bacterial adhesion to bovine submaxillary gland mucin and porcine stomach mucin. Differences between H6 and H7 flagella in the interactions with host molecules were also observed: Purified H6 flagella bound to extracellular matrix proteins collagen and laminin, whereas H7 flagella did not interact with these proteins. Also, EHEC O157:H7 cells agglutinated rabbit erythrocytes independently of mannoside-binding type 1 fimbriae typically expressed by E. coli. The hemagglutination (HA) was inhibited by anti-H7 antibodies and mucins (see above), whereas a number of characterized carbohydrates and glycoproteins had no inhibitory effect. Moreover, the FliC-deficient isogenic mutant lost the HA capacity, and similar HA properties were not observed with EPEC O127:H6 or purified H6 flagella. The authors speculate that the observed mucin-binding property of H6 and H7 flagella may favor intestinal colonization, whereas the adhesion to the extracellular matrix proteins by H6 flagellum may contribute to colonization at sites where the intestinal barrier is disrupted. The relevance of the flagella-mediated HA property of EHEC remains to be elucidated, but apparently it reflects the receptor specificity of H7 flagella.

The adhesive properties of H7 of EHEC to BREC, which is the colonization site in cattle, were studied by Mahajan and colleagues [26]. They demonstrated that wild-type EHEC O157:H7 adhered to BREC, that adhesion of an isogenic fliC deletion mutant was significantly decreased, and that adhesion of the fliC mutant was restored by trans complementation with $fliC_{H7}$ but not with $fliC_{H6}$. The wild-type adhesion was blocked by anti-$FliC_{H7}$ antiserum and by purified H7 flagella. Purified H7 flagella of EHEC bound directly to BREC, whereas EHEC flagella of serotypes H11 and H21 did not interact with bovine cells as assessed by indirect immunofluorescence microscopy. EHEC O157:H7 expressed flagella when applied onto BRECs, but indirect immunofluorescence microscopy and in-cell Western assay revealed a temporal expression of H7 flagella; the flagella were present during early colonization and on bacteria not associated with epithelial cells, but expression of flagella was repressed on bacteria in microcolonies or bacteria associated with the typical attaching and effacing lesions formed by EHEC on BRECs. The results are in good agreement with the study by Girón and coworkers [23] showing that EPEC, but not EHEC, express flagella when grown on cultured epithelial cells, and with the reports by Luck et al. [46] and Rogers et al. [47] that state that H21 flagella do not mediate adhesion to cultured epithelial cells. The results indicate that H7 flagella act as adhesins at the initiating phase of the infection, but are not required at later stages of EHEC infection. The preceding reports further demonstrate the relevance of the appropriate selection of host cells prior to the analysis of flagellar adhesion.

Flagellum-mediated ExPEC Adhesion

The major ExPEC pathovars are uropathogenic E. coli (UPEC) causing urinary tract infections and the newborn-meningitis causing E. coli (NMEC), and more recently, APEC that have been shown to closely resemble ExPEC [67,68]. In contrast to IPECs, NMEC strains are associated with high mortality rates, are genetically closely related, express a number of well-characterized fitness factors and belong to only a few serotypes of which O18:K1:H7 is widely spread and the most studied serotype [71]. Transcriptome analysis has revealed that the expression level of the fliI gene is significantly increased in NMEC associated with HBMEC cells compared to NMEC not associated with cultured cells [72]. Further analysis indicated that the H7 flagellum of E. coli O18:K1:H7 may be involved in NMEC pathogenesis. Flagella-deficient (ΔfliI, ΔflhDC, and ΔfliC) mutants associated to HBMECs and invaded them significantly less efficiently than the isogenic wild-type strain. Further, the mutants complemented with plasmid-cloned fliI, flhDC and fliC, and the mutants (ΔcheW, ΔmotAB) carrying non-motile flagella, associated with and invaded HBMECs equally well as the wild-type strain [31]. Also, the association rate of the wild-type NMEC strain was reduced to 40% after the treatment of HBMEC with purified recombinant flagellin, indicating that purified flagella block flagellum-mediated interaction of NMEC with HBMEC. Deletion of fliC did not affect the expression of Type 1 fimbriae, previously shown to interact with HBMEC [73]. Taken together, the results thus indicate that motility per se is not a key determinant in the association with and invasion into HBMECs by NMEC, but presence of the flagellar filament is apparently required for the interaction of NMEC with HBMEC, and FliC may, in concert with other virulence factors, play a role in meningitis caused by E. coli O18:K1:H7. Interestingly, according to the reports above, expression of H7 flagella of both EHEC and NMEC is upregulated upon contact with putative host cells.

Role of Flagellum in AIEC Adhesion

In the AIEC type strain LF82, the flagellar master operon flhDC reduces bacterial adhesion to Intestine-407 cells indirectly by down-regulating the expression of adhesive Type 1 fimbriae, but flagella also affect bacterial invasion in a manner not restricted to motility or the presence of a flagellar filament [55]. Similarly, the flhC gene, but not the flagellin gene, reduces in an unknown manner the colonization of cattle by Shiga-toxin-producing EHEC O157:H7 [56]. Claret and colleagues [57] have shown that the flagellar sigma factor FliA regulates the Intestine-407 adhesion and invasion by strain LF82 via a cyclic-di-GMP-dependent pathway that down-regulates expression of Type 1 fimbriae. In addition to the indirect effect, the flagellum can also directly be involved in AIEC adhesion and invasion. Two characterized AIEC strains, LF82 and O83:H1, isolated from Crohn's disease patients, weakly adhered to Caco-2BBe cells and invaded both Caco-2BBe and T-84 cells. On the other hand, two uncharacterized AIEC isolates, a laboratory E. coli HS strain used as a negative control, and an isogenic fliC mutant of AIEC O83:H1, were non-adhesive and non-invasive, whereas complementation of the fliC mutation fully restored the adhesiveness to intestinal epithelial cells and increased invasion of the complemented strain in comparison to the fliC mutant [25]. Conclusively, AIEC flagella contribute to bacterial adhesion and invasion indirectly by regulating expression of other adhesins but also directly by facilitating FliC-mediated adhesion in a strain-dependent manner.

Flagellum-mediated Adhesion of Porcine Pathogenic E. coli

Porcine pathogenic E. coli frequently express various types of fimbriae, such as F18 and F4 fimbriae [66], which mediate bacterial adhesion to host cells. Recently, the role of flagellum as a virulence factor has been studied in more detail in porcine isolates. Duan and colleagues [28,29,48] have reported that fliC deletion mutants of the swine edema-disease causing STEC (O139:H1:F18ab) and post-weaning diarrhea (PWD) causing E. coli (O157:H019:F18ac) show reduced adhesion and invasion into IPEC-J2 and IPEC-1 cells compared to the isogenic wild-type strains. The adhesive phenotype of the mutants was restored by complementation of the deletion with plasmid-cloned fliC, and specific inhibition of bacterial adhesion by purified flagella further demonstrated the adhesive role of flagella. As in AIEC, regulatory cross talk between adhesive organelles was also indicated in porcine E. coli isolates. Type 1 fimbria expression was upregulated in the F18 ΔfliC mutants and in ETEC O8:F4ac, another cause of PWD, the F4-fimbria appeared to be the main adhesin whereas flagella mainly affected expression of the fimbriae [30,48]. The results thus show that flagella of porcine-pathogenic E. coli possess similar adhesive, invasive and regulatory properties to flagella of IPEC and ExPEC.

Flagellum-mediated Adhesion of Probiotic E. coli

Bacterial adhesion to host surfaces is not always related to infection. In the human intestine, the microbiota and to some extent also probiotic bacterial strains protect the host surfaces against harmful intruders [74]. In order to successfully colonize the gut, probiotic bacteria should be able to adhere to intestinal surfaces. E. coli Nissle 1917 is a probiotic strain well-characterized regarding fimbrial genes and known to express H1 flagella, but other adhesive properties are not known [33]. A hyperflagellated variant of Nissle 1917 carrying up to 12 flagella per cell was assessed along with the wild-type strain, a ΔfliC mutant, and fimbriae-deficient mutants for adhesion to human intestinal cryosections. The hyperflagellated strain adhered efficiently, whereas the ΔfliC mutant showed no adhesion, and deletion of fimbriae genes had no impact on the adhesion capacity of Nissle 1917. More detailed analysis of the adhesive properties of Nissle 1917 indicated that the strain attached in a FliC-dependent manner to mucin-producing LS174T cells, but not to Caco-2 and T24 cells nor to immobilized murine mucus. Adhesion of wild-type Nissle 1917 and the hyperflagellated variant to intestinal cryosections was efficiently inhibited by mucin II, and flagella isolated from the wild-type strain bound to mucin II as well as human mucus. For identification of the receptor molecule of Nissle 1917 flagellum, isolated flagella were preincubated with carbohydrates known to be constituents of mucin II, and assessed for binding to immobilized mucin II and human mucus. In both cases, the binding was reduced only in the presence of gluconate. The adhesion of wild-type and hyperflagellated Nissle 1917 to human intestinal cryosections and immobilized mucin II was also inhibited by 100 mM gluconate, indicating that gluconate functions as a receptor for Nissle 1917 flagella in mucus. The results show that not only pathogenic but also probiotic E. coli possess adhesive flagella, and further studies are required to reveal the prevalence and biological significance of flagellum-mediated direct adhesion to host cell surfaces.

Flagellum-mediated Indirect E. coli Adhesion

Flagella can also mediate bacterial adhesion indirectly via other molecules. EtpA is a secreted ETEC adhesin shown to bind to Caco-2 cells and mucin-expressing regions of mouse small intestine [32]. Adherence of ETEC to Caco-2 cells requires EtpA as well as flagella, since deletion of etpA decreased and deletion of fliC abolished the adhesion. Complementation of the fliC deletion with plasmid-cloned fliCrestored the adhesive capacity to the level of the isogenic wild-type ETEC strain. Purification of EtpA revealed co-isolation of flagellin, regardless of flagellar serotype, indicating an interaction between the two types of molecules. Pull-down experiments showed that the toxin recognized the conserved N-terminus of monomeric FliC and bound to it and in situ immune electron microscopy revealed localization of EtpA to broken flagellar tips lacking FliD. Conclusively, EtpA and intact flagella together form a prerequisite for efficient adhesion of ETEC to intestinal cells and mouse small intestine.

As evident from the examples above, flagella of E. coli can be involved directly in bacterial adhesion and/or invasion or indirectly via another bacterial protein, such as EtpA in ETEC, or due to flagellar regulation of other bacterial adhesive/invasive factors. Frequently, the flagellum-associated adhesive/invasive property is closely related to specific E. coli pathovars.

Pseudomonas aeruginosa

P. aeruginosa is an opportunistic pathogen that can cause urinary tract infections and conditions such as pneumonia, particularly in cystic fibrosis patients. P. aeruginosa has a single polar flagellum that is crucial for motility and chemotaxis. The role of P. aeruginosa flagellum in adhesion has been investigated in several publications, and mucin is thought to be important target for P. aeruginosa adhesion as the lung epithelium contains a thick mucus layer.

P. aeruginosa ΔfliF, a mutant without flagellar membrane and supramembrane ring, has been reported to be defective in the adhesion to mucin [35]. However, as FliF is embedded in the outer membrane, it cannot act as an adhesin, and thus the adhesion-enabling function must be indirect rather than FliF-mediated. Later Arora et al. [36] reported that purified flagellar cap protein FliD inhibited the adhesion of non-piliated P. aeruginosa PAK to mucin from human sputum. They also showed that ΔfliD mutant did not bind mucin, and complementation with fliD restored the mucin binding ability, suggesting that FliD might mediate adhesion to mucin. Wild-type P. aeruginosa PAO1 has been shown to bind human airway mucins and Lewis x glycotypes on human respiratory mucins, while the ΔfliC and ΔfliD mutants do not [38]. Also in this study, complementation with fliD restored adherence while complementation with fliC was not performed. However, as flagellar biosynthesis requires FliD for the filament polymerization, it remains unclear whether FliD alone is responsible for adhesion or whether FliC or some other flagellar component has a role in adhesion as well. Direct interaction between FliD and mucins was not shown [36,38]. FliD of P. aeruginosa PAK mediated adhesion to mucin from human sputum [36], but seemed not to be involved in airway mucin binding [38], which may be due to the expression of distinct FliD types by the PAK and PAO1 strains [75].

Lillehoj et al. [37] showed that P. aeruginosa PAK adhered to MUC1 mucin expressed by CHO cells (CHO-Muc1), and that deletion of fliC abolished the adhesion. Purified flagellin and flagellin antiserum inhibited adhesion of P. aeruginosa to CHO-Muc1 cells. Surprisingly however, deletion of fliD did not affect binding, although flagellin filament cannot polymerize without FliD. It is therefore possible that P. aeruginosa also possess other mucin adhesins than flagellum. P.

aeruginosa PAK ΔfliC, a derivative of a clinical isolate, has also been shown to be impaired in adhesion to 1HAEo cells compared to wild-type strain [34] and the ΔfliC mutant showed reduced virulence in a mouse model of pneumonia. Purified flagellin of P. aeruginosa bound to monosialylated glycolipid GM1 that is common in cell membranes, disialylated ganglioside GD_{1a}—the major ganglioside in mammalian brain- and to asialo-GM1 present in several immune cells, for example [34].

In a recent study, P. aeruginosa PAO1 flagellum was shown to bind to heparan sulfate chains of heparan sulfate proteoglycans at the basolateral surface of polarized Calu-3 cells [39]. This was demonstrated with bacteria and flagella-coated fluorescent beads. The same study also showed evidence for the adherence of Type IV pili (Tfp) to N-glycans at the apical surface of the polarized lung epithelial cells. Subsequently, the adhesion of P. aeruginosa by flagellum or Tfp was shown to lead to changes in host signal transduction pathways by inducing phosphatidylinositol 3-kinase (PI3K)/Akt pathway. The activation of PI3K/Akt pathway led to bacterial entry into epithelial cells, and inhibition of PI3K decreased invasion but not adhesion at both apical and basolateral surfaces. Phosphorylation of Akt, a PI3K effector, was increased by flagellum- or Tfp-mediated binding. The flagellum-mediated adherence at the basolateral surface occurred via the epidermal growth factor receptor that is phosphorylated by flagellum, but Tfp-mediated adherence is independent of it. The ability to adhere and invade at both the basolateral and apical surfaces of epithelial cells may be crucial in the pathogenesis of P. aeruginosa. Flagellar component(s) responsible for the interaction with heparin sulfate remain(s) to be determined.

Clostridium difficile

The anaerobic spore-forming Gram-positive bacterium C. difficile is an opportunistic pathogen causing antibiotic-associated diarrhea especially in health-care environments. C. difficile carry peritrichous flagella, but the role of the flagellum in virulence has not been widely studied and reports on the adhesive properties of C. difficile flagella are partially contradictory. Thus, current results regarding the role of C. difficile flagella in virulence require further investigation. The early report by Tasteyre and colleagues [21] showed that C. difficile crude flagella extracts as well as purified recombinant FliD and FliC proteins bound in immunodot analyses to immobilized axenic mouse cecal mucus, but not to porcine stomach mucin. On the other hand, Dingle and colleagues [22] demonstrated, using wild-type C. difficile as well as isogenic fliC and fliD mutants, that the mutants adhered more efficiently to differentiated Caco-2 cells than the wild-type strain. In a hamster model FliC and FliD were nonessential for cecal colonization, but as bacterial growth curves indicated that the mutants grew more slowly and produced larger amounts of toxin than the wild-type strains, final conclusions regarding the role of C. difficile flagella in colonization cannot be drawn. Recent transcriptional analysis of the genome-sequenced C. difficile has demonstrated a differential expression of flagella genes during heat shock conditions; fliC was down-regulated, the level of fliD remained unchanged, and genes encoding the hook, rod, and basal body were up-regulated [76]. The authors speculated that motility may be down-regulated, but adhesion enhanced during infection. It is obvious that much of the data in the older papers is misleading.

Flagellar Adhesion of Additional Bacterial Species

In addition to the bacterial species already mentioned, the role of flagella in adhesion has been studied for example in Burkholderia pseudomallei that causes melioidosis, Burkholderia cepacia, an opportunistic pathogen, Bordetella bronchiseptica, which can colonize the respiratory tract and cause canine and porcine bronchitis, Bordetella pertussis, which causes whooping cough, the gastroenteritis-causing Vibrio vulnificusand Campylobacter jejuni, and the opportunistic pathogen Stenotrophomonas maltophilia.

B. pseudomallei is a facultative intracellular bacterium that is able to invade several cell types and survive inside macrophages and free-living amoebae [77,78]. The flagellum has been shown to mediate adhesion of B. pseudomallei to Acanthamoeba astronyxis, and the fliC-lacking mutant did not adhere or become internalized, even if the bacteria were centrifuged onto the amoebae [19]. Other studies have also been performed to investigate the adhesive properties of B. pseudomallei flagella, but in most cases, the flagellum does not seem to be the adhesin, but instead promotes motility toward targets. For example, flagella facilitated the invasion of B. pseudomallei into the mouse macrophage cell line RAW264.7 and human lung epithelial cell line A549, but flagella were not the only component responsible for bacterial invasion into epithelial cells and were not shown to contribute to adhesion [79]. The same study also compared the FliC of B. pseudomallei to FliC of a closely related but avirulent B. thailandensis, which has a 15 bp deletion in the variable region of fliC [80], but complementation of B. pseudomallei ΔfliC with B. thailandensis fliC improved invasion as much as complementation with fliCfrom B. pseudomallei [79]. No differences were observed between the ΔfliC mutant and the other strains, including B. thailandensis, when the bacteria were centrifuged to bring them in contact with the eukaryotic cells. Thus it must be noted that invasion ability seems to be correlated with motility, but other proteins than flagella probably mediate adhesion and invasion by B. pseudomallei, except in the case of amoeba adhesion. B. cepacia, which can cause disease in immunocompromised persons, particularly respiratory tract infections in cystic fibrosis patients, has polar flagella composed of flagellin that exhibits high sequence homology to B. pseudomallei fliC [81]. The invasiveness of B. cepacia into respiratory cells was dependent on functional flagella and motility, but flagella did not function as adhesins [82] thus resembling the pattern observed with B. pseudomallei.

Bordetella species are pathogens, which colonize the respiratory tract, and their virulence factors, including flagella, are regulated by the bvg locus. Flagella of B. bronchiseptica and B. pertussis may contribute to bacterial adhesion to HeLa cells, shown also with purified B. bronchiseptica flagella [18].

S. maltophilia is frequently associated with respiratory tract infections, and it has been found to adhere to mouse tracheal mucus via flagella, shown with flagella-expressing clinical isolates and flagellin isolated from S. maltophilia [41].

V. vulnificus has a polar flagellum, and a non-motile flagellum-deficient hook mutant ΔflgE was less adherent to Intestine-407 cells, formed less biofilm, and the lethal dose in mice was increased [83]. The results thus indicate that the flagellum plays a role in V. vulnificus virulence.

C. jejuni adhered to Intestine-407 cells with flagellum and LPS, and purified C. jejuni flagella bound to epithelial cells but not to intestinal mucus [20]. C. jejuni has two highly (95%) identical flagellin genes, flaA and flaB, but only flaA is required for flagella-mediated motility and adhesion [84]. Motility was suggested to be important for invasion into Intestine-407 cells, but direct adhesive function of flagellum was not observed [84]. According to Grant et al. [85], flagella of C. jejuni are not required for adhesion to epithelial cells but are involved in invasion. This might be explained by FlaC, a protein secreted via flagella and found to increase invasion of C. jejuni to Hep-2 cells [86], or by other secreted proteins. FlaC is homologous to flagellin proteins FlaA and FlaB at N- and C-terminal regions but lacks the central domain and is not required for the expression of the functional flagellum [86].

ADDITIONAL ADHESION-RELATED MECHANISMS OF FLAGELLA

The flagellum has an important role in biofilm formation, and biofilms on abiotic surfaces pose a remarkable health threat in cases such as clinical catheters and cooling systems. To explore in detail the adhesion and thereby formation of biofilms on catheters, the mechanism of flagellum binding was studied by Friedlander and colleagues [87]. UPEC were cultured on flat silicon surfaces as well as on patterned surfaces covered with bumps and submicrometer crevices too narrow to fit the bacterial cells, and the surfaces were analyzed for bacterial attachment and presence of flagella by scanning electron microscopy. During the first two hours after inoculation, adhesion to flat surfaces was more efficient, but after a longer incubation period, the colonization of the patterned surface was more efficient. Adherent bacterial cells were surrounded by flagella, as demonstrated by site-specific mutagenesis and scanning electron microscopy. The non-flagellated ΔflhDC and ΔfliC mutants showed reduced adhesion in comparison to the wild-type strain, whereas the flagellated paralyzed ΔmotB mutant colonized the patterned surface as efficiently as the wild-type strain. The results indicate that flagella reach crevices, grasp to improve bacterial adhesion, and are able to penetrate substructures not accessible to the bacterial cells, as was described also for EPEC adhesion to HeLa cells and EHEC interaction with BRECs [23,26].

As Salmonella enterica serovar Typhi is frequently associated with cholesterol-rich gallstones in chronic carriers, the adhesion of Salmonella and formation of biofilm on cholesterol have been studied in more detail [88]. When a library of random transposon mutants of S. Typhimurium was assessed for adherence to cholesterol, mutants carrying transposons in flagellum-related genes showed reduced adhesion to cholesterol in comparison to the wild-type strain. The importance of the flagella filament in cholesterol-binding and biofilm formation on cholesterol was demonstrated by the inability of the non-flagellated *flhC* mutant to form a biofilm, whereas the flagellated but non-motile motA deletion mutant formed biofilm on cholesterol as efficiently as the wild-type strain. Further, formalin-killed wild-type flagellated S. Typhimurium adhered to immobilized cholesterol as efficiently as live bacteria and functioned as a scaffold for biofilm formation of living flagellated as well as non-flagellated Salmonella. Adhesion to cholesterol by a ΔfliC mutant (expressing the FljB flagellin) was significantly reduced compared to the adhesion of ΔfljB mutant (expressing FliC flagellin) or the wild-type strain. Conclusively, flagella composed of FliC-type flagellin mediate adhesion to cholesterol in S. Typhimurium and promotes the early formation of biofilm.

Flagella composed of polymeric flagellin do not bind to TLR5, whereas monomeric flagellin induces a TLR5-mediated inflammatory response [12]. Interestingly, Subramanian and Quadri [89] reported that host cell-produced lysophospholipids induced the secretion of biologically active, monomeric flagellin in S. Typhi and S. Typhimurium. The results were further supported by the observations that externally added lysophospholipid triggered secretion of monomeric flagellin in salmonellae and flagellin secretion was reduced in Salmonella if host cells were pretreated with inhibitors of lysophospholipid synthesis. The secretion of monomeric flagellin was not due to depolymerization of flagella filaments but a result of eukaryotic cell-induced flagellin expression in S. Typhi, and the results demonstrated that secretion of monomeric flagellin was dependent on cAMP-dependent signaling. The authors speculate that during intestinal infection, Salmonella can sense lysophospholipids produced by host epithelial cells, activate the export of monomeric flagellin, and thereby modulate the TLR5-mediated innate immune response, which could promote bacterial dissemination.

Gangliosides have been reported to act as receptors for flagellin, but also appear to facilitate flagellin-mediated signaling in eukaryotic cells. Purified P. aeruginosa PAO flagellin has been shown to bind glycolipid receptors (including GM1 and asialo-GM1) on CHO Lec-2 cells with membrane-inserted GM1, trigger significant inflammatory response in murine lung, and via an ATP-dependent signaling pathway stimulate mucin production in MUC2-transfected human cells [34,90]. After addition of P. aeruginosa flagellin onto polarized airway epithelial cells (16HBE), FliC was observed to co-localize with asialo-GM1 on the apical surface. TLR5 is predominantly expressed basolaterally, but after a prolonged incubation time, TLR5 was found mainly at the apical surface of 16HBE cells co-localized with flagellin [91]. Gangliosides were also reported to act together with TLR5 as co-receptors for Salmonella flagellin and trigger expression of the human antimicrobial peptide β-defensin-2 in Caco-2 cells [92]. An analysis of the role of asialo-GM1 and TLR5 in binding of flagellin and in flagellin-mediated signaling suggests that signaling down-stream of gangliosides is TLR-dependent and that the two flagellin receptors co-operate in activation of signaling pathways in epithelial cells [93]. Thus, the adhesive capacity of flagellin in combination with host ganglioside and TLR5 appears as an efficient mechanism in host defense in the intestine as well as lungs but may also enable bacterial migration.

Flagella have also been reported to function in bacterial symbiosis. Bacteria persist in multifaceted environments containing variable and dense bacterial populations, and they frequently cooperate in different aspects, like in the degradation of organic matter. Earlier reports have indicated that unidentified extracellular filaments are involved in the initiation of symbiosis based on nutritional cooperation, also called syntrophy, between the fermentative bacterium Pelotomaculum thermopropionicumand the methanogenic archaea Methanothermobacter thermautotrophicus [94]. Shimoyama and colleagues [95] observed putative flagella and fimbriae gene clusters in the genomes of M. thermautotrophicus and P. thermopropionicum. Extracellular filaments found in a monoculture of P. thermopropionicum were partially purified, and the 55 kDa major protein of the filaments was shown to be flagellin. Flagellum was shown to connect M. thermautotrophicus and P. thermopropionicum in cocultures. Purified recombinant FliC and FliD of P. thermopropionicum bound to cells of M. thermautotrophicus and to Methanosaeta thermophile, a syntrophic-association forming methanogen. In M. thermautotrophicus, methanogenesis was accelerated by FliD but not by FliC. Transcriptome analysis verified the results and revealed that FliD up-regulated more than 50 genes encoding e.g., methanogenesis-related enzymes, ATP synthase, and hydrogenases. The authors speculate that flagellum of P. thermopropionicum has a dual role in maintaining proximity between the two species and in synchronizing their metabolism.

CONCLUSIONS

Flagella are involved in bacterial adhesion and invasion both indirectly, i.e., by providing motility towards target cells and receptors, and directly by adhering to these targets. However, only a few receptors for flagellar adhesion have so far been convincingly revealed: gangliosides GM1 and GD_{1a}, asialo-GM1, blood-group-antigen-related Lewis x glycotype, and heparan sulfate in the case of P. aeruginosa, and asialo-GM1, lipid rafts, gluconate, and EtpA in the case of E. coli. Flagellum-mediated adhesion and invasion have been mostly studied with whole flagella, or alternatively with purified flagellin (FliC) or flagellar cap (FliD) proteins. With the exception of the N- and C-terminal regions of FliC involved in innate immunity [12,13,14,15,16] and binding to the EtpA adhesin of ETEC [32], information on receptor-binding sites in flagella proteins is completely missing. Reports on other flagellar subunits, e.g., the hook or junction proteins, in adhesion, invasion, or innate immunity are totally lacking and would be important topics of further research. The existing knowledge on flagella adhesion is summarized in Figure 2.

Functions of flagella have mainly been studied in bacterial pathogens and from the viewpoint of bacterial virulence due to the various potential applications that exist for flagella and flagellin such as in vaccine development and diagnostics. For example, flagellin has been used for a generation of vaccines against Salmonella enterica serovar Paratyphi A, P. aeruginosa, and E. coli O157:H7 [96,97,98], and the colonization of EHEC in cattle was significantly decreased after vaccination with FliC [97]. Due to its high immunogenicity, flagellin has also been used as a vaccine adjuvant together with poorly immunogenic antigens, and FliC or anti-FliC antibodies have been used in diagnostics of e.g., melioidosis and inflammatory bowel disease [99]. Interestingly, flagellin has also been shown to suppress apoptosis and protect against radiation-related damage [100].

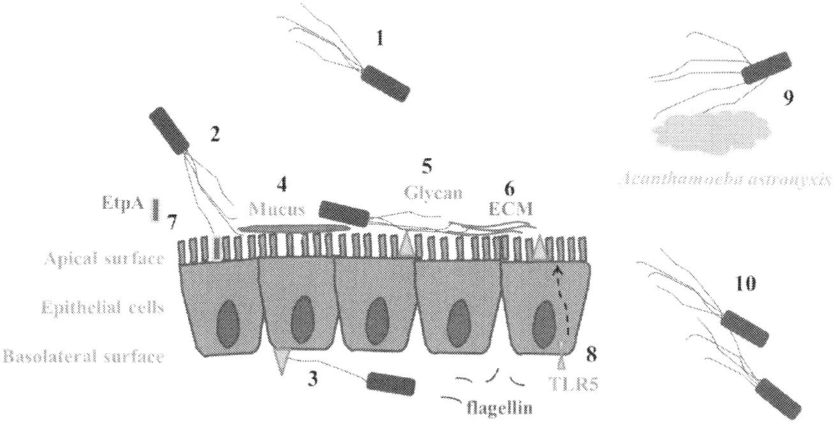

Figure 2. Summary of the bacterial flagellum as an adhesin. Flagellum can mediate bacterial adhesion to eukaryotic cells indirectly via motility (1), or by binding directly to epithelial cells either on apical (2) or basolateral surface (3). Flagellar target receptors include mucus and mucins (4), different glycans on cells or in mucus (gluconate, heparan sulfate proteoglycans, Lewis x glycotype, GM1, asialo-GM1, GD_{1a}) (5), extracellular matrix (ECM) proteins (6), or bacterial-secreted EtpA, which in turn adheres to host cells (7). Toll-like receptor 5 (TLR5) is expressed mostly at the basolateral cell surface and binds FliC, inducing the host's immune defense (8). FliC binding also induces TLR5 expression at the apical surface. In addition to various epithelia, flagella may also adhere to amoebae (9) or connect two bacterial species (10).

More recently, it has been noted that probiotic bacteria may also benefit from the flagellum, and that flagella can also be beneficial in symbiotic relationships between bacterial species. Thus many interesting aspects regarding the roles of flagella remain to be investigated in more detail.

ACKNOWLEDGMENTS

We thank Ladan Cockshut for help with language editing. The study was supported by University of Helsinki and the SalWe Research Program for IMO (Tekes—the Finnish Funding Agency for Technology and Innovation grant 648/10).

Conflicts of Interest

The authors declare no conflict of interest.

REFERENCES

1. Duan, Q.; Zhou, M.; Zhu, L.; Zhu, G. Flagella and bacterial pathogenicity. J. Basic Microb. 2013, 53, 1–8.
2. Prüss, B.M.; Besemann, C.; Denton, A.; Wolfe, A.J. A complex transcription network controls the early stages of biofilm development by Escherichia coli. J. Bacteriol. 2006, 188, 3731–3739.
3. Conrad, J.C. Physics of bacterial near-surface motility using flagella and type IV pili: Implications for biofilm formation. Res. Microbiol. 2012, 163, 619–629.
4. Chevance, F.F.; Hughes, K.T. Coordinating assembly of a bacterial macromolecular machine. Nat. Rev. Microbiol. 2008, 6, 455–465.
5. Sourjik, V.; Wingreen, N.S. Responding to chemical gradients: Bacterial chemotaxis. Curr. Opin. Cell Biol. 2012, 24, 262–268.
6. Erhardt, M.; Namba, K.; Hughes, K.T. Bacterial nanomachines: The flagellum and type III injectisome. Cold Spring Harbor Perspect. Biol. 2010, 2, a000299.
7. Büttner, D. Protein export according to schedule: Architecture, assembly, and regulation of type III secretion systems from plant- and animal-pathogenic bacteria. Microbiol. Mol. Biol. Rev. 2012, 76, 262–310.
8. Smith, T.G.; Hoover, T.R. Deciphering bacterial flagellar gene regulatory networks in the genomic era. Adv. Appl. Microbiol. 2009, 67, 257–295.
9. Klemm, P.; Vejborg, R.M.; Hancock, V. Prevention of bacterial adhesion. Appl. Microbiol. Biotechnol.2010, 88, 451–459.
10. Young, G.M.; Schmiel, D.H.; Miller, V.L. A new pathway for the secretion of virulence factors by bacteria: The flagellar export apparatus functions as a protein-secretion system. Proc. Natl. Acad. Sci. USA 1999, 96, 6456–6461.

11. Majander, K.; Anton, L.; Antikainen, J.; Lång, H.; Brummer, M.; Korhonen, T.K.; Westerlund-Wikström, B. Extracellular secretion of polypeptides using a modified Escherichia coli flagellar secretion apparatus. Nat. Biotechnol. 2005, 23, 475–481.
12. Hayashi, F.; Smith, K.D.; Ozinsky, A.; Hawn, T.R.; Yi, E.C.; Goodlett, D.R.; Eng, J.K.; Akira, S.; Underhill, D.M.; Aderem, A. The innate immune response to bacterial flagellin is mediated by Toll-like receptor 5. Nature 2001, 410, 1099–1103.
13. Gong, Y.N.; Shao, F. Sensing bacterial infections by NAIP receptors in NLRC4 inflammasome activation. Protein Cell 2012, 3, 98–105.
14. Halff, E.F.; Diebolder, C.A.; Versteeg, M.; Schouten, A.; Brondijk, T.H.; Huizinga, E.G. Formation and structure of a NAIP5-NLRC4 inflammasome induced by direct interactions with conserved N- and C-terminal regions of flagellin. J. Biol. Chem. 2012, 287, 38460–38472.
15. Yoon, S.I.; Kurnasov, O.; Natarajan, V.; Hong, M.; Gudkov, A.V.; Osterman, A.L.; Wilson, I.A. Structural basis of TLR5-flagellin recognition and signaling. Science 2012, 335, 859–864.
16. Yonekura, K.; Maki-Yonekura, S.; Namba, K. Complete atomic model of the bacterial flagellar filament by electron cryomicroscopy. Nature 2003, 424, 643–650.
17. Beatson, S.A.; Minamino, T.; Pallen, M.J. Variation in bacterial flagellins: From sequence to structure. Trends Microbiol. 2006, 14, 151–155.
18. Savelkoul, P.H.; de Kerf, D.P.; Willems, R.J.; Mooi, F.R.; van der Zeijst, B.A.; Gaastra, W. Characterization of the fim2 and fim3 fimbrial subunit genes of Bordetella bronchiseptica: Roles of Fim2 and Fim3 fimbriae and flagella in adhesion. Infect. Immun. 1996, 64, 5098–5105.
19. Inglis, T.J.; Robertson, T.; Woods, D.E.; Dutton, N.; Chang, B.J. Flagellum-mediated adhesion by Burkholderia pseudomallei precedes invasion of Acanthamoeba astronyxis. Infect. Immun. 2003, 71, 2280–2282.
20. McSweegan, E.; Walker, R.I. Identification and characterization of two Campylobacter jejuniadhesins for cellular and mucous substrates. Infect. Immun. 1986, 53, 141–148.

21. Tasteyre, A.; Barc, M.C.; Collignon, A.; Boureau, H.; Karjalainen, T. Role of FliC and FliD flagellar proteins of Clostridium difficile in adherence and gut colonization. Infect. Immun. 2001, 69, 7937–7940.
22. Dingle, T.C.; Mulvey, G.L.; Armstrong, G.D. Mutagenic analysis of the Clostridium difficile flagellar proteins, FliC and FliD, and their contribution to virulence in hamsters. Infect. Immun. 2011, 79, 4061–4067.
23. Girón, J.A.; Torres, A.G.; Freer, E.; Kaper, J.B. The flagella of enteropathogenic Escherichia colimediate adherence to epithelial cells. Mol. Microbiol. 2002, 44, 361–379.
24. Erdem, A.L.; Avelino, F.; Xicohtencatl-Cortes, J.; Girón, J.A. Host protein binding and adhesive properties of H6 and H7 flagella of attaching and effacing Escherichia coli. J. Bacteriol. 2007, 189, 7426–7435.
25. Eaves-Pyles, T.; Allen, C.A.; Taormina, J.; Swidsinski, A.; Tutt, C.B.; Jezek, G.E.; Islas-Islas, M.; Torres, A.G. Escherichia coli isolated from a Crohn's disease patient adheres, invades, and induces inflammatory responses in polarized intestinal epithelial cells. Int. J. Med. Microbiol. 2008, 298, 397–409.
26. Mahajan, A.; Currie, C.G.; Mackie, S.; Tree, J.; McAteer, S.; McKendrick, I.; McNeilly, T.N.; Roe, A.; La Ragione, R.M.; Woodward, M.J.; et al. An investigation of the expression and adhesin function of H7 flagella in the interaction of Escherichia coli O157:H7 with bovine intestinal epithelium. Cell. Microbiol. 2009, 11, 121–137.
27. Sampaio, S.C.; Gomes, T.A.; Pichon, C.; du Merle, L.; Guadagnini, S.; Abe, C.M.; Sampaio, J.L.; Le Bouguénec, C. The flagella of an atypical enteropathogenic Escherichia coli strain are required for efficient interaction with and stimulation of interleukin-8 production by enterocytes in vitro. Infect. Immun. 2009, 77, 4406–4413.
28. Duan, Q.; Zhou, M.; Zhu, X.; Bao, W.; Wu, S.; Ruan, X.; Zhang, W.; Yang, Y.; Zhu, J.; Zhu, G. The flagella of F18ab Escherichia coli is a virulence factor that contributes to infection in a IPEC-J2 cell model in vitro. Vet. Microbiol. 2012, 160, 132–140.
29. Duan, Q.; Zhou, M.; Zhu, X.; Yang, Y.; Zhu, J.; Bao, W.; Wu,

S.; Ruan, X.; Zhang, W.; Zhu, G. Flagella from F18+ Escherichia coli play a role in adhesion to pig epithelial cell lines. Microb. Pathog.2013, 55, 32–38.

30. Zhou, M.; Duan, Q.; Zhu, X.; Guo, Z.; Li, Y.; Hardwidge, P.R.; Zhu, G. Both flagella and F4 fimbriae from F4ac+ enterotoxigenic Escherichia coli contribute to attachment to IPEC-J2 cells in vitro. Vet. Res. 2013, 44, 30.
31. Parthasarathy, G.; Yao, Y.; Kim, K.S. Flagella promote Escherichia coli K1 association with and invasion of human brain microvascular endothelial cells. Infect. Immun. 2007, 75, 2937–2945.
32. Roy, K.; Hilliard, G.M.; Hamilton, D.J.; Luo, J.; Ostmann, M.M.; Fleckenstein, J.M. Enterotoxigenic Escherichia coli EtpA mediates adhesion between flagella and host cells. Nature 2009, 457, 594–598.
33. Troge, A.; Scheppach, W.; Schroeder, B.O.; Rund, S.A.; Heuner, K.; Wehkamp, J.; Stange, E.F.; Oelschlaeger, T.A. More than a marine propeller—The flagellum of the probiotic Escherichia colistrain Nissle 1917 is the major adhesin mediating binding to human mucus. Int. J. Med. Microbiol.2012, 302, 304–314.
34. Feldman, M.; Bryan, R.; Rajan, S.; Scheffler, L.; Brunnert, S.; Tang, H.; Prince, A. Role of flagella in pathogenesis of Pseudomonas aeruginosa pulmonary infection. Infect. Immun. 1998, 66, 43–51.
35. Arora, S.K.; Ritchings, B.W.; Almira, E.C.; Lory, S.; Ramphal, R. Cloning and characterization of Pseudomonas aeruginosa fliF, necessary for flagellar assembly and bacterial adherence to mucin. Infect. Immun. 1996, 64, 2130–2136.
36. Arora, S.K.; Ritchings, B.W.; Almira, E.C.; Lory, S.; Ramphal, R. The Pseudomonas aeruginosaflagellar cap protein, FliD, is responsible for mucin adhesion. Infect. Immun. 1998, 66, 1000–1007.
37. Lillehoj, E.P.; Kim, B.T.; Kim, K.C. Identification of Pseudomonas aeruginosa flagellin as an adhesin for Muc1 mucin. Am. J. Physiol. Lung C. 2002, 282, L751–L756.
38. Scharfman, A.; Arora, S.K.; Delmotte, P.; van Brussel, E.; Mazurier, J.; Ramphal, R.; Roussel, P. Recognition of Lewis x derivatives present on mucins by flagellar components of Pseudomonas

aeruginosa. Infect. Immun. 2001, 69, 5243–5248.

39. Bucior, I.; Pielage, J.F.; Engel, J.N. Pseudomonas aeruginosa pili and flagella mediate distinct binding and signaling events at the apical and basolateral surface of airway epithelium. PLoS Pathog. 2012, 8, e1002616.
40. Olsen, J.E.; Hoegh-Andersen, K.H.; Casadesus, J.; Rosenkranzt, J.; Chadfield, M.S.; Thomsen, L.E. The role of flagella and chemotaxis genes in host pathogen interaction of the host adapted Salmonella enterica serovar Dublin compared to the broad host range serovar S. Typhimurium. BMC Microbiol. 2013, 13, 67.
41. Zgair, A.K.; Chhibber, S. Adhesion of Stenotrophomonas maltophilia to mouse tracheal mucus is mediated through flagella. J. Med. Microbiol. 2011, 60, 1032–1037.
42. Hartmann, I.; Carranza, P.; Lehner, A.; Stephan, R.; Eberl, L.; Riedel, K. Genes involved in Cronobacter sakazakii biofilm formation. Appl. Environ. Microbiol. 2010, 76, 2251–2261.
43. La Ragione, R.M.; Cooley, W.A.; Woodward, M.J. The role of fimbriae and flagella in the adherence of avian strains of Escherichia coli O78:K80 to tissue culture cells and tracheal and gut explants. J. Med. Microbiol. 2000, 49, 327–338.
44. La Ragione, R.M.; Sayers, A.R.; Woodward, M.J. The role of fimbriae and flagella in the colonization, invasion and persistence of Escherichia coli O78:K80 in the day-old-chick model. Epidemiol. Infect. 2000, 124, 351–363.
45. Best, A.; La Ragione, R.M.; Sayers, A.R.; Woodward, M.J. Role for flagella but not intimin in the persistent infection of the gastrointestinal tissues of specific-pathogen-free chicks by shiga toxin-negative Escherichia coli O157:H7. Infect. Immun. 2005, 73, 1836–1846.
46. Luck, S.N.; Badea, L.; Bennett-Wood, V.; Robins-Browne, R.; Hartland, E.L. Contribution of FliC to epithelial cell invasion by enterohemorrhagic Escherichia coli O113:H21. Infect. Immun. 2006, 74, 6999–7004.
47. Rogers, T.J.; Thorpe, C.M.; Paton, A.W.; Paton, J.C. Role of lipid rafts and flagellin in invasion of colonic epithelial cells by Shiga-

toxigenic Escherichia coli O113:H21. Infect. Immun. 2012, 80, 2858–2867.

48. Duan, Q.; Zhou, M.; Liang, H.; Zhu, X.; Guo, Z.; Li, Y.; Hardwidge, P.R.; Zhu, G. Contribution of flagellin subunit FliC to piglet epithelial cells invasion by F18ab E. coli. Vet. Microbiol. 2013.
49. Rogers, T.J.; Paton, J.C.; Wang, H.; Talbot, U.M.; Paton, A.W. Reduced virulence of an fliC mutant of Shiga-toxigenic Escherichia coli O113:H21. Infect. Immun. 2006, 74, 1962–1966.
50. Allison, C.; Coleman, N.; Jones, P.L.; Hughes, C. Ability of Proteus mirabilis to invade human urothelial cells is coupled to motility and swarming differentiation. Infect. Immun. 1992, 60, 4740–4746.
51. Mobley, H.L.; Belas, R.; Lockatell, V.; Chippendale, G.; Trifillis, A.L.; Johnson, D.E.; Warren, J.W. Construction of a flagellum-negative mutant of Proteus mirabilis: Effect on internalization by human renal epithelial cells and virulence in a mouse model of ascending urinary tract infection. Infect. Immun. 1996, 64, 5332–5340.
52. Dibb-Fuller, M.P.; Allen-Vercoe, E.; Thorns, C.J.; Woodward, M.J. Fimbriae- and flagella-mediated association with and invasion of cultured epithelial cells by Salmonella enteritidis. Microbiology1999, 145, 1023–1031.
53. Van Asten, F.J.; Hendriks, H.G.; Koninkx, J.F.; Van Dijk, J.E. Flagella-mediated bacterial motility accelerates but is not required for Salmonella serotype Enteritidis invasion of differentiated Caco-2 cells. Int. J. Med. Microbiol. 2004, 294, 395–399.
54. La Ragione, R.M.; Cooley, W.A.; Velge, P.; Jepson, M.A.; Woodward, M.J. Membrane ruffling and invasion of human and avian cell lines is reduced for aflagellate mutants of Salmonella entericaserotype Enteritidis. Int. J. Med. Microbiol. 2003, 293, 261–272.
55. Barnich, N.; Boudeau, J.; Claret, L.; Darfeuille-Michaud, A. Regulatory and functional co-operation of flagella and type 1 pili in adhesive and invasive abilities of AIEC strain LF82 isolated from a patient with Crohn's disease. Mol. Microbiol. 2003, 48, 781–794.
56. Dobbin, H.S.; Hovde, C.J.; Williams, C.J.; Minnich, S.A. The Escherichia coli O157 flagellar regulatory gene flhC and

not the flagellin gene fliC impacts colonization of cattle. Infect. Immun.2006, 74, 2894–2905.

57. Claret, L.; Miquel, S.; Vieille, N.; Ryjenkov, D.A.; Gomelsky, M.; Darfeuille-Michaud, A. The flagellar sigma factor FliA regulates adhesion and invasion of Crohn disease-associated Escherichia coli via a cyclic dimeric GMP-dependent pathway. J. Biol. Chem. 2007, 282, 33275–33283.

58. Clyne, M.; Ocroinin, T.; Suerbaum, S.; Josenhans, C.; Drumm, B. Adherence of isogenic flagellum-negative mutants of Helicobacter pylori and Helicobacter mustelae to human and ferret gastric epithelial cells. Infect. Immun. 2000, 68, 4335–4339.

59. Guentzel, M.N.; Berry, L.J. Motility as a virulence factor for Vibrio cholerae. Infect. Immun. 1975, 11, 890–897.

60. Zunino, P.; Piccini, C.; Legnani-Fajardo, C. Flagellate and non-flagellate Proteus mirabilis in the development of experimental urinary tract infection. Microb. Pathog. 1994, 16, 379–385.

61. Legnani-Fajardo, C.; Zunino, P.; Piccini, C.; Allen, A.; Maskell, D. Defined mutants of Proteus mirabilis lacking flagella cause ascending urinary tract infection in mice. Microb. Pathogenesis 1996, 21, 395–405.

62. Armbruster, C.E.; Mobley, H.L. Merging mythology and morphology: The multifaceted lifestyle of Proteus mirabilis. Nat. Rev. Microbiol. 2012, 10, 743–754.

63. Sheu, B.S.; Yang, H.B.; Yeh, Y.C.; Wu, J.J. Helicobacter pylori colonization of the human gastric epithelium: A bug's first step is a novel target for us. J. Gastroen. Hepatol. 2010, 25, 26–32.

64. Kim, J.S.; Chang, J.H.; Chung, S.I.; Yum, J.S. Molecular cloning and characterization of the Helicobacter pylori fliD gene, an essential factor in flagellar structure and motility. J. Bacteriol. 1999, 181, 6969–6976.

65. Croxen, M.A.; Finlay, B.B. Molecular mechanisms of Escherichia coli pathogenicity. Nat. Rev. Microbiol. 2010, 8, 26–38.

66. Fairbrother, J.M.; Nadeau, E.; Gyles, C.L. Escherichia coli in postweaning diarrhea in pigs: An update on bacterial types, pathogenesis, and prevention strategies. Anim. Health Res. Rev. 2005, 6, 17–39.

67. Tivendale, K.A.; Logue, C.M.; Kariyawasam, S.; Jordan, D.; Hussein, A.; Li, G.; Wannemuehler, Y.; Nolan, L.K. Avian-pathogenic Escherichia coli strains are similar to neonatal meningitis E. colistrains and are able to cause meningitis in the rat model of human disease. Infect. Immun. 2010, 78, 3412–3419.
68. Köhler, C.D.; Dobrindt, U. What defines extraintestinal pathogenic Escherichia coli? Int. J. Med. Microbiol. 2011, 301, 642–647.
69. Clements, A.; Young, J.C.; Constantinou, N.; Frankel, G. Infection strategies of enteric pathogenic Escherichia coli. Gut Microbes 2012, 3, 71–87.
70. Wang, L.; Rothemund, D.; Curd, H.; Reeves, P.R. Species-wide variation in the Escherichia coliflagellin (H-antigen) gene. J. Bacteriol. 2003, 185, 2936–2943.
71. Bonacorsi, S.; Bingen, E. Molecular epidemiology of Escherichia coli causing neonatal meningitis. Int. J. Med. Microbiol. 2005, 295, 373–381.
72. Xie, Y.; Parthasarathy, G.; Di Cello, F.; Teng, C.H.; Paul-Satyaseela, M.; Kim, K.S. Transcriptome of Escherichia coli K1 bound to human brain microvascular endothelial cells. Biochem. Biophys. Res. Commun. 2008, 365, 201–206.
73. Teng, C.H.; Xie, Y.; Shin, S.; di Cello, F.; Paul-Satyaseela, M.; Cai, M.; Kim, K.S. Effects of ompAdeletion on expression of type 1 fimbriae in Escherichia coli K1 strain RS218 and on the association of E. coli with human brain microvascular endothelial cells. Infect. Immun. 2006, 74, 5609–5616.
74. Human Microbiome Project Consortium. Structure, function and diversity of the healthy human microbiome. Nature 2012, 486, 207–214.
75. Arora, S.K.; Dasgupta, N.; Lory, S.; Ramphal, R. Identification of two distinct types of flagellar cap proteins, FliD, in Pseudomonas aeruginosa. Infect. Immun. 2000, 68, 1474–1479.
76. Ternan, N.G.; Jain, S.; Srivastava, M.; McMullan, G. Comparative transcriptional analysis of clinically relevant heat stress response in Clostridium difficile strain 630. PLoS One 2012, 7, e42410.

77. Jones, A.L.; Beveridge, T.J.; Woods, D.E. Intracellular survival of Burkholderia pseudomallei. Infect. Immun. 1996, 64, 782–790.
78. Inglis, T.J.; Rigby, P.; Robertson, T.A.; Dutton, N.S.; Henderson, M.; Chang, B.J. Interaction between Burkholderia pseudomallei and Acanthamoeba species results in coiling phagocytosis, endamebic bacterial survival, and escape. Infect. Immun. 2000, 68, 1681–1686.
79. Chuaygud, T.; Tungpradabkul, S.; Sirisinha, S.; Chua, K.L.; Utaisincharoen, P. A role of Burkholderia pseudomallei flagella as a virulent factor. T. Roy. Soc. Trop. Med. H. 2008, 102, S140–S144.
80. Wajanarogana, S.; Sonthayanon, P.; Wuthiekanun, V.; Panyim, S.; Simpson, A.J.; Tungpradabkul, S. Stable marker on flagellin gene sequences related to arabinose non-assimilating pathogenic Burkholderia pseudomallei. Microbiol. Immunol. 1999, 43, 995–1001.
81. DeShazer, D.; Brett, P.J.; Carlyon, R.; Woods, D.E. Mutagenesis of Burkholderia pseudomallei with Tn5-OT182: Isolation of motility mutants and molecular characterization of the flagellin structural gene. J. Bacteriol. 1997, 179, 2116–2125.
82. Tomich, M.; Herfst, C.A.; Golden, J.W.; Mohr, C.D. Role of flagella in host cell invasion by Burkholderia cepacia. Infect. Immun. 2002, 70, 1799–1806.
83. Lee, J.H.; Rho, J.B.; Park, K.J.; Kim, C.B.; Han, Y.S.; Choi, S.H.; Lee, K.H.; Park, S.J. Role of flagellum and motility in pathogenesis of Vibrio vulnificus. Infect. Immun. 2004, 72, 4905–4910.
84. Wassenaar, T.M.; Bleumink-Pluym, N.M.; van der Zeijst, B.A. Inactivation of Campylobacter jejuniflagellin genes by homologous recombination demonstrates that flaA but not flaB is required for invasion. EMBO J. 1991, 10, 2055–2061.
85. Grant, C.C.; Konkel, M.E.; Cieplak, W., Jr.; Tompkins, L.S. Role of flagella in adherence, internalization, and translocation of Campylobacter jejuni in nonpolarized and polarized epithelial cell cultures. Infect. Immun. 1993, 61, 1764–1771.
86. Song, Y.C.; Jin, S.; Louie, H.; Ng, D.; Lau, R.; Zhang, Y.; Weerasekera, R.; Al Rashid, S.; Ward, L.A.; Der, S.D.; Chan, V.L. FlaC, a protein of Campylobacter jejuni TGH9011 (ATCC43431)

secreted through the flagellar apparatus, binds epithelial cells and influences cell invasion. Mol. Microbiol.2004, 53, 541–553.

87. Friedlander, R.S.; Vlamakis, H.; Kim, P.; Khan, M.; Kolter, R.; Aizenberg, J. Bacterial flagella explore microscale hummocks and hollows to increase adhesion. Proc. Natl. Acad. Sci. USA 2013, 110, 5624–5629.
88. Crawford, R.W.; Reeve, K.E.; Gunn, J.S. Flagellated but not hyperfimbriated Salmonella entericaserovar Typhimurium attaches to and forms biofilms on cholesterol-coated surfaces. J. Bacteriol.2010, 192, 2981–2990.
89. Subramanian, N.; Qadri, A. Lysophospholipid sensing triggers secretion of flagellin from pathogenic salmonella. Nat. Immunol. 2006, 7, 583–589.
90. McNamara, N.; Khong, A.; McKemy, D.; Caterina, M.; Boyer, J.; Julius, D.; Basbaum, C. ATP transduces signals from ASGM1, a glycolipid that functions as a bacterial receptor. Proc. Natl. Acad. Sci. USA 2001, 98, 9086–9091.
91. Adamo, R.; Sokol, S.; Soong, G.; Gomez, M.I.; Prince, A. Pseudomonas aeruginosa flagella activate airway epithelial cells through asialoGM1 and toll-like receptor 2 as well as toll-like receptor 5. Am. J. Respir. Cell Mol. 2004, 30, 627–634.
92. Ogushi, K.; Wada, A.; Niidome, T.; Okuda, T.; Llanes, R.; Nakayama, M.; Nishi, Y.; Kurazono, H.; Smith, K.D.; Aderem, A.; et al. Gangliosides act as co-receptors for Salmonella enteritidis FliC and promote FliC induction of human beta-defensin-2 expression in Caco-2 cells. J. Biol. Chem. 2004, 279, 12213–12219.
93. McNamara, N.; Gallup, M.; Sucher, A.; Maltseva, I.; McKemy, D.; Basbaum, C. AsialoGM1 and TLR5 cooperate in flagellin-induced nucleotide signaling to activate Erk1/2. Am. J. Respir. Cell Mol.2006, 34, 653–660.
94. Ishii, S.; Kosaka, T.; Hori, K.; Hotta, Y.; Watanabe, K. Coaggregation facilitates interspecies hydrogen transfer between Pelotomaculum thermopropionicum and Methanothermobacter thermautotrophicus. Appl. Environ. Microbiol. 2005, 71, 7838–7845.

95. Shimoyama, T.; Kato, S.; Ishii, S.; Watanabe, K. Flagellum mediates symbiosis. Science 2009, 323, 1574.
96. Döring, G.; Meisner, C.; Stern, M.; Flagella Vaccine Trial Study Group. A double-blind randomized placebo-controlled phase III study of a Pseudomonas aeruginosa flagella vaccine in cystic fibrosis patients. Proc. Natl. Acad. Sci. USA 2007, 104, 11020–11025.
97. McNeilly, T.N.; Naylor, S.W.; Mahajan, A.; Mitchell, M.C.; McAteer, S.; Deane, D.; Smith, D.G.E.; Low, J.C.; Gally, D.L.; Huntley, J.F. Escherichia coli O157:H7 colonization in cattle following systemic and mucosal immunization with purified H7 flagellin. Infect. Immun. 2008, 76, 2594–2602.
98. Gat, O.; Galen, J.E.; Tennant, S.; Simon, R.; Blackewelder, W.C.; Silverman, D.J.; Pasetti, M.F.; Levine, M.M. Cell-associated flagella enhance the protection conferred by mucosally-administered attenuated Salmonella Paratyphi A vaccines. PLoS Neglect. Trop. D. 2011, 5, e1373.
99. Wajanarogana, S.; Nimnuch, P.; Thongmee, A.; Kritsiriwuthinan, K. Potential of recombinant flagellin fragment from Burkholderia thailandensis as an antigen for melioidosis antibody detection by indirect ELISA. Mol. Cell. Probes 2013, 27, 98–102.
100. Burdelya, L.G.; Krivokrysenko, V.I.; Tallant, T.C.; Strom, E.; Gleiberman, A.S.; Gupta, D.; Kurnasov, O.V.; Fort, F.L.; Osterman, A.L.; Didonato, J.A.; et al. An agonist of Toll Like Receptor 5 has radioprotective activity in mouse and primate models. Science 2008, 320, 226–230.

PART III: MOLECULAR MACHINES

CHAPTER

9

MAN-MADE MOLECULAR MACHINES: MEMBRANE BOUND

Matthew A. Watson and Scott L. Cockroft

EaStCHEM School of Chemistry, University of Edinburgh, Joseph Black Building, David Brewster Road, Edinburgh, EH9 3FJ, UK.

ABSTRACT

Nature's molecular machines are a constant source of inspiration to the chemist. Many of these molecular machines function within lipid membranes, allowing them to exploit potential gradients between spatially close, but chemically distinct environments to fuel their work cycle. Indeed, the realisation of such principles in synthetic transmembrane systems remains a tantalising goal. This tutorial review opens by highlighting seminal examples of synthetic molecular machines. We illustrate the importance of surfaces for facilitating the extraction of work from molecular switches and motors. We chart the development of man-made transmembrane systems; from passive to machine-like stimuli-responsive channels, to fully autonomous transmembrane molecular machines. Finally, we highlight higher-order compartmentalised systems that exhibit emergent properties. We suggest that such higher-order architectures could serve as platforms for sophisticated devices that co-ordinate the activity of numerous transmembrane molecular machines.

INTRODUCTION

Life depends upon the interplay of countless molecular machines.[1] These natural molecular machines operate at length-scales where the dominant energetic factors are random thermal fluctuations. Indeed, proteins have been described as 'kicking, screaming, stochastic molecules›.[2] Membrane-spanning biological molecular machines such as adenosine triphosphate (ATP) synthase[3] and the bacterial flagellar motor[4] offer proven strategies for overcoming the constant buffeting of Brownian motion to perform work that facilitates the existence of life.[5,6] Many biologically inspired mechanically interlocked and light-responsive molecules have been synthesised,[7–9] some of which are capable of performing macroscopic work when arrayed on surfaces.[10–14] While the importance of synthetic nanotechnology continues to be highlighted,[5,15] significant advances have also been made using molecules of biological origin such as DNA.[16] The characteristics of many man-made transmembrane systems[17] are somewhat limited compared to the exquisite functionalities displayed by biological transmembrane machines.[18] Nonetheless, the bleeding-edge is starting to deliver man-made nanomechanical transmembrane assemblies that integrate lessons from both evolved and designed systems to operate as true molecular machines under non-equilibrium conditions. Such advances offer great promise for the realisation of man-made transmembrane molecular machines that perform useful work.[19]

Anchoring motor proteins within two-dimensional lipid membranes provides a means of countering the thermal forces that dominate behaviour at the molecular scale. One of the most studied class of biological transmembrane molecular machine are the ATP (synth)ase rotary motors (Fig. 1a).[3,20] ATP (synth)ases have a rotaxane-like architecture[21] in which a spindle (blue) is threaded through both an F_0 motor in the membrane (green) and an F_1 motor in the cytosol (red and grey). In addition, the motors are held in close association by a stator (orange and pink). This assembly operates in two reciprocal modes. In the ATP synthesis mode, the reaction between ADP (adenosine diphosphate) and P_i (phosphate) is catalysed at the active sites of the F_1motor. The catalytic cycle couples unidirectional rotation (driven by the flow of cations through the F_0 motor down a transmembrane gradient) to a cycle of conformational changes in the F_1 motor that involve sequential substrate binding, reaction, and ATP-dissociation.[20] Conversely, in the reciprocal mode, a transmembrane cation gradient is generated by the hydrolysis of ATP.[3]

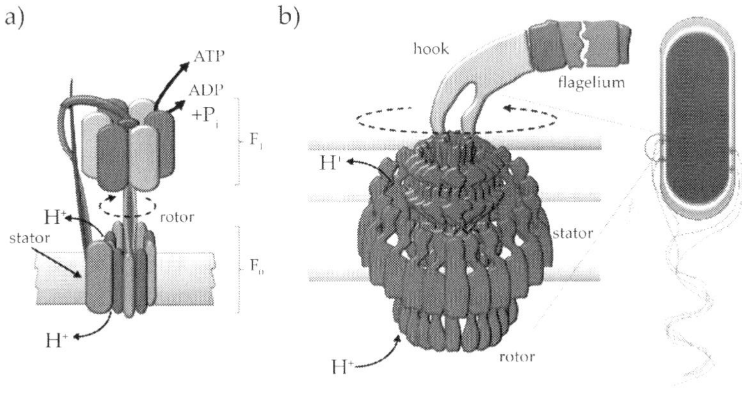

Figure 1. Natural transmembrane molecular machines. (a) Archetypal F1F0 rotary ATP (synth) ase.3,20 (b) Bacterial flagellar motor complex.

The equally impressive bacterial flagella motor (Fig. 1b)[22] also possesses a transmembrane rotaxane-like architecture and shares other mechanistic characteristics with ATP synthases. Flagellar motors utilise a ratchet mechanism to convert a transmembrane cation gradient into unidirectional motion that drives bacterial motility.[22] The flow of cations through the stator (green) transiently changes the electrostatic interaction between the stator and the rotor to generate a torque that drives unidirectional rotation of the flagellum.[22]

MAN-MADE MOLECULAR MACHINES; IN SOLUTION AND AT INTERFACES

Natural molecular machines exploit the principles of molecular self-assembly, and thereby provide lessons for the construction of synthetic molecular machines. The underlying operating principles of molecular machines have been thoroughly discussed.[5,15] However, the categorisation of a candidate system as a machine can be contentious due to differing interpretations of what constitutes "work" (particularly when energy is partitioned subjectively between different components of a system). Nonetheless, we have suggested one means of categorisation in Fig. 2. Briefly, molecular shuttles flip between multiple states under thermodynamic control as a result of Brownian motion. The result is a time-averaged Boltzmann distribution between the states at equilibrium, with the position of the equilibrium being governed by the relative free energies of those states under differing conditions.

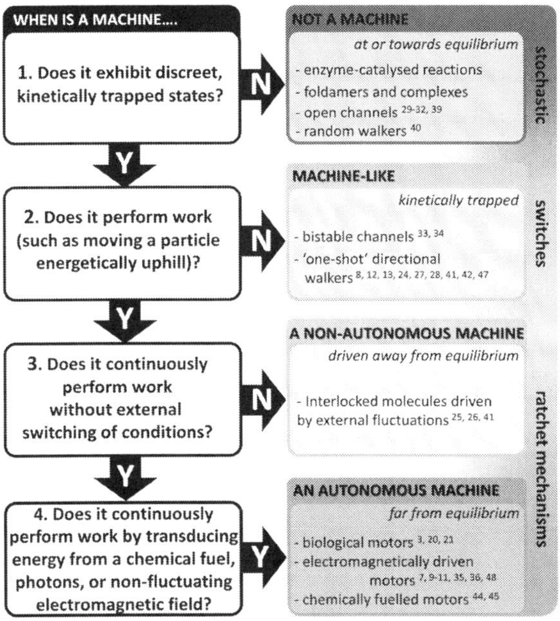

Figure 2. Flow chart for categorising machine and machine-like systems. Superscript numbers indicate examples cited in this review.

A switch differs from a stochastic shuttle and can be defined as having machine-like properties due to the possession of at least two kinetically trapped states that can only be interconverted by the application of a specific stimulus. Work can be extracted from such a system when switching allows the return from a kinetically trapped state to a lower energy equilibrium state. In contrast, a true molecular machine can repetitively and asymmetrically cycle through different nanomechanical states by transducing potential energy into work.[5,15] Notably, the changes between distinct states must not simply be the reverse of each other (as they are in a switch). Further distinctions can be drawn between autonomous and non-autonomous molecular machines. Non-autonomous machines rely on externally fluctuating conditions to ratchet asymmetrical cycles of work, while autonomous systems turn over chemical fuel, photons or other potential gradients to perform work continuously. In doing so, autonomous molecular machines generate a locally fluctuating potential, often manifested via ratcheted cycles of conformational change. The ratchet mechanisms that operate in both autonomous and non-autonomous machines can involve information[23] or energy.[24] Autonomous operation far from equilibrium is a defining characteristic of life exhibited by all biological molecular machines.[19] Indeed, achieving comparable autonomous operation in synthetic systems remains a goal of utmost importance.

A. Mechanically interlocked architectures

Synthetic mechanically interlocked structures such as rotaxanes (that share the same topology as ATP synthases and bacterial flagellar motors) have been designed that exhibit impressive and diverse machine-like behaviours. One recent example mimics the function of the ribosome (or perhaps more closely, biogenic non-ribosomal peptide synthesis).[8] The system shown in Fig. 3was based on a rotaxane in which a linear 'track' molecule (blue) was threaded through a macrocyclic host (green). The macrocycle could shuttle back and forth at random between the stoppered end of the track and the site of the first pendant amino acid group on the thread (orange) (Fig. 3i). The macrocycle was covalently linked to a peptide chain, the first residue of which was a cysteine protected by a trityl group (red, STrt). Following deprotection, the thiolate group of the cysteine residue underwent transacylation with the first amino-acid blocking group (orange), removing it from the track. A subsequent reaction between the terminal amine of the peptide chain on the macrocycle extended the peptide chain and regenerated the thiolate group (Fig. 3ii). This process was repeated until the final blocking amino acid was transferred to the growing peptide chain and the macrocycle dissociated from the track (Fig. 3iii and iv). Finally, the complete oligopeptide was cleaved from the macrocycle and the remaining protecting groups removed.

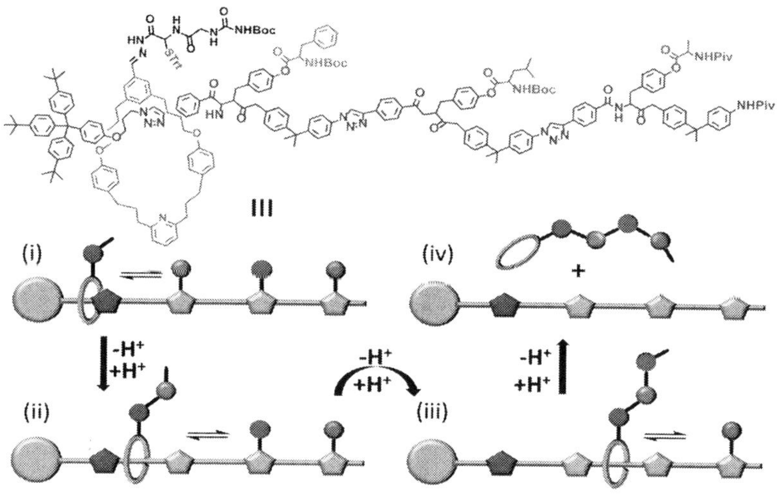

Figure 3. A rotaxane-based peptide synthesiser. Trt = CPh_3, Boc = CO_2t-Bu and Piv = OCOt-Bu.

A topologically interlocked molecule has recently been employed as a rudimentary molecular pump (Fig. 4).[25] The system was composed of a molecular axle that could accept up to two macrocycles (blue). When the positively charged viologen units on both the macrocycle and the axle were reduced to radical forms (purple) the repulsion between the macrocycle and the axle was decreased. The macrocycle was then able to thread onto the partially reduced viologen station on the axle. Following threading, the viologen units on the macrocycle and axle were re-oxidised using $NOPF_6$. The electrostatic repulsion between the macrocycle and the positively charged end of the axle was sufficient to overcome the steric barrier presented by the isopropylphenyl group (green) such that the macrocycle became trapped on the stoppered alkyl chain. Repetition of the cycle of reduction, oxidation and reduction allowed a second macrocycle to be 'pumped' onto the alkyl chain in an overall energetically uphill process.

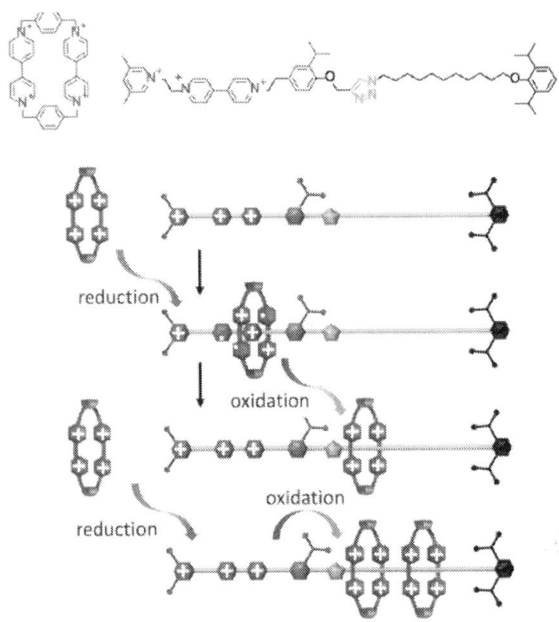

Figure 4. An artificial molecular pump driven by external switching of redox conditions.

A key attribute of biological molecular machines is autonomous nanomechanical operation in repetitive cycles that dissipate energy supplied by a chemical or electromagnetic potential. One recent synthetic example that accomplished such dynamism is shown in Fig. 5.[7] The system is composed of a 2,3-dinaphtho[24]crown-8 ether macrocycle (purple) and a molecular axle containing a photoisomerisable azobenzene moiety (E green, Z red), a cationic ammonium recognition site (blue) and a cyclopentyl terminus (grey). Under constant UV irradiation, the threading and dethreading proceeded with a directional bias (clockwise in Fig. 5). The macrocycle axle binding equilibria for both the E and the Z axle isomers lay towards the complex, but binding to the E-isomer was more favourable. Control experiments performed with double-stoppered compounds revealed that the barrier to macrocycle threading presented by the E-azobenzene moiety was lower than of the cyclopentyl group, while macrocycle unthreading over the Z-azobenzene moiety was less favoured compared to unthreading over the cyclopentyl group.

Moreover, the E-complex was more sensitive to E/Z-photoisomerisation than the free E-axle, while the lifetime of the Z-complex was long enough that dethreading preferentially occurred over the cyclopentyl unit. The net result of the differing energetic barriers to threading/unthreading and the differing rates of isomer interconversion meant that threading and dethreading proceeded with a directional bias under constant photo-irradiation (clockwise in Fig. 5). Although the system operated autonomously in a cyclical manner with molecular-level directionality, this directionality was lost when summed over the random orientations of the freely tumbling assemblies in solution. However, if it were possible to directionally orientate such a system across a lipid membrane then this would constitute a man-made molecular pump.

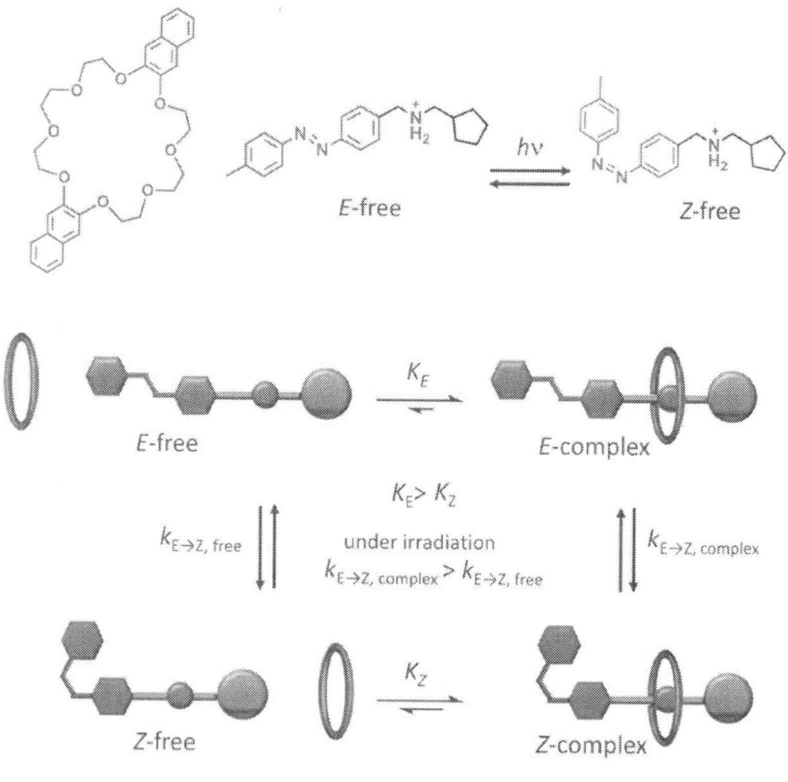

Figure 5. Light-driven autonomous cycle of unidirectional supramolecular assembly/disassembly.

The challenge of extracting work from the controlled motion of mechanically interlocked species has been addressed by attaching them to surfaces. This reduces the degrees of freedom such that changes in the nanomechanical states of a large number of individual molecules may be manifested as a coherent change in the macroscopic properties of the surface. One demonstration of the extraction of macroscopic work from the collective nanomechanical movements of rotaxanes anchored to a surface is shown in Fig. 6.[13] A bistable [3]rotaxane was prepared using two macrocycles (blue) that each featured a disulfide anchor (gold) threaded onto a stoppered thread containing two naphthalene stations (red) and two electroactive tetrathiofulvalene stations (green). The [3] rotaxanes formed a self-assembled monolayer on a gold cantilever via the disulfide anchors on the macrocycles. Oxidation of the tetrathiofulvalene stations to their cationic form (purple) using $Fe(ClO_4)_3$ resulted in electrostatic repulsion between the cationic tetrathiofulvalene stations and macrocycles. Lateral displacement of the cantilever beam of 35 nm was measured due to the force imparted by the collective contractions of the self-assembled monolayer.

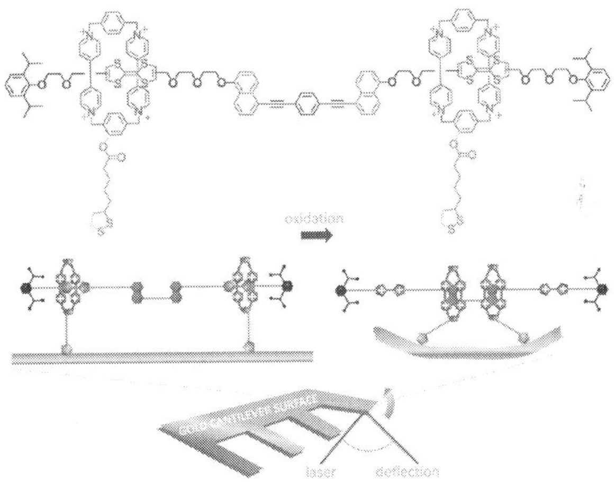

Figure 6. Cantilever actuation driven by redox switching of surface-bound rotaxane shuttles

B. Synthetic rotary motors

In contrast to switches, rotary motors operate by cycling through a number of distinct states in response to energetic stimulation or the availability of a suitable fuel. Such behaviour has been demonstrated in a fully synthetic catenated architecture (Fig. 7).[26] The system was composed of two identical macrocycles (blue and purple) that were both threaded on a larger macrocycle containing four distinct binding sites (A to D). Each binding site exhibited different association constants for the threaded macrocycles due to their differing H-bonding abilities. Photoirradiation at 350 nm converted station A from the E to the Z isomer (1), while photoirradiation at 254 nm converted station B from the E to the Z isomer (2). Both stations A and B could be re-isomerised from Z to E (3) by heating to 100 °C for 24 hours, or by treatment with ethylenediamine at 50 °C for 48 hours or Br_2 under ambient light at −78 °C for 10 minutes. Repeated, sequential application of stimuli (1), (2) then (3), resulted in the net unidirectional rotational movement of both small macrocycles with respect to the larger macrocycle. Although the system can be defined as performing local work, no work can be extracted as the rotors are freely tumbling in solution.

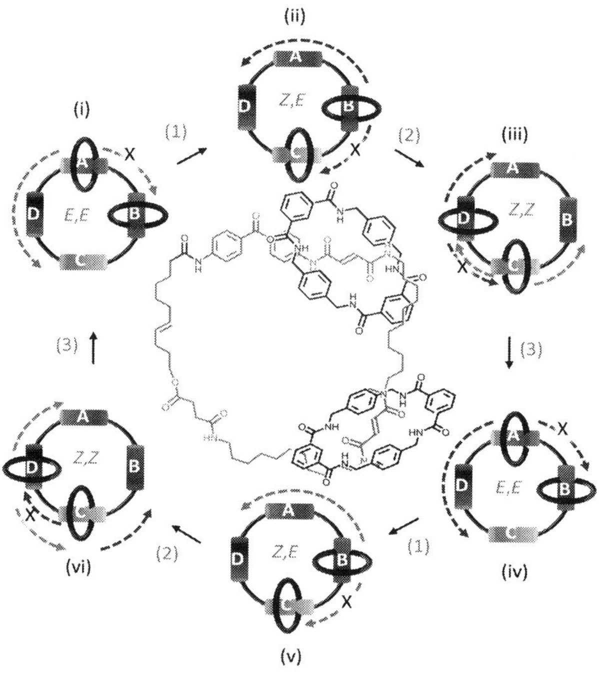

Figure 7. Unidirectional rotation of catenated rings driven by switching of external light and chemical conditions.

Synthetic rotary motors have also been developed that can be driven by light (Fig. 8).[9] These systems exploit repeated cycles of E/Z photoisomerisation of the central double bond and subsequent helicity inversion by thermal relaxation to convert between conformational states. Upon irradiation at 355 nm the central double bond photoisomerises from the most stable cisconformer (i) to a higher energy trans conformer (ii). Next, favourable helicity inversion facilitated by thermal fluctuations gives the stable trans conformer (iii). A second photo-excitation converts the compound to a higher energy cis conformer (iv) that undergoes thermal helicity inversion to return the system to the initial conformational state. Hence, under constant irradiation these motors rotate unidirectionally. Due to the diastereomeric relationship between each half of the molecule, retrograde motion does not occur because of the substantially higher barrier to re-isomerisation from the unstable state $(\Delta^{\ddagger}G_2^{\circ})$ compared to the barrier height for helix inversion $(\Delta^{\ddagger}G_1^{\circ})$. Variation of the X, Y and R groups has yielded motors that rotate autonomously under constant irradiation at remarkable rates of up to 3 MHz![9] Although each motor molecule generates a local torque, work cannot be extracted from an ensemble of freely tumbling, randomly orientated molecules in solution. Once again, the key to the extraction of work from these molecular rotors was the involvement of a surface. In one study such motors were used as dopants in a chiral liquid-crystal film which caused the unidirectional rotation of microscale objects when photoirradiated.[10]

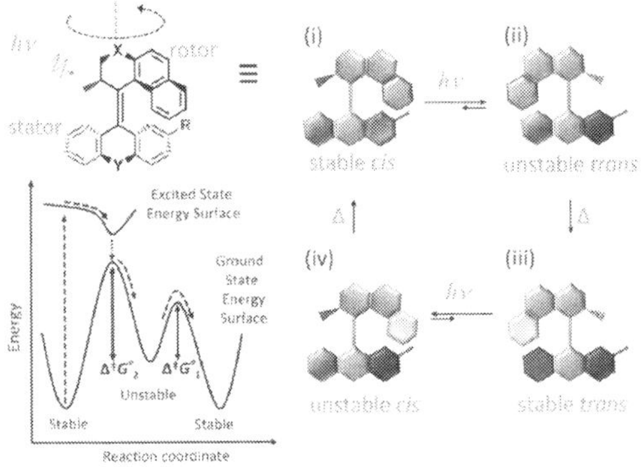

Figure 8. Light-driven unidirectional intramolecular rotation of a synthetic molecular motor.

The unidirectional rotation of bisarylidene rotors has also been transduced into directional linear motion (Fig. 9).[11] When a car-like molecule containing four bisarylidene rotors was stimulated with electrons from an STM tip, the concerted motion of the rotors caused it to move along the surface. The net movement of each molecule was determined by the relative stereochemistries of the four rotors. The R,R,S,S meso-isomer shown in Fig. 9 gave unidirectional translation, as did the S,S,R,Rconfiguration. The R,R,R,R and S,S,S,S isomers moved randomly while R,S,R,S and S,R,S,R did not move due to the cancellation of directionally opposed motions.

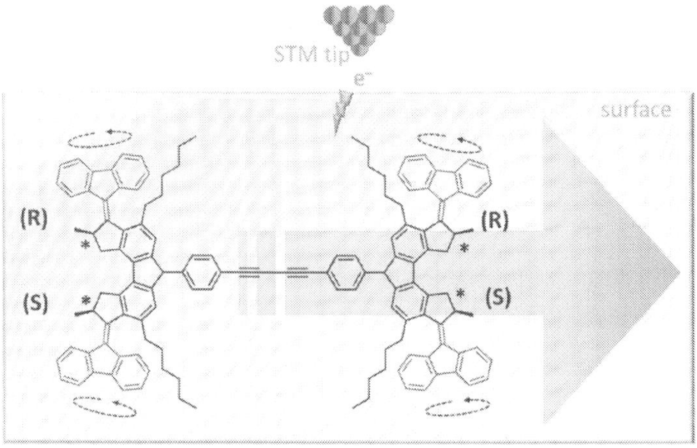

Figure 9. Unidirectional linear translocation of a molecule containing bisarylidene rotors.

C. DNA-based molecular machines

Nature's molecular machines certainly exceed our own in terms of elegance, modularity and functionality. Bionanotechnological research has sought to exploit the principles of molecular recognition to repurpose biomolecules such as DNA and proteins as machine elements.[16] Many of the achievements of wholly synthetic molecular machines, including programmable walkers[27] and molecular synthesisers[28] have also been accomplished by synthetic molecular machines based on DNA (sometimes years previously). Such DNA nanomachines exploit the programmable hydrogen-bonding interactions of canonical base pairing. One of the most impressive demonstrations of a DNA-based molecular machine is a peptide synthesiser based on a DNA walker (Fig. 10).[28] The system operated in a stepwise fashion reminiscent of the naturally occurring ribosome.

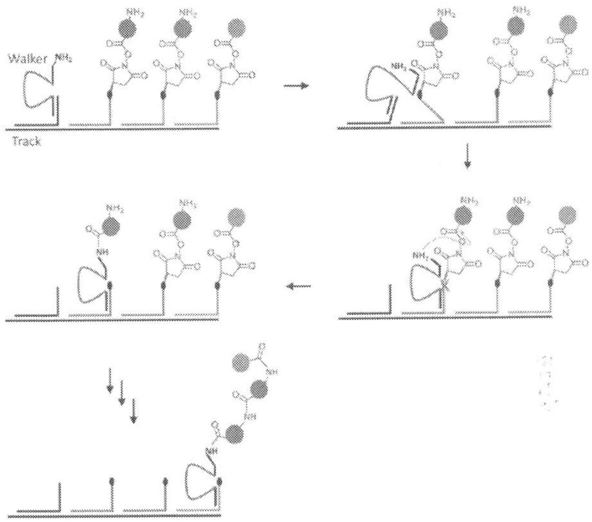

Figure 10. A peptide synthesised based on a DNA walker

The DNA track (black) and the strands bearing the amino acid monomers (blue, red and green) were first assembled in solution, and the walker (purple) subsequently added. Next, the walker could hybridise to the first DNA foothold (blue) bringing the free amine on the walker and the succinyl ester of the first foothold into close enough proximity to initiate an acylation reaction. The reaction caused the walker to fully migrate onto the blue foothold to reduce strain in the system. In this configuration, a DNAzyme region of the walker (purple) was able to catalytically cleave an RNA linker sequence (black dot). Since the first foothold (blue) had now been shortened by cleavage, binding to the subsequent (red) foothold was favoured. Repetition of this sequence of events facilitated the autonomous synthesis of a tripeptide. Overall, the system compares favourably to the more recent fully synthetic peptide synthesiser introduced in Fig. 3; it is easily constructed and operates under isothermal conditions. However, in common with the synthetic peptide synthesiser (and unlike the biogenic ribosome) processivity is limited because the effective molarity of the reacting groups decreases with each amino-acid addition step. Furthermore, the fidelities of the man-made systems are limited by the increased flexibility of longer chains, which might allow monomers to be added in an unintended order.

MAN-MADE TRANSMEMBRANE MOLECULAR MACHINES

In biology, lipid bilayers provide a route to system complexity. Self-assembly facilitates the spatial organisation of discrete entities with specific orientations within a membrane. This includes selective, stimuli-responsive channels that operate at equilibrium, as well as active transmembrane pumps that operate far from equilibrium to transduce energy into useful work. Efforts to mimic the functionality of natural transmembrane systems have mirrored this progression in complexity; from passive (Section 3A) to selective channels (Section 3B), to light-driven (Section 3C) and nanomechanical transmembrane assemblies that can operate far from equilibrium (Section 3D).

A. Artificial transmembrane channels

Lipid membranes provide an attractive platform for the construction of man-made molecular machines; bilayers are easily prepared and they are inherently biocompatible. Furthermore, they permit the spatial organisation of self-assembling biological and synthetic components in the realisation of man-made transmembrane devices. For example, both cyclic peptides (Fig. 11, left)[29] and DNA (Fig. 11, right)[30] have been used to create stable artificial nanopores across lipid bilayers. Both of these assemblies have dimensions close to those of natural protein channels and permit the passage of ions down a transmembrane potential or concentration gradient. The cyclic peptides (green) self-assemble into membrane-spanning tubes as they partition into the membrane. The assembly process was reasoned to occur by the stacking of aromatic side chains within the hydrophobic membrane interior, while the inside of the tubes present the polar amide backbone to the solvent. Similarly, the DNA origami nanopore was designed such that it was anchored into the bilayer via cholesterol-modified helix bundles (grey and red). The machine-like, voltage-sensitive gating (opening and closing) of the DNA-origami channels was also investigated. A channel variant bearing a strand that could readily adopt a channel-blocking conformation in response to high transmembrane potentials was observed to gate more than the original channel. Similarly, DNA origami tiles have been designed that modulate the flow of ions and biomolecules across channels at the ends of glass capillaries.[31] Given the maturity of DNA nanotechnology,[16,27] it can be envisioned that such approaches will soon develop beyond switchable channels to yield true transmembrane molecular machines capable of performing work (such as active transport).

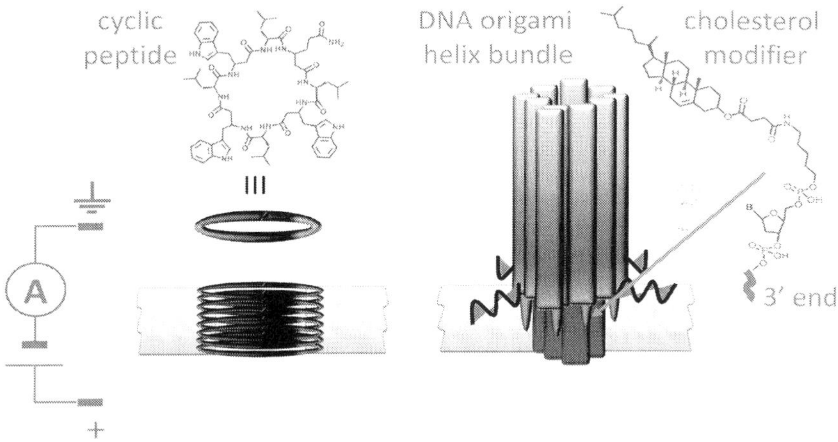

Figure 11. Artificial nanopores in lipid bilayer membranes made using cyclic peptides (left) and DNA origami (right).

B. Ligand-responsive transmembrane channels

Biological nanopores have also inspired the development of systems in which micro- to nano-scale apertures in solid-state wafers separate fluid-filled compartments. These artificial apertures can be functionalised to give tuneable systems that respond to stimuli such as voltage, pH, temperature, light, and even specific ions and enantiomers.[18] Furthermore, selective transport across a partition can be achieved by exploiting the interactions between an analyte and the molecules used to functionalise the aperture. Although this machine-like behaviour mimics cellular function, most approaches developed to date act as simple gates or switches that are incapable of performing work such as establishing a chemical gradient.

The most sophisticated systems employing this top-down approach facilitate the simultaneous translocation and analysis of macromolecules. One such system mimics the nuclear pore complex (Fig. 12)[32] and consisted of a nanopore formed in a polycarbonate wafer (grey) coated with gold and further modified with phenylalanine–glycine–nucleoporin proteins (blue). The membrane-like wafer acted as a selective filter that facilitated the diffusion of specific, free and cargo-bound transport factors (red) through the pore via favourable non-covalent interactions with the nucleoporins. In contrast, passive transmembrane diffusion of similarly sized molecules (green) that were not recognised by the nucleoporins was hindered. One obvious difference between many solid-state and biological channels is their size. This difference arises from the difficulty in reproducibly obtaining nanometre-scale features using top-down engineering techniques. Channel size is important because the interactions between an aperture and an analyte become more pronounced when they have similar dimensions, and hence the likelihood of departure from simple diffusive behaviour increases.

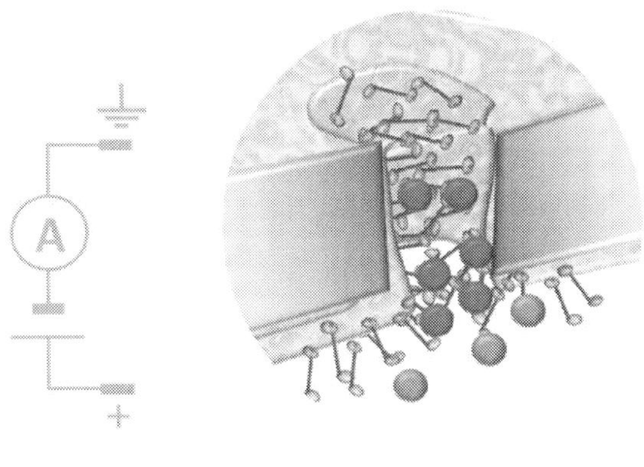

Figure 12. Artificial nanopore system that mimics the nuclear pore complex.

Molecularly reproducible membrane-spanning devices can be obtained by repurposing biological pore proteins and modifying them as needed. For example, a protein chimera (Fig. 13) that combined a K^+ channel protein (yellow) with a G-protein-coupled receptor (red) has been engineered.[33] Analysis of ion currents flowing through the chimeric protein channels indicated that the ion channel opened upon binding of a known ligand (green) to the G-coupled-receptor. Thereby, artificial transmembrane signal transduction was achieved from a chemical signal without the molecule itself crossing the membrane.

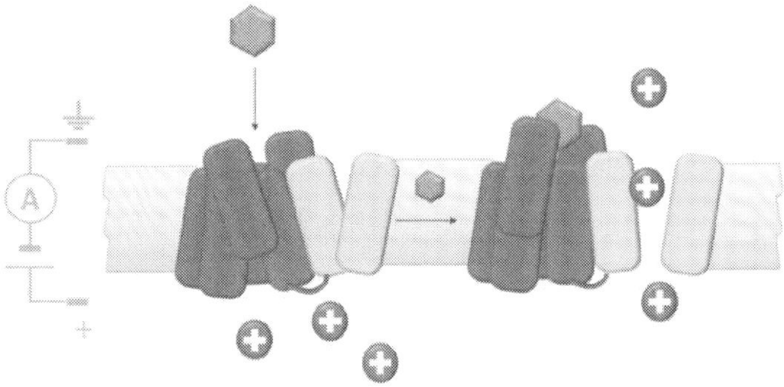

Figure 13. A man-made ligand-gated transmembrane ion channel formed from the chimera of a K^+ ion channel (yellow) and G-protein-coupled receptor (red).

C. Light-driven transmembrane devices

A number of devices that use light either as an external stimulus or as an energy source to drive transmembrane processes have been demonstrated. Synthetic photoactive functionalities that undergo conformational changes have been used in conjunction with biological channel proteins to facilitate optical control over membrane transport. One example of this approach involved engineering the mechanosensitive channel of large conduction (MscL) from E. coli (Fig. 14).[34] The protein (yellow) was modified to have one cysteine residue per monomer allowing the incorporation of a photo-responsive cyanine group that could be reversibly switched between a neutral spiropyran form (red) and a charged zwitterionic merocyanine form (purple). When 366 nm UV light was shone on a membrane containing the chemically modified channels, a large-scale conformational change caused the channel to open, and permitting the flow of ions across the membrane under an applied transmembrane potential. The conformational change was fully reversed by ambient light at wavelengths >460 nm. Hence, light-actuated gating could be performed repeatedly by turning the UV light on and off.

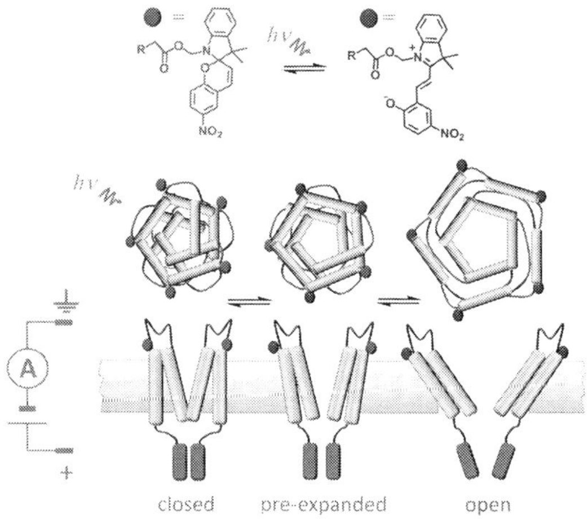

Figure 14. Semi-synthetic light-gated ion channel prepared from a chemically modified mechanosensitive channel protein.

ATP synthase has also been used as a component in a variety of artificially assembled devices. The key to achieving non-equilibrium properties in the example shown in Fig. 15 was the addition of a diphenyl quinone (Q_S) and a photoactive carotene–porphyrin–naphthoquinone (C–P–Q) (Fig. 15).[36] The synthetically functionalised membrane was shown to generate a transmembrane proton gradient in response to visible light, which could be exploited to drive the rotation of ATP synthase and the associated synthesis of ATP (as explained in Fig. 1a).

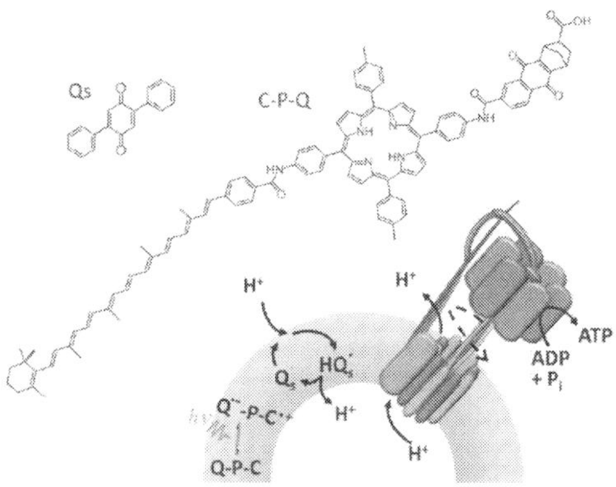

Figure 15. An artificial photosynthetic vesicle.

A system that combines both ligand-mediated channel gating and the mimicry of transmembrane photosynthesis has been developed from wholly synthetic molecules (Fig. 16).[35] The system comprised four p-octiphenyl chains (maroon), each bearing eight naphthalene-diimides (green). The complex self-assembled in lipid bilayer vesicles such that the π-surfaces of the naphthalene-diimides stacked to give a canted-barrel structure. In this geometry, irradiation of the complex at 638 nm caused electronic excitation that resulted in a transmembrane charge-separated state. The generated electrons and holes were used to drive a redox couple (reduction of quinones and oxidation of EDTA on opposite sides of the bilayer), maintaining the charge-separated state and generating a transmembrane proton gradient. Finally, the addition of an intercalating naphthalene derivative (yellow) switched the barrel into an open conformation, thereby permitting neutralisation of the accumulated proton gradient by the flow of hydroxide ions from outside the vesicle.

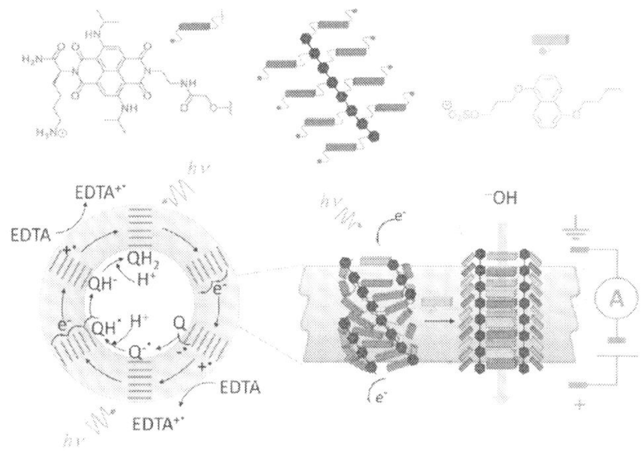

Figure 16. A self-assembled synthetic transmembrane electron pump that can also open in response to ligand binding. EDTA = ethylene diamine tetraacetate, Q = quinone.

D. Nanomechanical motions in transmembrane devices

Although macroscopic mechanistic principles do not describe nanoscale processes,[15] machine behaviour on the molecular level is most easily recognised where well-defined nanomechanical motions occur. Nanomechanical processes have been widely explored using biological nanopores. In particular, protein nanopores have found applications in single-molecule biophysics[37]and in single-molecule DNA sequencing.[38] In many es nanopore-based detection methods are facilitated viananomechanical processes such as the translocation of single-stranded DNA. For example, the connector protein associated with the φ29 DNA-packaging motor has been modified with a band of hydrophobic residues (yellow) such that it could form channels across lipid bilayers (Fig. 17).[39] In this study, the DNA-packaging motor protein was not present, and the translocation of DNA through the channel was driven by the application of a transmembrane voltage.

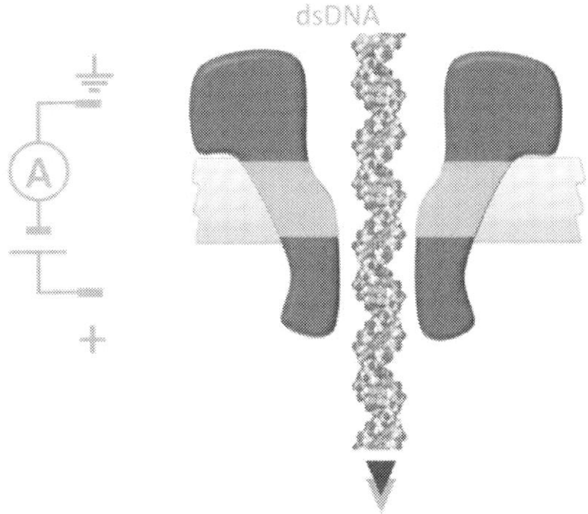

Figure 17. A membrane-adapted φ29 connector protein that permits the voltage-driven passage of double-stranded DNA across a membrane.

Although open protein channels are not molecular machines, they can provide useful chassis for the construction of transmembrane nanomechanical devices. Mirroring earlier advances in the field of synthetic molecular walkers, the random walk of a small molecule inside a protein nanopore has been observed (Fig. 18).[40] The approach employed a mutant α-haemolysin (α-HL) nanopore that contained five cysteine residues within the transmembrane stem. The molecular walker was formed in situ by the reaction of 4-sulfophenyl arsenous acid and 2-(2-methoxyethoxy)ethanethiol. Changes in the transmembrane ion current flowing through the pore revealed that the arsenous molecule walked stochastically by reversible formation of covalent bonds between the arsenic of the walker and the cysteine residues in the transmembrane stem. The motion of the walker was biased by the direction of the applied potential to move towards a thermodynamic sink.

Figure 18. Stochastic walking of a small molecule within a transmembrane nanopore.

Another example of machine-like behaviour was recently demonstrated using a DNA-modified cytolysin A nanopore (Cly A) that facilitated the selective transport of DNA across a membrane (Fig. 19).[41] The mutant pore was covalently modified with single-stranded DNA 'arms' (black) that recognised a specific target DNA complex (yellow and purple) that was present on one side of the bilayer. Complexation to the arm increased both the negative charge density and the effective molarity in the vicinity of the pore compared to the free target. The combination of these two effects meant that the barrier to DNA translocation could be overcome at an applied voltage of +50 mV. Next, complexation of streptavidin (green circle) to the biotin group on the target sequence trapped the translocated DNA on that side of the membrane. A DNA strand (pink) was then able to displace the cargoes into solution via a strand displacement process. The free DNA arm was then able to diffuse back through the pore, allowing further cycles to occur. This system has several characteristics of a machine; it operates autonomously, cycling directionally through multiple nanomechanical states via an information ratchet mechanism,[23] while also consuming a strand-displacing DNA fuel. Although DNA transport proceeded at a voltage that would normally be too low to overcome the energy barrier to translocation, DNA transport occurred in the same direction as the applied field (down the potential energy landscape). Thus, the translocation of DNA by this mechanism does not constitute work performed by the nanomechanical cycle (Qn. 2 in Fig. 2). Hence, the system can be considered as a catalytic transloe that lowers the activation barrier to a thermodynamically downhill process, rather than being an active transporter, pump or motor.

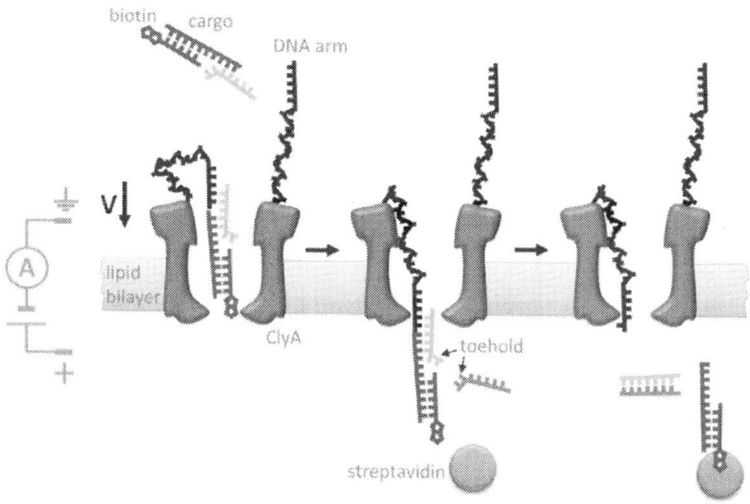

Figure 19. A transmembrane DNA transporter based on the cytolysin A pore protein.

Indeed, motor proteins such as DNA polymerases have been employed in conjunction with biological pores to create transmembrane nanomechanical devices with machine-like properties. For example, the replication of DNA hairpins by φ29 DNA polymerase has been monitored at the single-molecule level using nanopores (Fig. 20).[42] In this approach, a DNA-hairpin (black) was captured in an α-HL channel (blue) under an applied potential. The turn-over of dNTP fuel during extension of the self-primed hairpin by the polymerase (yellow) resulted in the captured DNA being ratcheted out of the pore against the applied transmembrane potential. The DNA hairpin contained an abasic region (red) that allowed the relative position of the strand in the pore to be determined from changes in the ion current flowing through a single channel.

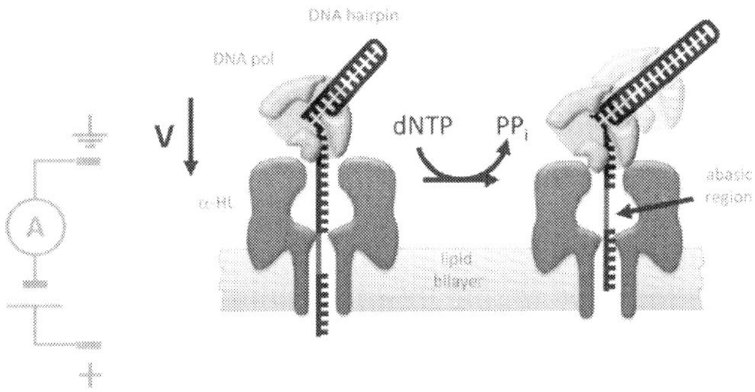

Figure 20. Polymerase-catalysed replication of DNA monitored at the single-molecule level using a nanopore.

A similar approach has been used to study the ratcheting of a protein through an α-HL nanopore by the motor protein AAA+ unfoldase (ClpX in Fig. 21).[43] The protein Smt3 (purple) was modified with a charged polypeptide tail (green) terminating in an ssrA recognition sequence (yellow). This modified protein was threaded through and captured within an α-HL nanopore by an applied transmembrane potential. The ssrA sequence was then recognised by the unfoldase (red) that was present on the opposite side of the lipid bilayer. Cycles of conformational change in the unfoldase driven by the turn-over of ATP allowed the captured protein to be unfolded and pulled through the pore (green to purple). The unfolding process was monitored viaratcheted changes in the ion current flowing through the pore under an applied potential.

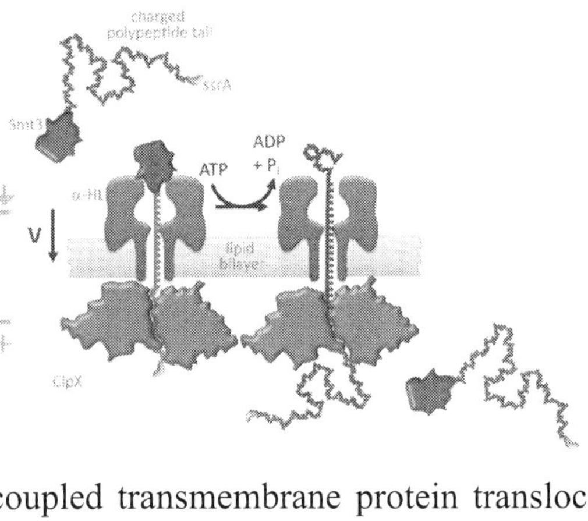

Figure 21. ATP-coupled transmembrane protein translocation through α-HL.

Further efforts to create synthetic transmembrane molecular machines have employed protein chimeras. A recent, impressive example appended loops of the cochaperonin protein GroES (green) onto the α-HL nanopore (blue) (Fig. 22).[44] Natively, GroES forms a complex with GroEL to yield a catalytically active chaperonin that turns over ATP to drive a series of conformational changes that mediate the folding of proteins (yellow) within the internal cavity of the GroES–GroEL complex. Since both the GroES–GroEL complex and the α-HL share a seven-fold symmetry, the α-HL–GroES chimera was able to template the assembly of an active GroEL heptamer (pink). The ATP-driven chaperonin function of this engineered transmembrane assembly was characterised using gel-shift experiments in the bulk, while dynamic activity was monitored at the single-molecule level using transmembrane ion currents.

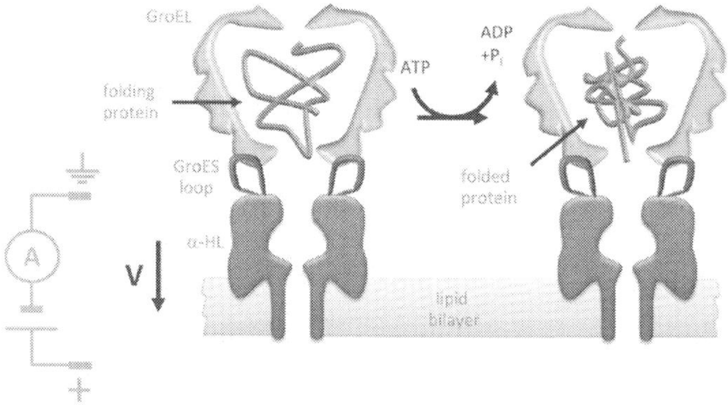

Figure 22. Single-molecule observation of the ATP-fuelled folding of a protein using an engineered transmembrane co-chaperonin.

The assemblies described in Fig. 20–22 were used to monitor the machine behaviour of known motor proteins; a further milestone has been the realisation of systems in which machine behaviour is an emergent quality possessed by the entire assembly. This goal was recently attained in the form of a transmembrane nanoactuator that turned over DNA molecules to fuel cycles of autonomous reciprocating (back-and-forth) nanomechanical motion (Fig. 23).[45] The approach built on the precedent set by a nanopore-based system capable of monitoring DNA polymerase activity with single-nucleotide resolution.[46,47] The device was assembled by threading a DNA–PEG copolymer (black-red) capped with a streptavidin–biotin stopper (green circle) through an α-HL nanopore under an applied potential. The captured co-polymer was then trapped in place by hybridisation to a fully complementary DNA primer (purple-green) to form a rotaxane-like transmembrane architecture. In the presence of DNA polymerase, pyrovanadate, Mn^{2+} and DNA fuel molecules, ion-current recordings revealed asymmetric saw-tooth patterns. These patterns corresponded to an emergent cycle of back and forth movement in the threaded co-polymer at rates of up to one cycle per minute. Stepwise 'downwards' movement of the threaded co-polymer (left to right in Fig. 23) arose due to DNA polymerase-catalysed degradation of the primer fuel, while displacement of the degraded primer by the invasion of a full-length primer pulled the threaded co-polymer 'upwards' in a single step (against the applied potential). The asymmetric conformational cycle was co-ordinated via an information ratchet mechanism since strand displacement could only occur once the primer had been sufficiently degraded.

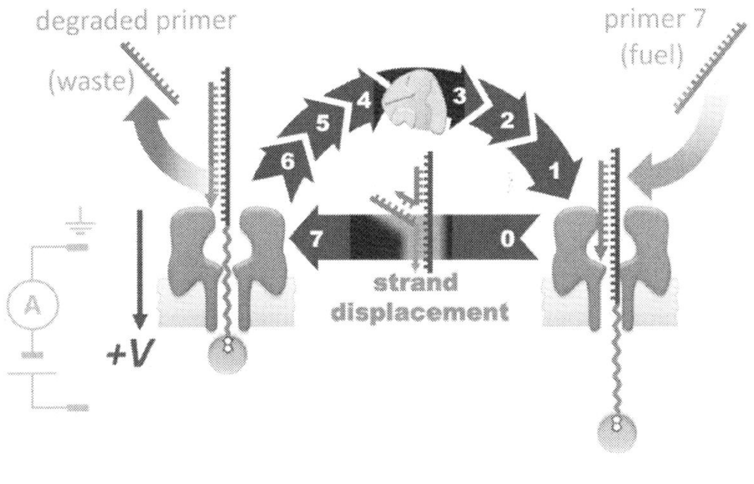

Figure 23. A transmembrane reciprocating nanoactuator that operates with a temporally asymmetric nanomechanical cycle.

E. Multi-compartment transmembrane devices

Contrasting with the isolated devices described above, higher-order living organisms are hierarchically organised into multiple chemically distinct environments by lipid bilayers. This partitioning allows the generation of transmembrane potentials by molecular machines working in different compartments, giving rise to emergent collective properties. Artificial systems have been developed that mimic the multi-compartmental nature of living organisms. For example, aqueous droplets can be submerged in oil and arranged into networks in which bilayers form at the interfaces between the drops (Fig. 24). The addition of proteins that insert into the bilayers connecting these droplet allows ions and analytes to pass between the droplets in a designed manner. In addition, the approach allows these processes to be monitored by both electrical and optical methods. Applications have included a rudimentary light sensor that mimics optical, to chemical, to electrical signal transduction in animal sight (Fig. 24).[48] The sensor consisted of three droplets containing bacteriorhodopsin that were each connected to a common electrode. These three droplets were interfaced with a central drop containing only buffer. A fifth drop containing α-HL (green) was also interfaced with the central drop and connected to a counter electrode. The bacteriorhodopsin and the α-HL were then able to insert into the bilayer interfaces with the central droplet. Photoirradiation caused the bacteriorhodopsin to pump protons across the bilayer, leading to a measurable electrical current arising from the flow of ions through the droplet circuit.

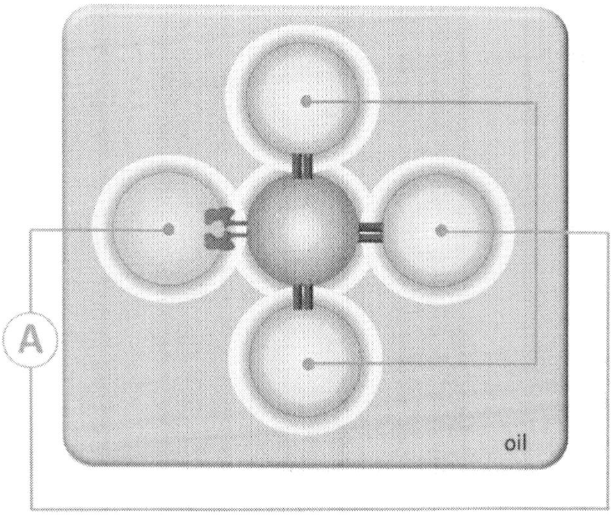

Figure 24. A simple circuit constructed from a droplet interface bilayer network connected by an α-haemolysin pore and bacteriorhodopsin.

The droplet network approach has been further expanded by the generation of higher-order 'multisomes' that can operate in aqueous environments rather than in oil baths (Fig. 25).[49] The assemblies were formed from multiple aqueous compartments that were organised, but not completely encapsulated within a larger droplet of oil. The aqueous droplets had distinct internal compositions (indicated by the different colours in Fig. 25) enabling distinct functionalities to be presented to the surrounding environment. In this manner, different droplets could respond to different stimuli in distinct ways. In addition, the aqueous droplets could be connected at their internal interfaces by the inclusion of transmembrane pores such as α-HL. Thus, these multisomes were able to engage with their environment by sensing, processing and responding.

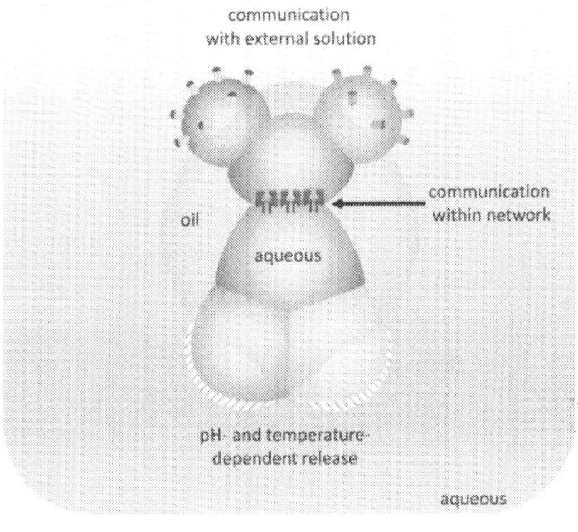

Figure 25. Droplet multisomes that operate in aqueous environments.

The droplet bilayer approach has been combined with 3D-printing technology to produce heterogeneous droplet networks that mimic some of the properties of living tissues (Fig. 26).[50] Sub-microlitre droplets were 'printed' in three dimensions within an oil droplet submerged in water (Fig. 26a). The resulting jelly-like material was held in place by a metal frame. Heterogeneous multi-compartment tissues could be printed that contained different concentrations (red and blue in Fig. 26a and b). The osmotic movement of water between droplets containing different salt concentrations in one- and two-dimensional tissues gave rise to assemblies with emergent three-dimensional morphologies (Fig. 26b). Furthermore, the inclusion of α-HL in specific droplets enabled the path of solute diffusion to be programmed in three dimensions (Fig. 26c). While these droplet-based systems do not constitute transmembrane molecular machines in their own right, they might serve as hosts for collectively operating transmembrane molecular machines in the future, paving the way to sophisticated multi-compartment devices with emergent properties.

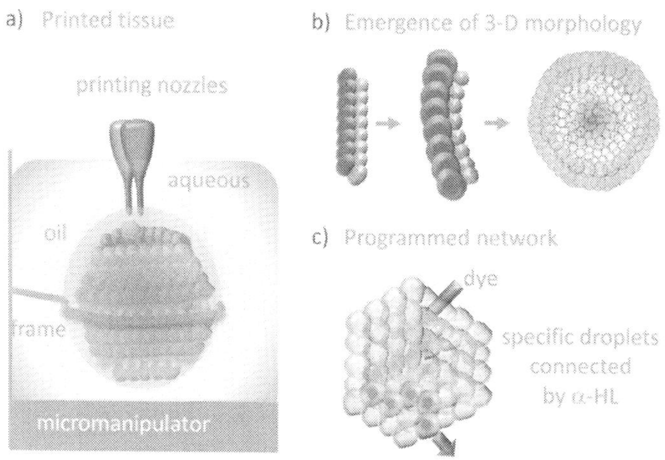

Figure 26. Tissue-like 3-D printed droplet networks.

CONCLUSIONS AND FUTURE DIRECTIONS

Many avenues are being pursued in the design of molecular machines. Biology has not only inspired the development of man-made molecular machines, but also provided insights into the mechanisms of energy transduction on the molecular level. Implementation of fundamental mechanisms such as ratcheting[23,24] have led to the construction of minimal synthetic designs that are free of the molecular 'baggage' acquired over billions of years of biological evolution.[5,15] Despite the simplicity and elegance of many synthetic systems,[7,9,25,26] no macroscopic work can be extracted from freely tumbling, randomly oriented molecules.[2] This fundamental Brownian limitation has been overcome by interfacing synthetic molecular machines with surfaces.[10,11,13] In biology, similar entropic constraints are afforded to molecular machines that work along one-dimensional biopolymers or across two-dimensional membranes. Indeed, it is hard to imagine life without the nucleic acid polymerases,[42,46,47] transmembrane photosynthesis,[35] or ATP synthase.[3] Substantial progress has been made since the first transmembrane channels were prepared,[29] including the realisation of stimuli-responsive channels that mimic some biological functions.[30–32] However, few artificial transmembrane systems are as robust or reproducible as their biological counterparts, or display the autonomous operation ubiquitous to the machinery of life. Thus, many researchers have repurposed biological molecules such as stable pore proteins[33,34,41] and DNA[30] to construct artificial transmembrane devices. Some of the most effective demonstrations of transmembrane machine behaviour have utilised motor proteins to attain assemblies that operate with autonomous nanomechanical cycles.[44,45] While lessons continue to be learned from biology, principles from synthetic nanotechnology[15] must be better integrated into the next generation of transmembrane architectures. Looking further ahead, multi-compartment devices[49,50] might deliver hybrid synthetic/biological systems that coordinate the functions of numerous transmembrane machines to operate far from equilibrium,[19] blurring the lines between life and the artificial.

ACKNOWLEDGEMENTS

We thank the Engineering and Physical Sciences Research Council (EP/H021620-1) and ERC Starting Grant No. 336935, 'Transmembrane molecular machines' for financial support.

REFERENCES

1. M. Schliwa and G. Woehlke, Nature, 2003, 422, 759–765.
2. R. D. Astumian, Phys. Chem. Chem. Phys., 2007, 9, 5067 .
3. S. Arai, S. Saijo, K. Suzuki, K. Mizutani, Y. Kakinuma, Y. Ishizuka-Katsura, N. Ohsawa, T. Terada, M. Shirouzu, S. Yokoyama, S. Iwata, I. Yamato and T. Murata, Nature, 2013, 493, 703–707.
4. X. Zhao, K. Zhang, T. Boquoi, B. Hu, M. A. Motaleb, K. A. Miller, M. E. James, N. W. Charon, M. D. Manson, S. J. Norris, C. Li and J. Liu, Proc. Natl. Acad. Sci. U. S. A., 2013, 110, 14390–14395.

5. E. R. Kay, D. A. Leigh and F. Zerbetto, Angew. Chem., Int. Ed., 2007, 46, 72–191.
6. A. Coskun, M. Banaszak, R. D. Astumian, J. F. Stoddart and B. A. Grzybowski, Chem. Soc. Rev., 2012, 41, 19–30 .
7. G. Ragazzon, M. Baroncini, S. Silvi, M. Venturi and A. Credi, Nat. Nanotechnol., 2014, 10, 70–75 .
8. B. Lewandowski, G. De Bo, J. W. Ward, M. Papmeyer, S. Kuschel, M. J. Aldegunde, P. M. E. Gramlich, D. Heckmann, S. M. Goldup, D. M. D'Souza, A. E. Fernandes and D. A. Leigh, Science, 2013, 339, 189–193.
9. M. Klok, N. Boyle, M. T. Pryce, A. Meetsma, W. R. Browne and B. L. Feringa, J. Am. Chem. Soc., 2008, 130, 10484–10485.
10. R. Eelkema, M. M. Pollard, J. Vicario, N. Katsonis, B. S. Ramon, C. W. M. Bastiaansen, D. J. Broer and B. L. Feringa, Nature, 2006, 440, 163.
11. T. Kudernac, N. Ruangsupapichat, M. Pahau, B. Macia, N. Katsonis, S. R. Harutyunyan, K.-H. Ernst and B. L. Feringa,Nature, 2011, 479, 208–211.
12. J. Berna, D. A. Leigh, M. Lubomska, S. M. Mendoza, E. M. Perez, P. Rudolf, G. Teobaldi and F. Zerbetto, Nat. Mater., 2005, 4, 704–710.
13. T. J. Huang, B. Brough, C.-M. Ho, Y. Liu, A. H. Flood, P. A. Bonvallet, H.-R. Tseng, J. F. Stoddart, M. Baller and S. Magonov,Appl. Phys. Lett., 2004, 85, 5391–5393 .
14. V. Balzani, A. Credi and M. Venturi, ChemPhysChem, 2008, 9, 202–220.
15. S. Erbas-Cakmak, D. A. Leigh, C. T. McTernan and A. L. Nussbaumer, Chem. Rev., 2015, 115, 10081–10206.
16. J. Bath and A. J. Turberfield, Nat. Nanotechnol., 2007, 2, 275–284.
17. S. Matile, A. Vargas Jentzsch, J. Montenegro and A. Fin, Chem. Soc. Rev., 2011, 40, 2453–2474 .
18. X. Hou, W. Guo and L. Jiang, Chem. Soc. Rev., 2011, 40, 2385–2401 .
19. E. Mattia and S. Otto, Nat. Nanotechnol., 2015, 10, 111–119.
20. J. Weber, Nat. Chem. Biol., 2010, 6, 794–795.

21. C. A. Schalley, K. Beizai and F. Vögtle, Acc. Chem. Res., 2001, 34, 465–476.
22. Y. Sowa and R. M. Berry, Q. Rev. Biophys., 2008, 41, 103–132.
23. V. Serreli, C.-F. Lee, E. R. Kay and D. A. Leigh, Nature, 2007, 445, 523–527.
24. A. M. Brouwer, C. Frochot, F. G. Gatti, D. A. Leigh, L. c. Mottier, F. Paolucci, S. Roffia and G. W. H. Wurpel, Science, 2001, 291, 2124–2128.
25. C. Cheng, P. R. McGonigal, S. T. Schneebeli, H. Li, N. A. Vermeulen, C. Ke and J. F. Stoddart, Nat. Nanotechnol., 2015, 10, 547–553.
26. D. A. Leigh, J. K. Y. Wong, F. Dehez and F. Zerbetto, Nature, 2003, 424, 174–179.
27. R. A. Muscat, J. Bath and A. J. Turberfield, Nano Lett., 2011, 11, 982–987.
28. Y. He and D. R. Liu, Nat. Nanotechnol., 2010, 5, 778–782.
29. M. R. Ghadiri, J. R. Granja and L. K. Buehler, Nature, 1994, 369, 301–304.
30. M. Langecker, V. Arnaut, T. G. Martin, J. List, S. Renner, M. Mayer, H. Dietz and F. C. Simmel, Science, 2012, 338, 932–936.
31. S. Hernández-Ainsa, K. Misiunas, V. V. Thacker, E. A. Hemmig and U. F. Keyser, Nano Lett., 2014, 14, 1270–1274 .
32. T. Jovanovic-Talisman, J. Tetenbaum-Novatt, A. S. McKenney, A. Zilman, R. Peters, M. P. Rout and B. T. Chait, Nature, 2009,457, 1023–1027.
33. C. J. Moreau, J. P. Dupuis, J. Revilloud, K. Arumugam and M. Vivaudou, Nat. Nanotechnol., 2008, 3, 620–625.
34. A. Koçer, M. Walko, W. Meijberg and B. L. Feringa, Science, 2005, 309, 755–758 .
35. S. Bhosale, A. L. Sisson, P. Talukdar, A. Fürstenberg, N. Banerji, E. Vauthey, G. Bollot, J. Mareda, C. Röger, F. Würthner, N. Sakai and S. Matile, Science, 2006, 313, 84–86.
36. G. Steinberg-Yfrach, J.-L. Rigaud, E. N. Durantini, A. L. Moore, D. Gust and T. A. Moore, Nature, 1998, 392, 479–482.
37. L. Ma and S. L. Cockroft, ChemBioChem, 2010, 11, 25–34.

38. D. Branton, D. W. Deamer, A. Marziali, H. Bayley, S. A. Benner, T. Butler, M. Di Ventra, S. Garaj, A. Hibbs, X. Huang, S. B. Jovanovich, P. S. Krstic, S. Lindsay, X. S. Ling, C. H. Mastrangelo, A. Meller, J. S. Oliver, Y. V. Pershin, J. M. Ramsey, R. Riehn, G. V. Soni, V. Tabard-Cossa, M. Wanunu, M. Wiggin and J. A. Schloss, Nat. Biotechnol., 2008, 26, 1146–1153.
39. D. Wendell, P. Jing, J. Geng, V. Subramaniam, T. J. Lee, C. Montemagno and P. Guo, Nat. Nanotechnol., 2009, 4, 765–772.
40. G. S. Pulcu, E. Mikhailova, L.-S. Choi and H. Bayley, Nat. Nanotechnol., 2014, 10, 76–83 .
41. L. Franceschini, M. Soskine, A. Biesemans and G. Maglia, Nat. Commun., 2013, 4, 2415 Search .
42. F. Olasagasti, K. R. Lieberman, S. Benner, G. M. Cherf, J. M. Dahl, D. W. Deamer and M. Akeson, Nat. Nanotechnol., 2010, 5, 798–806.
43. J. Nivala, D. B. Marks and M. Akeson, Nat. Biotechnol., 2013, 31, 247–250.
44. C.-W. Ho, V. van Meervelt, K.-C. Tsai, P.-J. de Temmerman, J. Mast and G. Maglia, Sci. Adv., 2015, 1, e1500905 Search .
45. M. A. Watson and S. L. Cockroft, Angew. Chem., Int. Ed., 2016, 55, 1345–1349.
46. S. L. Cockroft, J. Chu, M. Amorin and M. R. Ghadiri, J. Am. Chem. Soc., 2008, 130, 818–820.
47. J. Chu, M. González-López, S. L. Cockroft, M. Amorin and M. R. Ghadiri, Angew. Chem., Int. Ed., 2010, 49, 10106–10109.
48. M. A. Holden, D. Needham and H. Bayley, J. Am. Chem. Soc., 2007, 129, 8650–8655.
49. G. Villar, A. J. Heron and H. Bayley, Nat. Nanotechnol., 2011, 6, 803–808.
50. G. Villar, A. D. Graham and H. Bayley, Science, 2013, 340, 48–52.

CHAPTER

10

MOLECULAR MACHINES IN THE SYNAPSE: OVERLAPPING PROTEIN SETS CONTROL DISTINCT STEPS IN NEUROSECRETION

L. Niels Cornelisse, Evgeni Tsivtsivadze, Marieke Meijer, Tjeerd M. H. Dijkstra, Tom Heskes, Matthijs Verhage

[1]Functional Genomics, Center for Neurogenomics and Cognitive Research (CNCR), Department of Clinical Genetics, VUmc, Neuroscience Campus Amsterdam (NCA), VU University (VUA) and VU University Medical Center (VUmc), Amsterdam, The Netherlands

[2]Machine Learning Group, Intelligent Systems, Institute for Computing and Information Sciences, Radboud University, Nijmegen, The Netherlands,

[3]Microbiology and Systems Biology Group, TNO, Zeist, The Netherlands

[4]Signal Processing Systems, Department of Electrical Engineering, Technical University Eindhoven, Eindhoven, The Netherlands

ABSTRACT

Activity regulated neurotransmission shapes the computational properties of a neuron and involves the concerted action of many proteins. Classical, intuitive working models often assign specific proteins to specific steps in such complex cellular processes, whereas modern systems theories emphasize more integrated functions of proteins. To test how often synaptic proteins participate in multiple steps in neurotransmission we present a novel probabilistic method to analyze complex functional data from genetic perturbation studies on neuronal secretion. Our method uses a mixture of probabilistic principal component analyzers to cluster genetic perturbations on two distinct steps in synaptic secretion, vesicle priming and fusion, and accounts for the poor standardization between different studies. Clustering data from 121 perturbations revealed that different perturbations of a given protein are often assigned to different steps in the release process. Furthermore, vesicle priming and fusion are inversely correlated for most of those perturbations where a specific protein domain was mutated to create a gain-of-function variant. Finally, two different modes of vesicle release, spontaneous and action potential evoked release, were affected similarly by most perturbations. This data suggests that the presynaptic protein network has evolved as a highly integrated supramolecular machine, which is responsible for both spontaneous and activity induced release, with a group of core proteins using different domains to act on multiple steps in the release process.

AUTHOR SUMMARY

Synapses, which are small structures where information is transmitted between neurons, are the functional units of computation in the brain implicated in information processing, learning and memory. In the last few years several genes that are expressed in synapses have been linked to brain disorders such as autism, depression and schizophrenia. However, good understanding of the molecular basis of these diseases requires more insight in the functional organization of the protein network. In the classic view proteins are involved in a specific step of a cellular process, whereas modern system theories suggest a more integrated and complex network where proteins can be involved in several steps. We have developed a method that allows comparison of high-end but low-throughput functional data from different genetic perturbation studies despite the poor standardization between studies. Our analysis shows that synaptic proteins are not restricted to a single function but that they are part of a highly integrated supramolecular machine, with overlapping protein sets being involved in distinct steps in neurotransmitter release. Further characterization of these complex protein network relations will be important for new drug design strategies.

INTRODUCTION

Synapses are complex biological structures, which evolved into highly specialized computational units that play an important role in learning, memory formation and information processing in the brain [1], [2], [3]. A key process in the synapse is the remarkably fast and precisely timed secretion of neurotransmitters from small synaptic vesicles in the nerve terminal upon arrival of an action potential (AP) [4]. In general, such complex cellular processes exist by virtue of the concerted action of many proteins, which are often found to sequester into multi-protein complexes, also referred to as molecular machines. It is common practice to assign specific proteins or sub-complexes to specific steps in a cascade of events. However, this intuitive idea has not been tested sufficiently. Gene knockout technology in mice and gene overexpression by viral constructs have been used in many studies for detailed functional analysis of individual mammalian proteins. However, systematic comparative analysis based on these studies is hampered by the fact that functional data from different studies are poorly standardized and therefore do not allow direct comparison of the observed perturbation effects as in less complex model organisms [5], [6], [7], [8], [9].

Over the past 15 years, the proteins that make up the synaptic release machinery have been largely identified [4], [10], [11]. Many of these were functionally studied in genetic perturbation studies using high-end but low-throughput assays. Here we present a novel probabilistic method to compare functional data from different studies and cluster perturbations of presynaptic genes according to their effect on different synaptic release parameters. The method addresses some fundamental problems in meta-analysis of poorly standardized functional data, such as large variation between different studies, incomplete reporting of strongly co-varying variables, and different effect sizes of perturbations, and was designed to be generally applicable to similar data sets in other experimental settings. We show that synaptic release is governed by a highly integrated molecular machine with a set of core proteins controlling both the steps of priming and fusion, implying that functionally distinct steps are not always regulated by distinct sets of proteins.

RESULTS/DISCUSSION

MPPCA allows clustering of functional data from different perturbation studies

Vesicle priming and fusion, two important steps in the life cycle of synaptic vesicles, are the main determinants of synaptic strength. According to the quantal hypothesis of neurotransmitter release synaptic strength is given by

$$EPSC = RRP\, P_v\, q, \qquad (1)$$

with EPSC the excitatory post-synaptic current, RRP the size of the readily releasable pool, which contains all the vesicles that are primed to be released, P_v the vesicular release probability, which is the probability that a vesicle will fuse upon arrival of an action potential at the terminal, and q the postsynaptic quantal size, which is defined as the current response to the release of a single vesicle [12]. It is shown in several studies that manipulation of different presynaptic genes affect RRP size and P_v. independently [13], [14], [15], [16]. To systematically compare the effect of presynaptic genes on vesicle priming and fusion we collected functional data from genetic perturbation studies in hippocampal island cultures (autapses) to reduce variation between datasets [17], [18] (Figure 1A). All data was obtained from genes with a known presynaptic gene product, transcription factors not included. We selected synaptic variables that were frequently enough reported among different studies to allow systematic comparison and retrieved data for 378 experiments (perturbations and control experiments) from 56 published studies. Included variables were the number of primed vesicles (RRP), the fusion probability (P_v), the amount of evoked release (A), and the amount of spontaneous release (F) (Figure 1B–C). Since the variables RRP and P_v are expected to co-vary with A according to Eq. 1 most papers only report a subset of these. Wild-type data from different studies showed a large variation, with only A and P_v, but not A and RRP, showing the linear correlation predicted by the release model. The variance could not be attributed to specific experimental conditions in the different studies, like the external calcium concentration or number of days in vitro (DIV). Altogether, this shows that large variation prevents direct comparison between functional data of different studies. To overcome this problem and in addition allow comparison between different data formats (e.g. normalized vs. absolute values, different units) we normalized each perturbation to a control group from the same experiment and log2-transformed the data, yielding 121 normalized perturbations of 27 genes from 39 studies with two or more variables reported (Figure 1D).

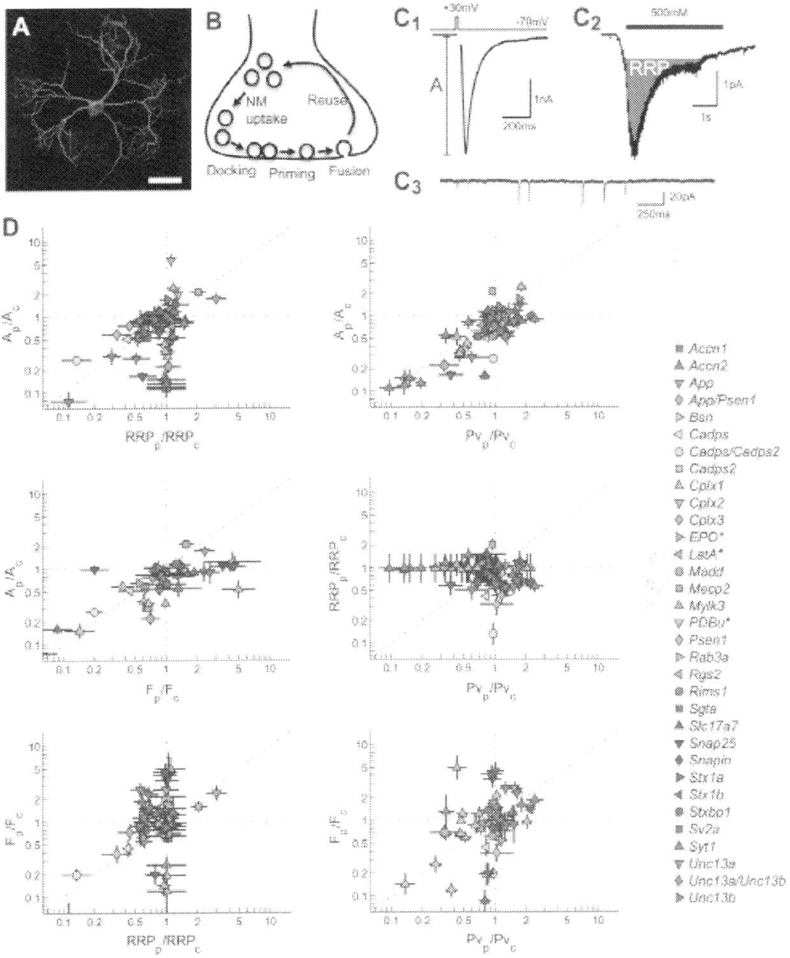

Figure 1. Functional data from genetic perturbation studies in autapses. A) Confocal image of an autaptic neuron, scale bar is 50 μm. B) Schematic representation of a synapse. In each synapse only a limited number of vesicles, referred to as the readily releasable pool (RRP), is in a primed

state that allows direct fusion during calcium triggering. The probability that a primed vesicle fuses during a calcium trigger is given by the vesicular release probability (P_v). C) Example traces of synaptic release variables. C1) Evoked EPSC (lower trace), upper trace shows somatic stimulation to trigger an action potential. C2) Postsynaptic response to a 500 mOsmol hypertonic stimulation for RRP estimation. C3) Spontaneous release events. D) Scatter plots of normalized and log-2 transformed synaptic variables in 2D projections of the 4D variable space for 121 perturbations (13 perturbations with >10-fold effect are not shown). *Non-genetic perturbations using a biochemical compound.

We designed a novel method to cluster perturbations with similar qualitative effects on vesicle release variables A, RRP, and P_v, irrespective of different effect sizes for different perturbation types (e.g. heterozygous vs homozygous knock-out, mild vs strong overexpression), different experimental conditions, or missing variables. To this end we used a mixture of probabilistic principal component analyzers (MPPCA), which clusters the orientations of the normalized Log-2 transformed data in 3D variable space and assigns probabilities to the perturbations to belong to each of the detected clusters (for schematic overview see Figure 2, for a detailed description see Suppl. Info). Running the cluster algorithm revealed three clusters based on the structure of the average co-occurrence matrix, with 87, 21 and 13 perturbations assigned with the highest probability to cluster 1, 2 and 3, respectively (Figure 3).

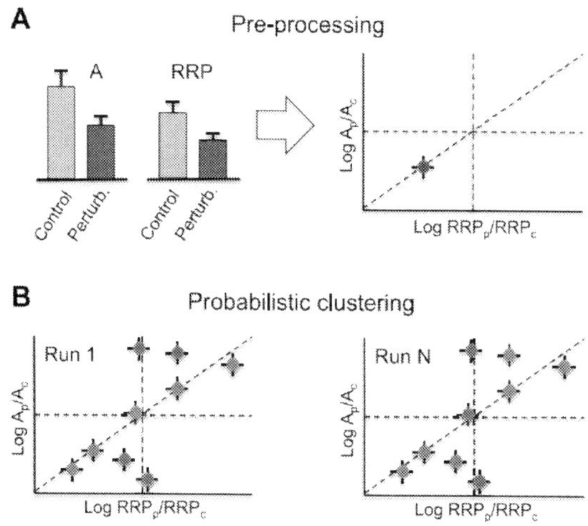

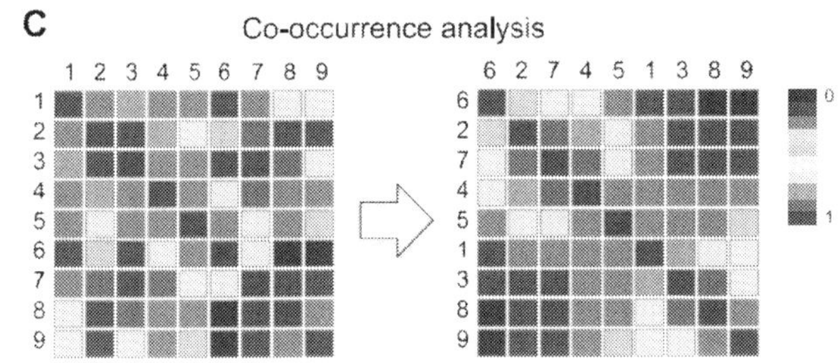

Figure 2. Schematic representation of data processing steps. A) Perturbations are extracted from publications then log-transformed and standardized relative to a control condition. B) Two exemplary runs of the probabilistic clustering algorithm for K=2 clusters are shown in the left and right panels. Perturbations are colored red or blue depending on the cluster to which they are assigned with the highest probability. Note that some of the perturbations change color (cluster) for different runs of the probabilistic clustering algorithm. C) Co-occurrence analysis. By observing the co-occurrence of pairs of perturbations in the same cluster, we built a co-occurrence matrix with color indicating how often each pair co-occurs in the same cluster. In the right panel we have reordered the rows and columns to highlight the two dominant clusters in the data.

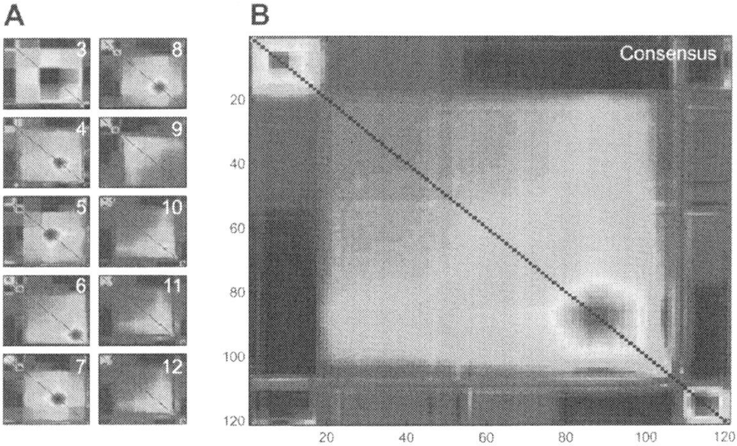

Figure 3. Co-occurrence matrices. A) Average over ten restarts of the co-occurrence matrices for clustering with a fixed number of clusters. Number of clusters K is indicated in each plot. Matrices were ordered for visualization using optimal leaf ordering. B) Consensus co-occurrence matrix obtained by averaging the individual co-occurrence matrices presented in A. Red color denotes high co-occurrence, and blue low co-occurrence.

Perturbation clusters reflect distinct synaptic function

To relate each cluster of perturbations with the priming and/or fusion step in synaptic release we analyzed the orientation of the cluster unit vectors in 3D variable space. This revealed that perturbations in cluster 1 were affecting P_v, whereas perturbations in cluster 2 were involved in RRP. Furthermore, an interesting negative correlation between P_v and RRP was found for perturbations in cluster 3. These findings were confirmed by analysis of the orientation of the individual perturbations in 2D. In Figure 4A all individual perturbations are plotted per cluster, with marker sizes scaled with the probability to belong to that cluster. We calculated for all variable subspaces the best fits out of four possible regression curves, which gave a significant 51% reduction of the total sum of weighted orthogonal errors compared to the non-clustered case (Figure 4A). Hence, perturbation data is not randomly scattered in variable space but clustered in at least three classes with specific orientations, indicative of different synaptic functions. As a validation of our method we included in our analysis specific perturbations of *Mecp2*, *Psen* and *App* that were reported to affect RRP by changing the number of synapses, whereas other perturbations in the database did not affect synapse number. Although these perturbations had either positive or negative effects on synapse number, all were assigned correctly to the RRP cluster.

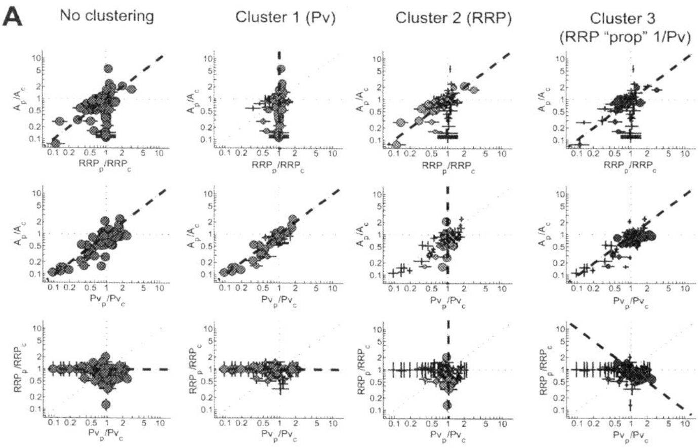

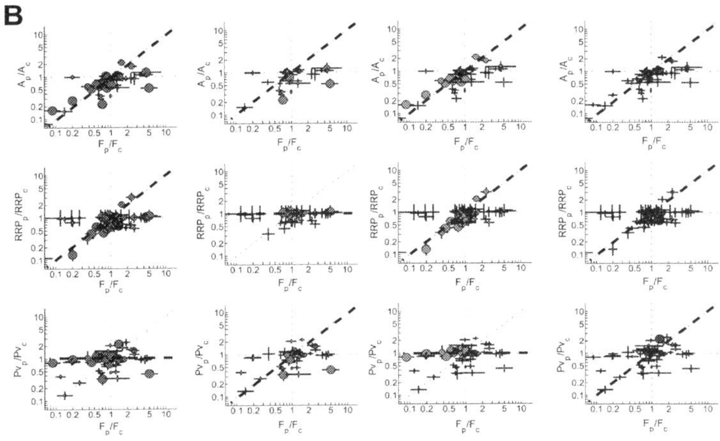

Figure 4. Perturbation clusters reflect distinct synaptic function. A) Functional data on synaptic vesicle release from all perturbations is plotted for the three clusters and the non-clustered case. RRP "prop" $1/P_v$ denotes the cluster with the RRP being proportionally related to the inverse of P_v. Marker sizes indicate the cluster probability. Dotted lines indicate the best fit of the data for four possible regression curves of 0,45, 90 and 135 degrees. B) Idem as in A but now for the correlation of evoked release variables A, RRP, P_v with the frequency F of spontaneous release. To avoid bias towards the A–F relation of genes with many perturbations in the database, cluster probabilities are corrected for the number of perturbations per gene plotted in the graph. Final weighted probabilities are indicated by the marker sizes.

Perturbations of the same gene associate with multiple release steps

Next we addressed the question to what extent presynaptic genes are involved in multiple steps in release. We analyzed how different perturbations of the same gene were distributed over the three clusters for 9 genes that had one or more perturbations reported in the database, including those from isoforms (Figure 5). Three genes had all their perturbations assigned to one cluster (*Cplx* and *Snap25* to the P_v cluster, *Mecp2* to the RRP cluster). However, the other 6 genes (*Cadps*, *Psen*, *Snapin*, *Stxbp*, *Syt* and *Unc13*) had their perturbations distributed over different clusters. This was most prominent for *Snapin*, *Syt1*, and *Unc13*, with 4 or more perturbations in the database, of which more than one third was assigned to a different cluster than their preferred cluster. Close inspection showed that most of the individual perturbations for these genes were highly significant in the original papers and tended to cluster with high specificity (P>0.8). Additional entropy analysis showed that indeed their distribution over different clusters was not due to weak classification (i.e. aspecific clustering). This confirms the idea that genes are involved in different steps in release, with individual perturbations having distinct functional effects. Homeostatic mechanisms, in which disturbance of synaptic transmission by genetic perturbations lead to the expression of other synaptic gene products to normalize the release process, could be an alternative explanation for these findings. However, studies in our database that measured protein levels reported these to be unchanged or mildly reduced (interpreted as decreased stability of direct binding partners) [13], [15], [16], [19], [20]. Most likely synaptic protein complexes operate at locations too remote from the cell nucleus to enable such direct feedback loops.

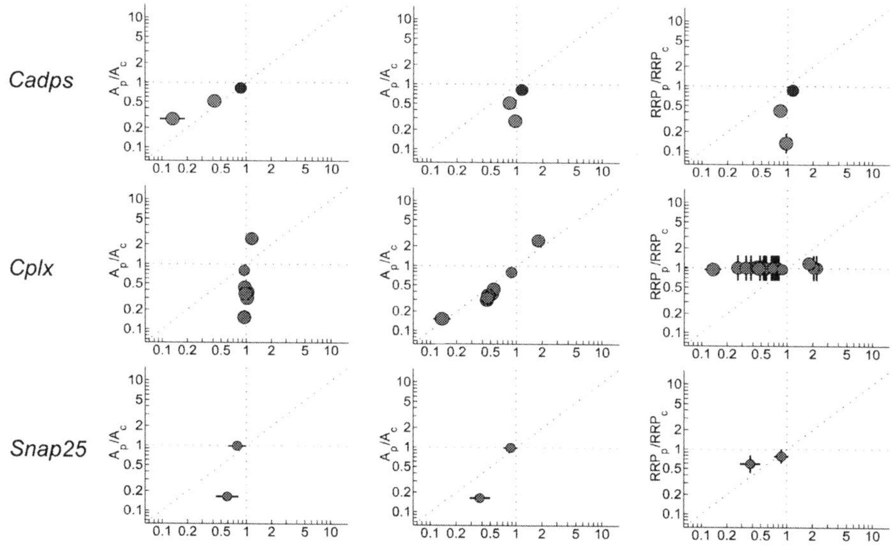

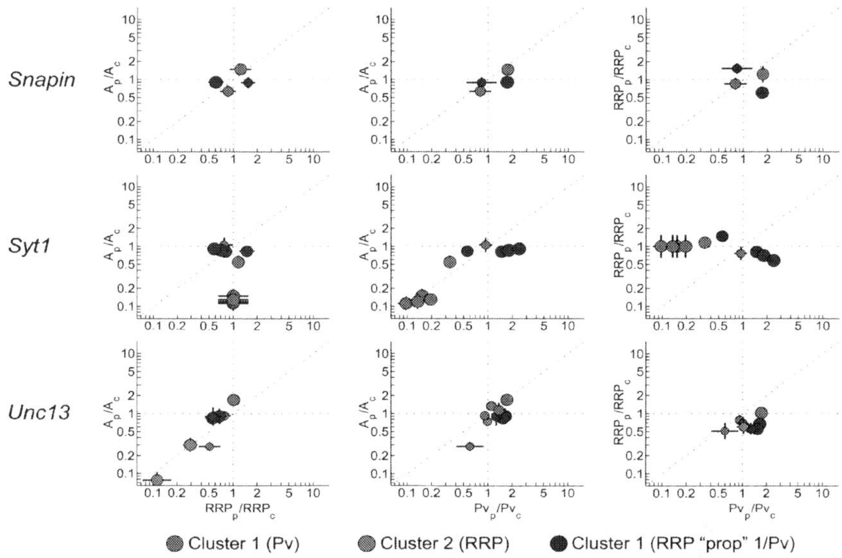

Figure 5. Perturbations from the same gene are assigned to different clusters. Perturbations are plotted per gene with marker color indicating to which cluster the perturbation is assigned and marker size indicating its cluster probability.

Studies in other model systems on individual genes seem to support this finding with reported effects of *Syt1* on RRP, P_v, positional priming, and docking, *Snapin* on RRP and synchronous fusion, *Stxbp1* on docking, priming and PDBU dependent potentiation, and *Unc13a* on priming and PDBU dependent potentiation [21], [22], [23], [24], [25], [26]. A picture emerges where some genes have a central role in the release machinery and are involved in multiple steps, whereas others, such as *Cplx*, act as modifiers of a specific step in the release cycle. These modifying genes have possibly evolved to fine-tune the specific behavior of different types of synapses. In this context it is interesting to see that in the case of *Cplx* different isoforms across species can have an opposite effect on release [27], [28].

Gain-of-function perturbations affect RRP inversely with P_v

We noted that a relatively large number of the perturbations in the RRP "prop" $1/P_v$ cluster was designed to enhance neurotransmission by introducing gain-of-function (GOF) mutants (for instance enhanced calcium sensitivity mutants of *Syt1* and PDBU activation mimicking mutants of *Unc13a*). To test if GOF perturbations were significantly more assigned to the RRP "prop" $1/P_v$ cluster we used a weighted contingency analysis to compare how GOF perturbations and loss-of-function (LOF) perturbations were distributed over the three clusters. Indeed, GOF perturbations did significantly cluster more in the RRP "prop" $1/P_v$ cluster, with an average probability of 0.86, which was corrected for a possible bias from multiple GOF perturbations per gene. In contrast, the LOF perturbations had no significant preference for any cluster. The inverse relation between the change in P_v and RRP for GOF mutants is in line with the idea that these proteins are involved both in priming and fusion. Mutating a protein domain to enhance one function might change the protein structure such that another function becomes impaired. Similarly, LOF mutations targeted to one function might be not specific as well. This would explain why some LOF perturbations with large effect size do not have a clear preference for one cluster, reflecting a mixed P_v and RRP phenotype. This suggests that the presynaptic protein network has evolved as a highly integrated molecular machine, with a group of core proteins that is involved in multiple steps of release. Indeed, biochemical evidence shows a strongly connected network with protein interactions between many presynaptic core proteins [29], [30]. Furthermore, it was shown that PKC-dependent and -independent pathways in the synapse are interdependent and need to be activated simultaneously to induce presynaptic plasticity [31].

Same set of presynaptic genes controls spontaneous and evoked release

Finally, we investigated whether genes that control AP induced release are also involved in spontaneous release which is relevant in a long-standing debate on whether or not evoked and spontaneous release share the same vesicle pools and synaptic pathways [32], [33], [34], [35], [36], [37]. In all 3 clusters and the non-clustered situation, the 45° regression curve yields the best fit of the A–F correlation indicating that the majority of the perturbations affecting AP induced release also affect spontaneous release proportionally (Figure 4B). This is confirmed by the fact that in the RRP cluster, F correlates well with RRP, but not with P_v, implying that spontaneous release and evoked release are to a large extent driven by the same RRP-modifying genes, whereas for the P_v cluster the opposite holds. The fact that the same set of presynaptic genes controls spontaneous and evoked release favors a model where both modes of release are governed by the same molecular release machinery instead of being recruited from different vesicle pools. Until now the only exception seems to be *Doc2* that most likely competes with *Syt1* for snare binding but exclusively regulates spontaneous release in hippocampal neurons [38]. It should be noted however, that by some unknown mechanism some genetic perturbations affect spontaneous release differentially in autaptic cultures compared to other model systems, especially for *Syt1* and *Cplx*, which seem to act as a clamp on spontaneous release in these systems [16], [22], [39], [40], [41].

Molecular machines in the synapse

Our novel approach to cluster non-standardized functional data from different perturbation studies allowed us to study the role of presynaptic genes in vesicle release at the network level, which could not be addressed in the original papers. We show that individual presynaptic proteins are not restricted to a single step in the release process but have evolved to contribute to multiple steps. Currently there is too little physical interaction data available, for instance from immunoprecipitation or yeast-two-hybrid experiments, to get a complete picture of the presynaptic protein interactome. This makes it difficult to interpret our results in terms of molecular interactions in the network, although we do know that protein complexes are formed and play an important role in the vesicle release machinery [10], [29], [30], [42]. We would like to discuss three network scenarios for the presynapse, and to what extend they can explain our findings. The first option is that the proteins in presynaptic terminal form a highly connected network, with a high number of cross-reactions, and many proteins involved in many steps of the vesicle release cycle. This seems unlikely since most protein networks have a scale-free organization, which implies that clusters of locally connected proteins are connected with each other through a few hub nodes in the network [43]. A second possibility is that there is a single macro-molecular complex, with many presynaptic proteins directly bound to each other, which is involved in the steps of docking, priming and fusion. However, there is no biochemical evidence for such a large complex in the presynapse and it would be difficult to reconcile with the observation that many genetic perturbations affect exclusively one step only. Finally, a third possibility is that the presynaptic release machinery comprises multiple smaller and dynamic protein complexes, which adapt their molecular configuration and function as the vesicles progress in the vesicle cycle. In this view, the stereotypical presynaptic protein complex does not exist. Instead, a whole range of complexes can be formed around a core set of proteins, with their specific function determined by different combinations of proteins or isoforms. For instance, different C2 domain proteins could compete for binding to the SNARE complex and make the vesicles more prone for spontaneous or action potential evoked release [44]. Likewise, a particular protein can be part of different complex configurations associated with different steps. For example, binding of Syt1 to Stx1/SNAP25 complexes is important for vesicle docking whereas Syt1 also functions as a calcium sensor for synchronous release when bound to the fully assembled SNARE complex [25], [45]. The fact that some Syt1 gain-of-function mutants both affect priming and fusion, whereas the Syt1 KO only shows a strong reduction of P_v, could be explained by partial redundancy of the system: *i.e.* other proteins can take over the role of Syt1 in docking/priming, but not in vesicular release probability, while the Syt1 mutants have a dominant negative effect.

Our method for probabilistic clustering of genetic perturbation data can in principle be applied to any high dimensional functional dataset obtained by medium or low-throughput experiments, including imaging or behavioral data. The success of this method will be enhanced through the availability of standardized and well-annotated functional data in publicly accessible databases, which are expected to become more and more available in the coming years.

METHODS

See the supplementary material for a detailed description of the materials and methods.

ACKNOWLEDGMENTS

We thank R. Wijntjes and K. Iffland for their help on data collection.

AUTHOR CONTRIBUTIONS

Conceived and designed the experiments: LNC ET TMHD TH. Performed the experiments: ET TMHD. Analyzed the data: LNC ET TMHD. Contributed reagents/materials/analysis tools: ET. Wrote the paper: LNC TMHD MV. Data retrieval from literature: LNC MM.

REFERENCES

1. Abbott LF, Regehr WG (2004) Synaptic computation. Nature 431: 796–803.LF AbbottWG Regehr2004Synaptic computation. Nature431796803
2. Chua JJ, Kindler S, Boyken J, Jahn R (2010) The architecture of an excitatory synapse. J Cell Sci 123: 819–823.JJ ChuaS. KindlerJ. BoykenR. Jahn2010The architecture of an excitatory synapse.J Cell Sci123819823
3. Emes RD, Pocklington AJ, Anderson CN, Bayes A, Collins MO, et al. (2008) Evolutionary expansion and anatomical specialization of synapse proteome complexity. Nat Neurosci 11: 799–806.RD EmesAJ PocklingtonCN AndersonA. BayesMO Collins2008Evolutionary expansion and anatomical specialization of synapse proteome complexity.Nat Neurosci11799806
4. Sudhof TC (2004) The synaptic vesicle cycle. Annu Rev Neurosci 27: 509–547.TC Sudhof2004The synaptic vesicle cycle.Annu Rev Neurosci27509547
5. Giaever G, Chu AM, Ni L, Connelly C, Riles L, et al. (2002) Functional profiling of the Saccharomyces cerevisiae genome. Nature 418: 387–391.G. GiaeverAM ChuL. NiC. ConnellyL. Riles2002Functional profiling of the Saccharomyces cerevisiae genome.Nature418387391
6. Ohya Y, Sese J, Yukawa M, Sano F, Nakatani Y, et al. (2005) High-dimensional and large-scale phenotyping of yeast mutants. Proc Natl Acad Sci U S A 102: 19015–19020.Y. OhyaJ. SeseM. YukawaF. SanoY. Nakatani2005High-dimensional and large-scale phenotyping of yeast mutants.Proc Natl Acad Sci U S A1021901519020
7. Fraser AG, Kamath RS, Zipperlen P, Martinez-Campos M, Sohrmann M, et al. (2000) Functional genomic analysis of C.

elegans chromosome I by systematic RNA interference. Nature 408: 325–330.AG FraserRS KamathP. ZipperlenM. Martinez-CamposM. Sohrmann2000Functional genomic analysis of C. elegans chromosome I by systematic RNA interference.Nature408325330

8. Lee SS, Lee RY, Fraser AG, Kamath RS, Ahringer J, et al. (2003) A systematic RNAi screen identifies a critical role for mitochondria in C. elegans longevity. Nat Genet 33: 40–48.SS LeeRY LeeAG FraserRS KamathJ. Ahringer2003A systematic RNAi screen identifies a critical role for mitochondria in C. elegans longevity. Nat Genet334048
9. Boutros M, Kiger AA, Armknecht S, Kerr K, Hild M, et al. (2004) Genome-wide RNAi analysis of growth and viability in Drosophila cells. Science 303: 832–835.M. BoutrosAA KigerS. ArmknechtK. KerrM. Hild2004Genome-wide RNAi analysis of growth and viability in Drosophila cells.Science303832835
10. Rizo J, Rosenmund C (2008) Synaptic vesicle fusion. Nat Struct Mol Biol 15: 665–674.J. RizoC. Rosenmund2008Synaptic vesicle fusion.Nat Struct Mol Biol15665674
11. Takamori S, Holt M, Stenius K, Lemke EA, Gronborg M, et al. (2006) Molecular anatomy of a trafficking organelle. Cell 127: 831–846.S. TakamoriM. HoltK. SteniusEA LemkeM. Gronborg2006Molecular anatomy of a trafficking organelle.Cell127831846
12. Schneggenburger R, Sakaba T, Neher E (2002) Vesicle pools and short-term synaptic depression: lessons from a large synapse. Trends Neurosci 25: 206–212.R. SchneggenburgerT. SakabaE. Neher2002Vesicle pools and short-term synaptic depression: lessons from a large synapse.Trends Neurosci25206212
13. Augustin I, Rosenmund C, Sudhof TC, Brose N (1999) Munc13-1 is essential for fusion competence of glutamatergic synaptic vesicles. Nature 400: 457–461.I. AugustinC. RosenmundTC SudhofN. Brose1999Munc13-1 is essential for fusion competence of glutamatergic synaptic vesicles.Nature400457461
14. Fernandez-Chacon R, Konigstorfer A, Gerber SH, Garcia J, Matos MF, et al. (2001) Synaptotagmin I functions as a calcium regulator of release probability. Nature 410: 41–49.R. Fernandez-ChaconA.

KonigstorferSH GerberJ. GarciaMF Matos2001Synaptotagmin I functions as a calcium regulator of release probability.Nature4104149

15. Jockusch WJ, Speidel D, Sigler A, Sorensen JB, Varoqueaux F, et al. (2007) CAPS-1 and CAPS-2 are essential synaptic vesicle priming proteins. Cell 131: 796–808.WJ JockuschD. SpeidelA. SiglerJB SorensenF. Varoqueaux2007CAPS-1 and CAPS-2 are essential synaptic vesicle priming proteins.Cell131796808
16. Reim K, Mansour M, Varoqueaux F, McMahon HT, Sudhof TC, et al. (2001) Complexins regulate a late step in Ca2+-dependent neurotransmitter release. Cell 104: 71–81.K. ReimM. MansourF. VaroqueauxHT McMahonTC Sudhof2001Complexins regulate a late step in Ca2+-dependent neurotransmitter release.Cell1047181
17. Bekkers JM, Stevens CF (1991) Excitatory and inhibitory autaptic currents in isolated hippocampal neurons maintained in cell culture. Proc Natl Acad Sci U S A 88: 7834–7838.JM BekkersCF Stevens1991Excitatory and inhibitory autaptic currents in isolated hippocampal neurons maintained in cell culture.Proc Natl Acad Sci U S A8878347838
18. Ikeda K, Bekkers JM (2006) Autapses. Curr Biol 16: R308.K. IkedaJM Bekkers2006Autapses.Curr Biol16R308
19. Toonen RF, Wierda K, Sons MS, de Wit H, Cornelisse LN, et al. (2006) Munc18-1 expression levels control synapse recovery by regulating readily releasable pool size. Proc Natl Acad Sci U S A 103: 18332–18337.RF ToonenK. WierdaMS SonsH. de WitLN Cornelisse2006Munc18-1 expression levels control synapse recovery by regulating readily releasable pool size.Proc Natl Acad Sci U S A1031833218337
20. Geppert M, Goda Y, Hammer RE, Li C, Rosahl TW, et al. (1994) Synaptotagmin I: a major Ca2+ sensor for transmitter release at a central synapse. Cell 79: 717–727.M. GeppertY. GodaRE HammerC. LiTW Rosahl1994Synaptotagmin I: a major Ca2+ sensor for transmitter release at a central synapse.Cell79717727
21. Dulubova I, Lou X, Lu J, Huryeva I, Alam A, et al. (2005) A Munc13/RIM/Rab3 tripartite complex: from priming to plasticity? EMBO J 24: 2839–2850.I. DulubovaX. LouJ. LuI. HuryevaA. Alam2005A Munc13/RIM/Rab3 tripartite complex: from priming

to plasticity?EMBO J2428392850

22. Liu H, Dean C, Arthur CP, Dong M, Chapman ER (2009) Autapses and networks of hippocampal neurons exhibit distinct synaptic transmission phenotypes in the absence of synaptotagmin I. J Neurosci 29: 7395–7403.H. LiuC. DeanCP ArthurM. DongER Chapman2009Autapses and networks of hippocampal neurons exhibit distinct synaptic transmission phenotypes in the absence of synaptotagmin I.J Neurosci2973957403
23. Lou X, Korogod N, Brose N, Schneggenburger R (2008) Phorbol esters modulate spontaneous and Ca2+-evoked transmitter release via acting on both Munc13 and protein kinase C. J Neurosci 28: 8257–8267.X. LouN. KorogodN. BroseR. Schneggenburger2008Phorbol esters modulate spontaneous and Ca2+-evoked transmitter release via acting on both Munc13 and protein kinase C.J Neurosci2882578267
24. Young SM Jr, Neher E (2009) Synaptotagmin has an essential function in synaptic vesicle positioning for synchronous release in addition to its role as a calcium sensor. Neuron 63: 482–496.SM Young JrE. Neher2009Synaptotagmin has an essential function in synaptic vesicle positioning for synchronous release in addition to its role as a calcium sensor.Neuron63482496
25. de Wit H, Walter AM, Milosevic I, Gulyas-Kovacs A, Riedel D, et al. (2009) Synaptotagmin-1 docks secretory vesicles to syntaxin-1/SNAP-25 acceptor complexes. Cell 138: 935–946.H. de WitAM WalterI. MilosevicA. Gulyas-KovacsD. Riedel2009Synaptotagmin-1 docks secretory vesicles to syntaxin-1/SNAP-25 acceptor complexes.Cell138935946
26. Gulyas-Kovacs A, de Wit H, Milosevic I, Kochubey O, Toonen R, et al. (2007) Munc18-1: sequential interactions with the fusion machinery stimulate vesicle docking and priming. J Neurosci 27: 8676–8686.A. Gulyas-KovacsH. de WitI. MilosevicO. KochubeyR. Toonen2007Munc18-1: sequential interactions with the fusion machinery stimulate vesicle docking and priming.J Neurosci2786768686
27. Cho RW, Song Y, Littleton JT (2010) Comparative analysis of Drosophila and mammalian complexins as fusion clamps and facilitators of neurotransmitter release. Mol Cell Neurosci 45:

389–397.RW ChoY. SongJT Littleton2010Comparative analysis of Drosophila and mammalian complexins as fusion clamps and facilitators of neurotransmitter release.Mol Cell Neurosci45389397

28. Xue M, Reim K, Chen X, Chao HT, Deng H, et al. (2007) Distinct domains of complexin I differentially regulate neurotransmitter release. Nat Struct Mol Biol 14: 949–958.M. XueK. ReimX. ChenHT ChaoH. Deng2007Distinct domains of complexin I differentially regulate neurotransmitter release.Nat Struct Mol Biol14949958
29. Schoch S, Gundelfinger ED (2006) Molecular organization of the presynaptic active zone. Cell Tissue Res 326: 379–391.S. SchochED Gundelfinger2006Molecular organization of the presynaptic active zone.Cell Tissue Res326379391
30. Wang X, Hu B, Zieba A, Neumann NG, Kasper-Sonnenberg M, et al. (2009) A protein interaction node at the neurotransmitter release site: domains of Aczonin/Piccolo, Bassoon, CAST, and rim converge on the N-terminal domain of Munc13-1. J Neurosci 29: 12584–12596.X. WangB. HuA. ZiebaNG NeumannM. Kasper-Sonnenberg2009A protein interaction node at the neurotransmitter release site: domains of Aczonin/Piccolo, Bassoon, CAST, and rim converge on the N-terminal domain of Munc13-1.J Neurosci291258412596
31. Wierda KD, Toonen RF, de Wit H, Brussaard AB, Verhage M (2007) Interdependence of PKC-dependent and PKC-independent pathways for presynaptic plasticity. Neuron 54: 275–290.KD WierdaRF ToonenH. de WitAB BrussaardM. Verhage2007Interdependence of PKC-dependent and PKC-independent pathways for presynaptic plasticity.Neuron54275290
32. Fredj NB, Burrone J (2009) A resting pool of vesicles is responsible for spontaneous vesicle fusion at the synapse. Nat Neurosci 12: 751–758.NB FredjJ. Burrone2009A resting pool of vesicles is responsible for spontaneous vesicle fusion at the synapse.Nat Neurosci12751758
33. Groemer TW, Klingauf J (2007) Synaptic vesicles recycling spontaneously and during activity belong to the same vesicle pool. Nat Neurosci 10: 145–147.TW GroemerJ. Klingauf2007Synaptic vesicles recycling spontaneously and during activity belong to the same vesicle pool.Nat Neurosci10145147

34. Ikeda K, Bekkers JM (2009) Counting the number of releasable synaptic vesicles in a presynaptic terminal. Proc Natl Acad Sci U S A 106: 2945–2950.K. IkedaJM Bekkers2009Counting the number of releasable synaptic vesicles in a presynaptic terminal.Proc Natl Acad Sci U S A10629452950
35. Mathew SS, Pozzo-Miller L, Hablitz JJ (2008) Kainate modulates presynaptic GABA release from two vesicle pools. J Neurosci 28: 725–731.SS MathewL. Pozzo-MillerJJ Hablitz2008Kainate modulates presynaptic GABA release from two vesicle pools.J Neurosci28725731
36. Sara Y, Virmani T, Deak F, Liu X, Kavalali ET (2005) An isolated pool of vesicles recycles at rest and drives spontaneous neurotransmission. Neuron 45: 563–573.Y. SaraT. VirmaniF. DeakX. LiuET Kavalali2005An isolated pool of vesicles recycles at rest and drives spontaneous neurotransmission.Neuron45563573
37. Hua Z, Leal-Ortiz S, Foss SM, Waites CL, Garner CC, et al. (2011) v-SNARE composition distinguishes synaptic vesicle pools. Neuron 71: 474–487.Z. HuaS. Leal-OrtizSM FossCL WaitesCC Garner2011v-SNARE composition distinguishes synaptic vesicle pools.Neuron71474487
38. Groffen AJ, Martens S, Diez Arazola R, Cornelisse LN, Lozovaya N, et al. (2010) Doc2b is a high-affinity Ca2+ sensor for spontaneous neurotransmitter release. Science 327: 1614–1618.AJ GroffenS. MartensR. Diez ArazolaLN CornelisseN. Lozovaya2010Doc2b is a high-affinity Ca2+ sensor for spontaneous neurotransmitter release. Science32716141618
39. Xue M, Lin YQ, Pan H, Reim K, Deng H, et al. (2009) Tilting the balance between facilitatory and inhibitory functions of mammalian and Drosophila Complexins orchestrates synaptic vesicle exocytosis. Neuron 64: 367–380.M. XueYQ LinH. PanK. ReimH. Deng2009Tilting the balance between facilitatory and inhibitory functions of mammalian and Drosophila Complexins orchestrates synaptic vesicle exocytosis.Neuron64367380
40. Maximov A, Tang J, Yang X, Pang ZP, Sudhof TC (2009) Complexin controls the force transfer from SNARE complexes to membranes in fusion. Science 323: 516–521.A. MaximovJ. TangX. YangZP

PangTC Sudhof2009Complexin controls the force transfer from SNARE complexes to membranes in fusion.Science323516521

41. Huntwork S, Littleton JT (2007) A complexin fusion clamp regulates spontaneous neurotransmitter release and synaptic growth. Nat Neurosci 10: 1235–1237.S. HuntworkJT Littleton2007A complexin fusion clamp regulates spontaneous neurotransmitter release and synaptic growth.Nat Neurosci1012351237
42. Sorensen JB (2009) Conflicting views on the membrane fusion machinery and the fusion pore. Annu Rev Cell Dev Biol 25: 513–537.JB Sorensen2009Conflicting views on the membrane fusion machinery and the fusion pore.Annu Rev Cell Dev Biol25513537
43. Vazquez A (2010) Protein Interaction Networks. In: Alzate O, editor. Neuroproteomics. Boca Raton, FL.A. Vazquez2010Protein Interaction Networks.O. AlzateNeuroproteomicsBoca Raton, FL
44. Walter AM, Groffen AJ, Sorensen JB, Verhage M (2011) Multiple Ca2+ sensors in secretion: teammates, competitors or autocrats? Trends Neurosci 34: 487–497.AM WalterAJ GroffenJB SorensenM. Verhage2011Multiple Ca2+ sensors in secretion: teammates, competitors or autocrats?Trends Neurosci34487497
45. Kochubey O, Schneggenburger R (2011) Synaptotagmin increases the dynamic range of synapses by driving Ca(2)+-evoked release and by clamping a near-linear remaining Ca(2)+ sensor. Neuron 69: 736–748.O. KochubeyR. Schneggenburger2011Synaptotagmin increases the dynamic range of synapses by driving Ca(2)+-evoked release and by clamping a near-linear remaining Ca(2)+ sensor. Neuron69736748

CHAPTER

11

MECHANICAL TUNING OF MOLECULAR MACHINES FOR NUCLEOTIDE RECOGNITION AT THE AIR-WATER INTERFACE

Taizo Mori[1], Ken Okamoto[1], Hiroshi Endo[1], Keita Sakakibara[1,2], Jonathan P Hill[1, 2], Satoshi Shinoda[2, 3], Miki Matsukura[3], Hiroshi Tsukube[2, 3], Yasumasa Suzuki[4], Yasumasa Kanekiyo[4] and Katsuhiko Ariga[1, 2]

[1]World Premier International (WPI) Research Center for Materials Nanoarchitectonics (MANA), National Institute for Materials Science (NIMS)

[2]JST, CREST

[3]Osaka City University

[4]Kitami Institute of Technology

ABSTRACT

Molecular machines embedded in a Langmuir monolayer at the air-water interface can be operated by application of lateral pressure. As part of the challenge associated with versatile sensing of biologically important substances, we here demonstrate discrimination of nucleotides by applying a cholesterol-armed-triazacyclononane host molecule. This molecular machine can discriminate ribonucleotides based on a twofold to tenfold difference in binding constants under optimized conditions including accompanying ions in the subphase and lateral surface pressures of its Langmuir monolayer. The concept of mechanical tuning of the host structure for optimization of molecular recognition should become a novel methodology in bio-related nanotechnology as an alternative to traditional strategies based on increasingly complex and inconvenient molecular design strategies.

INTRODUCTION

Supramolecular structures constructed through bottom-up processes play crucial roles in nanoscience and nanotechnology [1, 2]. In particular, those structures can be applied in bio-related nanotechnologies such as drug discrimination. Molecular assemblies immobilized at the air-water interface are appropriate media for incorporation of the sensing and diagnostic modules of aqueous biological molecules, since they provide great opportunities for molecular recognition of water-soluble guests by designer hosts in an insoluble floating monolayer [3]. Enhanced binding efficiencies of host-guest recognition at the air-water interface are in accord with theoretical simulations [4, 5] and are supported experimentally as seen in selective sensing of aqueous peptides [6, 7, 8]. We have recently applied the concept of nanotechnology to these interfacial molecular recognition systems by embedding molecular machines in a Langmuir monolayer at the air-water interface where their mechanical operation can be operated by compressive surface pressure applied laterally [9]. The morphologies of the molecular machines can be controlled by macroscopic mechanical forces, resulting in optimization of structure for molecular sensing. We have previously demonstrated the (i) capture and release of fluorescent molecules upon cavity closure-opening motions of molecular machines [10, 11, 12, 13], (ii) control of enantioselective binding of amino acids upon twisting motion of molecular machines [14, 15], and (iii) discrimination of single-methyl-group difference between nucleobases (thymine and uracil) by control of macroscopic lateral pressures [16]. In our next demonstration of the utility of host molecules at the air-water interface, we show discrimination of some naturally occurring nucleotides, which are important in biological activities such as energy storage and signal transduction, using cholesterol-armed-triazacyclononane (**1**) as a molecular machine (see Figure 1 for recognition system). Using this strategy, we were able to discriminate between several ribonucleotides based on the twofold to tenfold difference in their binding constants under optimized conditions.

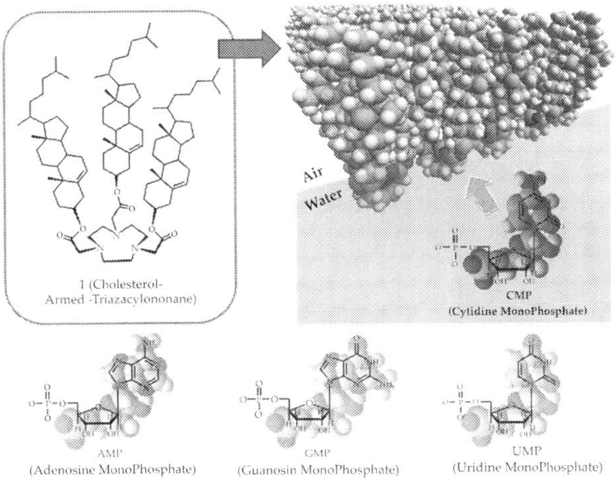

Figure 1. Structures and schematic drawing of host 1 and guest nucleotides.

EXPERIMENTAL

Water used for the subphase was distilled using an Autostill WG220 (Yamato) and deionized using a Milli-Q Lab (Millipore). Its specific resistance was greater than 18 MΩ · cm. Spectroscopic grade chloroform (Wako Pure Chemical Co., Osaka, Japan) was used as the spreading solvent. Ribonucleotides [adenosine 5'-monophosphate disodium salt (AMP), cytidine 5'-monophosphate disodium salt (CMP), guanosine 5'-monophosphate disodium salt (GMP), and uridine 5'-monophosphate disodium salt (UMP)] and lithium chloride were purchased from Wako Pure Chemical Co. (Osaka, Japan). The synthesis of the molecular machine, cholesterol-armed-triazacyclononane (**1**), was described previously [16]. Isotherms of surface pressure and molecular area (ϖ-A isotherm) were measured at 20.0°C using an FSD-300 computer-controlled film balance (USI System, Fukuoka, Japan). A period of 15 min was allowed for spreading solvent evaporation, compression was commenced at a rate of 0.2 mm s^{-1}. Fluctuation of the subphase temperature was within ± 0.2°C.

RESULTS AND DISCUSSION

ϖ-*A* isotherms of the molecular machine **1** with four different ribonucleotides (AMP, CMP, GMP, and UMP) in the subphase are shown in Figure 2 (on pure water) and Figure 3 (on aqueous solution of [LiCl] = 10 mM). In general, isotherms of **1** under each condition exhibit monotonic increases without phase transitions. Increase in the nucleotide concentration in the subphase shifted the isotherms to larger molecular areas, suggesting that the molecular packing of **1** was disturbed by interaction between the nucleotides and **1** at the air-water interface. According to a reported method [14, 16], the shifts in molecular areas at various guest concentrations can be converted into the binding constants (*K*) of nucleotides to the monolayer of **1** at each surface pressure. The calculated values are summarized in Figure 4. In all the cases, assumption of an equimolecular binding gave the best fitting of the binding curves.

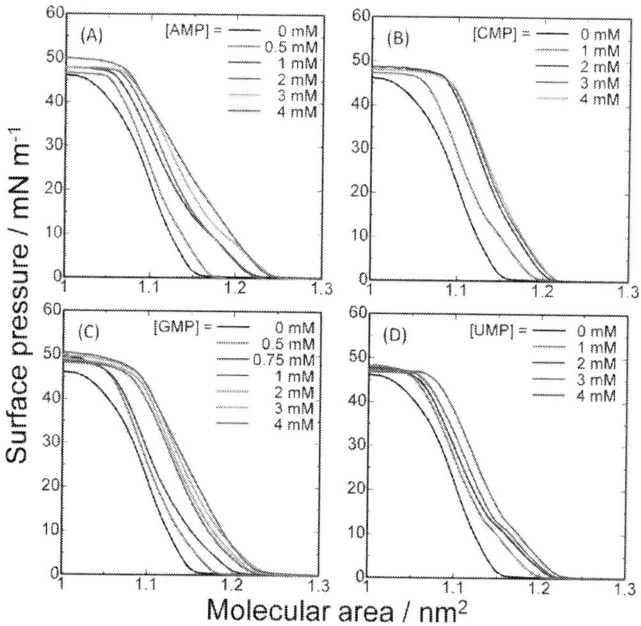

Figure 2. ***π-A*** Isotherms of 1 with guests at 20°C without addition of LiCl into subphase: (A) AMP; (B) CMP; (C) GMP; (D) UMP.

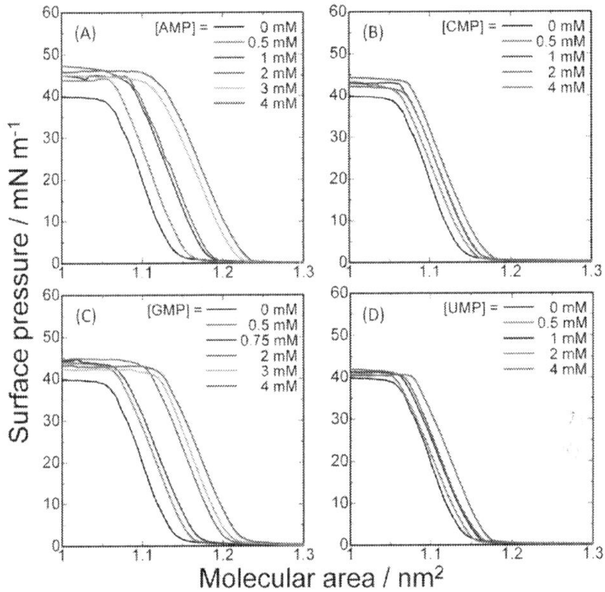

Figure 3. ***π-A*** Isotherms of 1 with guests at 20°C with 10 mM of LiCl: (A) AMP; (B) CMP; (C) GMP; (D) UMP.

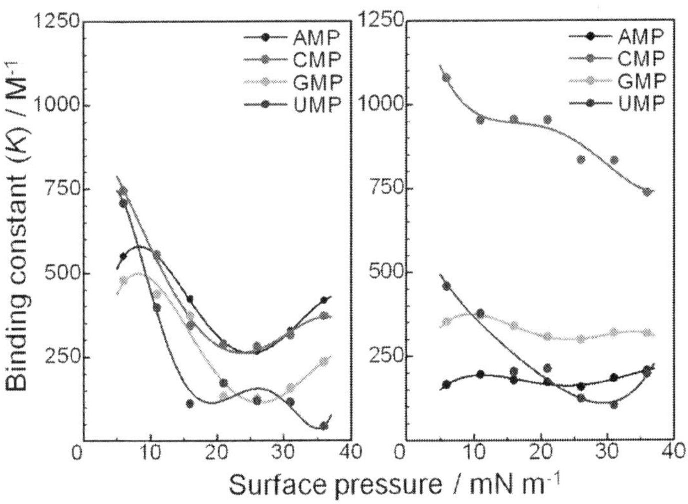

Figure 4. Binding constant (K) of AMP, CMP, GMP, and UMP to the monolayer of 1 at various surface pressures at 20°C: (A) without LiCl and (B) with 10 mM of LiCl.

As shown in Figure 4A, the binding constants of the nucleotides to the monolayer of **1** gradually decreased as the surface pressure increased. This is because expansion of the molecular area of **1** by binding to the nucleotides is thermodynamically unfavourable at higher pressures. As will be described later, when the triazacyclononane moiety is not complexed with a central Li^+ ion electrostatic interaction between **1** and the phosphate group within the nucleotide becomes less important. Hence, on the surface of pure water, there exists a rather ambiguous interaction between **1** and the base portion of the nucleotides, and this interaction is quite sensitive to other factors. Although the absolute value of binding constants decreased drastically, differences in the binding efficiencies amongst the nucleotides became obvious at higher surface pressures. For example, ratios of binding constants, *K*(AMP/UMP), *K*(CMP/UMP), and *K*(GMP/UMP), are 0.78, 1.05, and 0.68, respectively, at a surface pressure of 5 mN m^{-1}, whereas *K*(AMP/UMP), *K*(CMP/UMP), and *K*(GMP/UMP) values become 9.89, 8.77, and 5.52, respectively, when compressed to 35 mN m^{-1}. Thus, discrimination of GMP and UMP from AMP and CMP is possible as well as between GMP and UMP, although differentiation between AMP and CMP is rather difficult even at greater surface pressures.

Complexation of Li^+ ion by the triazacyclononane ring causes two variations in the characteristics of the recognition system. The presence of Li^+ ion at the core of **1** ensures strong electrostatic interaction between the monolayer and the nucleotides. In addition, the complexation of Li^+ ion stabilizes the conformation of the cyclononane ring of **1**, resulting in a rather simple situation of discrimination amongst the nucleotides (Figure 4B). Although the binding constants of UMP to the monolayer exhibit a distinct dependence on surface pressure, an order of binding constant (*K*(CMP) > *K*(GMP) > *K* (AMP)) is maintained over the entire pressure range. An apparent advantage in the Li^+-containing system is due to a significant increase in the binding constant of CMP. As seen in Figure 3B, binding of CMP to the monolayer of **1** does not require large expansion of the monolayer in contrast to AMP and GMP (Figure 3A, C). This binding mode should provide more favourable binding to the molecular assembly. On the other hand, the binding curve for UMP is unusual when compared with the other nucleotides. As has been suggested in previous research [16], the uridine moiety of UMP probably interacts with the cyclononane ring, thus competing with the major interactions between the phosphate and the Li^+ ion.

These results clearly indicate that the recognition of aqueous nucleotides can be tuned both by the surface pressure and the presence of Li^+ ion, although the same recognition element (**1**) was used throughout this investigation. The optimum discrimination between nucleotides can be obtained as follows, where the maximum ratio of binding constants and the conditions applied are summarized: *K*(CMP/AMP) = 6.5 ([Li^+] = 10 mM and ϖ = 5 mN · m^{-1}); *K*(CMP/GMP) = 3.11 ([Li^+] = 10 mM and ϖ = 20 mN · m^{-1}); *K*(CMP/UMP) = 8.77 ([Li^+] = 0 mM and ϖ = 35 mN · m^{-1}); *K*(AMP/GMP) = 2.22 ([Li^+] = 0 mM and ϖ = 20 mN · m^{-1}); *K*(AMP/UMP) = 9.89 ([Li^+] = 0 mM and ϖ = 35 mN · m^{-1}); *K*(GMP/UMP) = 5.52 ([Li^+] = 0 mM and ϖ = 35 mN · m^{-1}). On the other hand, the maximum binding constants for individual nucleotides are: *K*(CMP) = 1080 M^{-1} ([Li^+] = 10 mM and ϖ = 5 mN · m^{-1}); *K*(AMP) = 550 M^{-1} ([Li^+] = 0 mM and ϖ = 5 mN · m^{-1}); *K*(GMP) = 480 M^{-1} ([Li^+] = 0 mM and ϖ = 5 mN · m^{-1}); *K*(UMP) = 710 M^{-1} ([Li^+] = 0 mM and ϖ = 5 mN · m^{-1}). Therefore, conditions suitable for discrimination of the nucleotides and for most efficient binding of a single nucleotide component can be selected. The molecular recognition system presented here is therefore distinct different from more conventional ones where the structure of recognition components primarily defines binding efficiency of guest molecules.

CONCLUSIONS

Prior to this and our other preliminary reports, discrimination of nucleotides has not been easy to achieve because of their structural similarity, and despite its importance in biological and pharmaceutical fields. This research strikingly demonstrates a method for molecular discrimination amongst structurally similar nucleotides by mechanical tuning of a simple host at a dynamic interfacial medium. Recognition and discrimination of ribonucleotides can also be optimized. The concept of mechanical tuning for optimization of molecular recognition should become a novel methodology in bio-related nanotechnology as an alternative to traditional strategies based on increasingly complex and inconvenient molecular design strategies.

DECLARATIONS

Acknowledgements

This work was partly supported by World Premier International Research Center Initiative (WPI Initiative), MEXT, Japan and Core Research for Evolutional Science and Technology (CREST) program of Japan Science and Technology Agency (JST), Japan.

Competing interests

The authors declare that they have no competing interests.

Authors' contributions

SS, MM, and HT carried out syntheses of molecular machines. TM, KO, HE, KS, JPH, YS, YK, and KA evaluate molecular recognition at the air-water interface. All authors read and approved the final manuscript.

Authors' Affiliations

(1) World Premier International (WPI) Research Center for Materials Nanoarchitectonics (MANA), National Institute for Materials Science (NIMS)

(2) JST, CREST

(3) Osaka City University

(4) Kitami Institute of Technology

REFERENCES

1. Ariga K, Hill JP, Lee MV, Vinu A, Charvet R, Acharya S: Challenges and breakthroughs in recent research on self-assembly. Sci Technol Adv Mater 2008, 9: 014109. 10.1088/1468-6996/9/1/014109
2. Ariga K, Ji Q, Hill JP, Kawazoe N, Chen G: Supramolecular approaches to biological therapy. Expert Opin Biol Ther 2009, 9: 307–320. 10.1517/14712590802715772
3. Ariga K, Kunitake T: Molecular Recognition at Air-Water and Related Interfaces:Complementary Hydrogen Bonding and Multisite Interaction. Acc Chem Res 1998, 31: 371–378. 10.1021/ar970014i

4. Sakurai M, Tamagawa H, Inoue Y, Ariga K, Kunitake T: Theoretical Study of Intermolecular Interaction at the Lipid-Water Interface. 1. Quantum Chemical Analysis Using a Reaction Field Theory. J Phys Chem B 1997, 101: 4810–4816. 10.1021/jp9700591
5. Tamagawa H, Sakurai M, Inoue Y, Ariga K, Kunitake T: Theoretical Study of Intermolecular Interaction at the Lipid-Water Interface. 2. Analysis Based on the Poisson-Boltzmann Equation. J Phys Chem B 1997, 101: 4817–4825. 10.1021/jp9700600
6. Cha X, Ariga K, Onda M, Kunitake T: Molecular Recognition of Aqueous Dipeptides by Noncovalently Aligned Oligoglycine Units at the Air/Water Interface. J Am Chem Soc 1995, 117: 11833–11838. 10.1021/ja00153a003
7. Cha X, Ariga K, Kunitake T: Molecular Recognition of Aqueous Dipeptides at Multiple Hydrogen-Bonding Sites of Mixed Peptide Monolayers. J Am Chem Soc 1996, 118: 9545–9551. 10.1021/ja961526f
8. Ariga K, Kamino A, Cha X, Kunitake T: Multisite Recognition of Aqueous Dipeptides by Oligoglycine Arrays Mixed with Guanidinium and Other Receptor Units at the Air-Water Interface. Langmuir 1999, 15: 3875–3885. 10.1021/la981047p
9. Ariga K, Lee MV, Mori T, Yu X-Y, Hill JP: Two-dimensional nanoarchitectonics based on self-assembly. Adv Colloid Interface Sci 2010, 154: 20–29. 10.1016/j.cis.2010.01.005
10. Ariga K, Terasaka Y, Sakai D, Tsuji H, Kikuchi J: Piezoluminescence Based on Molecular Recognition by Dynamic Cavity Array of Steroid Cyclophanes at the Air-Water Interface. J Am Chem Soc 2000, 122: 7835–7836. 10.1021/ja000924m
11. Ariga K, Nakanishi T, Terasaka Y, Tsuji H, Sakai D, Kikuchi J: Piezoluminescence at the Air-Water Interface through Dynamic Molecular Recognition Driven by Lateral Pressure Application. Langmuir 2005, 21: 976–981. 10.1021/la0477845
12. Ariga K, Nakanishi T, Hill JP: A paradigm shift in the field of molecular recognition at the air-water. Soft Matter 2006, 2: 465–477. 10.1039/b602732f
13. Ariga K, Nakanishi T, Terasaka Y, Kikuchi J: Catching a Molecule at the Air-Water Interface: Dynamic Pore Array for Molecular

Recognition. J Porous Mater 2006, 13: 427–430. 10.1007/s10934-006-8041-2

14. Michinobu T, Shinoda S, Nakanishi T, Hill JP, Fujii K, Player TN, Tsukube H, Ariga K: Mechanical Control of Enantioselectivity of Amino Acid Recognition by Cholesterol-Armed Cyclen Monolayer at the Air-Water Interface. J Am Chem Soc 2006, 128: 14478–14479. 10.1021/ja066429t
15. Ariga K, Michinobu T, Nakanishi T, Hill JP: Chiral Recognition at Air-Water Interfaces. Curr Opin Colloid Interface Sci 2008, 13: 23–30. 10.1016/j.cocis.2007.08.010
16. Mori T, Okamoto K, Endo H, Hill JP, Shinoda S, Matsukura M, Tsukube H, Suzuki Y, Kanekiyo Y, Ariga K: Mechanical Tuning of Molecular Recognition To Discriminate the Single-Methyl-Group Difference between Thymine and Uracil. J Am Chem Soc2010, 132: 12868–12870. 10.1021/ja106653a

CHAPTER

12

ACTIVE RNA POLYMERASES: MOBILE OR IMMOBILE MOLECULAR MACHINES?

Argyris Papantonis[1], Joshua D. Larkin[1], Youichiro Wada[2], Yoshihiro Ohta[2], Sigeo Ihara[2], Tatsuhiko Kodama[2], Peter R. Cook[1]

[1]Sir William Dunn School of Pathology, Medical Sciences Division, University of Oxford, Oxford, United Kingdom,

[2]Laboratory for Systems Biology and Medicine, Research Center for Advanced Science and Technology, The University of Tokyo, Tokyo, Japan

ABSTRACT

It is widely assumed that active RNA polymerases track along their templates to produce a transcript. We test this using chromosome conformation capture and human genes switched on rapidly and synchronously by tumour necrosis factor alpha (TNFα); one is 221 kbp SAMD4A, which a polymerase takes more than 1 h to transcribe. Ten minutes after stimulation, the SAMD4A promoter comes together with other TNFα-responsive promoters. Subsequently, these contacts are lost as new downstream ones appear; contacts are invariably between sequences being transcribed. Super-resolution microscopy confirms that nascent transcripts (detected by RNA fluorescence in situ hybridization) co-localize at relevant times. Results are consistent with an alternative view of transcription: polymerases fixed in factories reel in their respective templates, so different parts of the templates transiently lie together.

AUTHOR SUMMARY

We were all taught that an RNA polymerase becomes active by diffusing to a promoter, initiating transcription, and then tracking like a locomotive down the DNA template. We test this using tumour necrosis factor alpha (TNFα) to switch on transcription of two human genes which lie far apart on the genetic map and then measure how close the two are in 3D nuclear space. If what we were taught were true, there is no reason to expect the two genes to lie together. What we find—using two different techniques (cutting/ligating nearby sequences, and super-resolution microscopy)—is that the two genes are initially apart; then the parts of the genes being transcribed at a particular moment transiently come into close proximity. Our results are consistent with a model in which genes diffuse to a cluster of polymerases—a transcription factory—with transcripts being made as immobile polymerases reel in their respective templates. The DNA moves, not the polymerase.

INTRODUCTION

It is widely assumed that an RNA polymerase transcribes by diffusing to a promoter, binding, and then tracking down the template as it makes its transcript [1]. Accumulating evidence, however, is consistent with an alternative: a promoter diffuses to a transcription factory where it binds to a transiently immobilized polymerase, which then reels in its template as it extrudes a transcript [2]–[6]. Here, we address the question: Are transcribing enzymes mobile or immobile?

Our strategy involves switching on transcription of two genes rapidly and synchronously using tumour necrosis factor alpha (TNFα). This cytokine orchestrates the inflammatory response in human umbilical vein endothelial cells (HUVECs) by signalling through nuclear factor kappa B (NF-κB) to activate a sub-set of genes [7]–[8]. SAMD4A—a 221 kbp-long gene that encodes a regulator of this pathway—is amongst the first few to respond. Microarray analysis reveals that a synchronous wave of transcription initiates within 15 min, before sweeping down the gene (at ~3 kbp/min) to reach the terminus ~70 min later; no transcripts from the non-coding strand are detected [9]. RNA FISH using intronic probes confirms that almost half the cells in the population respond; essentially no nascent RNA can be detected prior to stimulation, no transcription occurs from the antisense strand, and probes targeting successive introns only yield signal as the wave passes by [9].

TNFAIP2—a short 11 kbp gene that lies ~50 Mbp away from SAMD4A on chromosome 14—encodes another regulator. It is switched on as rapidly and then repeatedly transcribed over the next 90 min. We use it as an external reference point (or "anchor") and analyze the contacts it makes with different parts of SAMD4A using chromosome conformation capture (3C)—a powerful tool for detecting proximity of two DNA sequences in 3D nuclear space [10]–[12]. If the conventional model for transcription applies, we would not expect the anchor to lie close to any part of SAMD4A either before or after adding TNFα, as it lies so far away on the chromosome (Figure 1, left). Even if polymerases on the two genes happened to lie together (for whatever reason), tracking of one down the long gene should increase the distance between transcribed parts of the two genes. But if both genes were transcribed by polymerases transiently immobilized in one factory, the short gene—which would repeatedly attach to (and detach from) the factory as it initiates (and terminates)—should always lie close to just the part of SAMD4Abeing transcribed at that particular moment (Figure 1, right). Thus, as the polymerase reels in SAMD4A, introns 1, 2, 3, etc. should successively be brought into the factory to lie transiently next to the anchor. Results using TNFAIP2 (and other anchors) are impossible to reconcile with the widely held assumption that polymerases track; rather they are consistent with active polymerases being immobilized in factories.

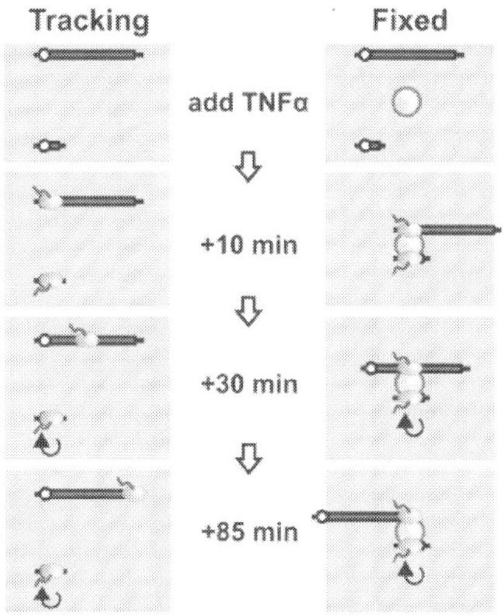

Figure 1. Distinguishing between tracking and fixed RNA polymerases. Before adding TNFα, both the long and short gene are not transcribed. Assuming they lie far apart on the same chromosome, they are unlikely to yield detectable 3C products. Ten min after adding TNFα, RNA polymerases (ovals) initiate on both genes. If active polymerases track, it remains unlikely that any part of the two genes will contact each other. However, if the two genes diffuse to one factory (sphere) and are then transcribed by transiently immobilized polymerases, the two promoters will lie close together. After 30 min, the pioneering polymerase on the short gene has terminated and been replaced by others that continuously transcribe it, while the pioneering polymerase on the long gene has only transcribed one-third of the gene. If polymerases track, the two genes are still unlikely to be together. But if polymerases are immobilized in a factory, the parts of the two genes currently being transcribed will lie together and yield a 3C product. After 85 min, the pioneering polymerase reaches the terminus of the long gene. If polymerases track, the two genes will still be apart; if immobilized, the terminus should now contact the short gene.

RESULTS

Some Interacting Partners of *SAMD4A* and Their Transcriptional Activation

As our strategy requires one gene to be used as an anchor, we applied 3C and a variant of "associated chromosome trap" (ACT) [13]–[14] to search for genes that interacted with SAMD4A. A number were found, and we chose four that were detected in independent experiments and which were relatively short (<60 kbp): TNFAIP2, GCH1, PTRF, and SLC6A5.

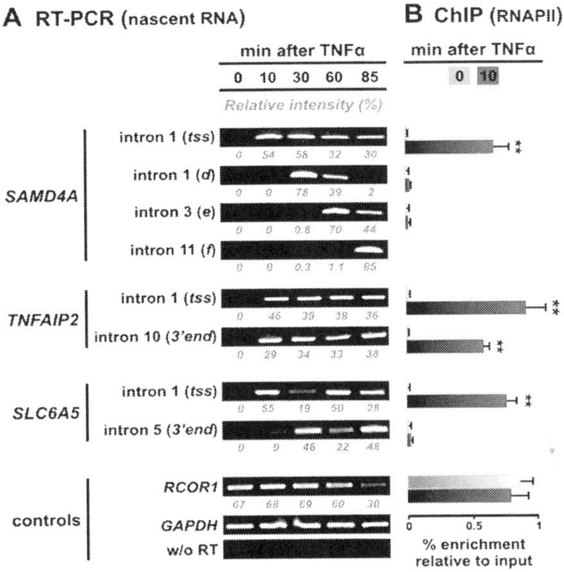

Figure 2. Polymerases initiate rapidly and synchronously on responding genes and elongate at expected rates. (A) Nascent RNA detected using reverse transcriptase PCR (RT-PCR). Total RNA was isolated from HUVECs 0–85 min after adding TNFα, treated with DNase, and intronic RNA detected. No nascent RNA copied from SAMD4A, TNFAIP2, or SLC6A5 is detected at 0 min. For 221 kbp SAMD4A, nascent RNA appears at the tss after 10 min. As polymerases continue to initiate thereafter (albeit at declining rates), signal is seen at the tss between 15 and 85 min; however, many of these polymerases abort within ~10 kbp of the tss . Nascent RNA from region d of intron 1 (34 kbp into the gene) is seen only after 30 min. Pioneering polymerases reach this region after 30 min and the slowest by 60 min; after 85 min all have passed by. Similarly, pioneering polymerases only reach introns 3 and 11 after 60 and 85 min, respectively. TNFAIP2 is 11 kbp, and polymerases in the population can generate intronic RNA from both 5′ and 3′ ends within 10 min (it is then transcribed throughout the 85 min). SLC6A5—a 56 kbp gene—yields an intermediate pattern. No signal is again seen at 0 min, and pioneering polymerases generate maximal levels of intronic RNA at the tss after 10 min, and the 3′ end after 30 min; then, the cycle repeats between 60 and 85 min. Controls show levels of intronic RNA from two non-responsive, active genes (GAP-

DH and RCOR1), and that amplimers do not result from contaminating genomic DNA (using GAPDH probes, but replacing RT by Taq polymerase). Numbers under each panel (orange) correspond to the relative intensity of bands, corrected for background, and normalized to GAPDHlevels. (B) Bound RNA polymerase II detected by chromatin immunoprecipitation (ChIP). Enrichments relative to input DNA (normalised to GAPDH levels) are shown 0 and 10 min after stimulation (light and dark grey bars, respectively). Almost no polymerase is bound to any part of SAMD4A, TNFAIP2, or SLC6A5 at 0 min. The tss of SAMD4A is occupied by polymerases within 10 min of induction; however, levels further 3′ remain low as pioneering polymerases have not yet reached these regions. For TNFAIP2, some polymerases bind after 10 min, while others have reached the 3′ end as the gene is so short. For SLC6A5, polymerases bind by 10 min to the tss but have not yet reached the 3′ end. RCOR1 was analyzed as it was used as a control in RNA FISH experiments. Error bars show standard deviations from two independent experiments. $^{**}p<0.01$, Student's t test compared to 0 min.

We initially verified that all five genes responded to TNFα by reverse-transcriptase PCR (RT-PCR). No intronic RNA (or only low levels in the case of PTRF) copied from the five genes was detected before induction, but higher levels were seen within 10 min of TNFα treatment (Figures 2A and S3F). Intronic RNA copied from further downstream in SAMD4A then appeared consistent with pioneering polymerases transcribing its 221 kbp at ~3 kbp/min. Thus, RNA copied immediately downstream of the transcription start site (tss) appeared after 10 min, from ~34 kbp into intron 1 after 30 min, from intron 3 after 60 min, and from the terminus after 85 min. In contrast, signal from each end of TNFAIP2 is seen by 10 min. This 11 kbp gene is so short, and synchrony sufficiently poor, that some polymerases in the population are initiating as others are terminating (Figure 2A). GCH1 and SLC6A5—both genes of ~60 kbp—present intermediate patterns; pioneering polymerases reach termini after ~30 min, before a second (reasonably synchronous) transcription cycle begins (Figures 2A and S3F). Such cycling has now been seen on various mammalian genes (e.g., [15]). Chromatin immunoprecipitation (ChIP) showed an enrichment of RNA polymerase II bound to the tss of all five genes within 10 min (Figures 2B and S3G). It also showed that NF-κB bound to promoters within 10 min, as might be expected [16]. RNA fluorescence in situ hybridization (FISH) also shows that intronic RNA copied from the relevant parts of the genes is present at the appropriate times. Therefore, results obtained with four independent methods (i.e., microarrays, RT-PCR, ChIP, RNA FISH) are in agreement and provide data on when polymerizing complexes are actively transcribing the sequences to be analyzed. These data are summarized in cartoons that accompany the results.

Changing Contacts Between Two TNFα-Responsive Genes on Chromosome 14

Contacts between selected regions of SAMD4A and TNFAIP2 were monitored by 3C, where the presence of a band after 34 PCR cycles reflects a high contact frequency (Figure 3). Essentially no contacts are seen between the tss of TNFAIP2 (the anchor) and regions ~25 kbp upstream or downstream of SAMD4A (a, h) at any time, or between the anchor and any region of SAMD4A (b–g) at 0 min—when no polymerases are engaged on either gene (Figure 3B, cartoon). By 10 min (when polymerases are first found on both genes; cartoon), contacts appear between the anchor and SAMD4A regions b, c (Figure 3B). Such contacts are soon lost, as new ones appear further 3′ on SAMD4A; they seem to steadily "slide" down the long gene. Thus, by 30 min, contacts are with regions c and d, by 60 min with region e, and by 85 min with regions e, f, and g. (The presence of more than one contact at certain times is consistent with imperfect synchrony amongst the ~10^6 cells assayed.) Treatment with DRB (5,6-dichloro-1-β-D-ribofuranosylbenzimidazole)—a reagent that inhibits transcription and releases polymerases from the template reduces contacts (Figure 3B, grey box). Similar changing contacts were seen using (i) real-time PCR to quantify selected interactions, (ii) the 3′ end of TNFAIP2 as an anchor (Figure 3C, D; the gene is short enough for polymerases to be found at the same times on promoter and terminus in different cells in the population), and (iii) if HindIII replaced SacI as the restriction enzyme used for 3C. In every case, contacts are only seen at times when active polymerases are transcribing contacting sequences. Note that several genes lying within 50 Mbp on either side of SAMD4A do not interact with it (e.g., responsive NFKBIA, SAV1, IRF9, GPR68, and PAPLN; non-responsive GMFB, YY1, HIF1A, and C14orf2; and constitutive RCOR1). As a whole, these results are inconsistent with the model involving tracking polymerases (Figure 1, left) but are simply explained if the two contacting templates are transiently tethered to polymerases fixed in one factory (Figure 1, right).

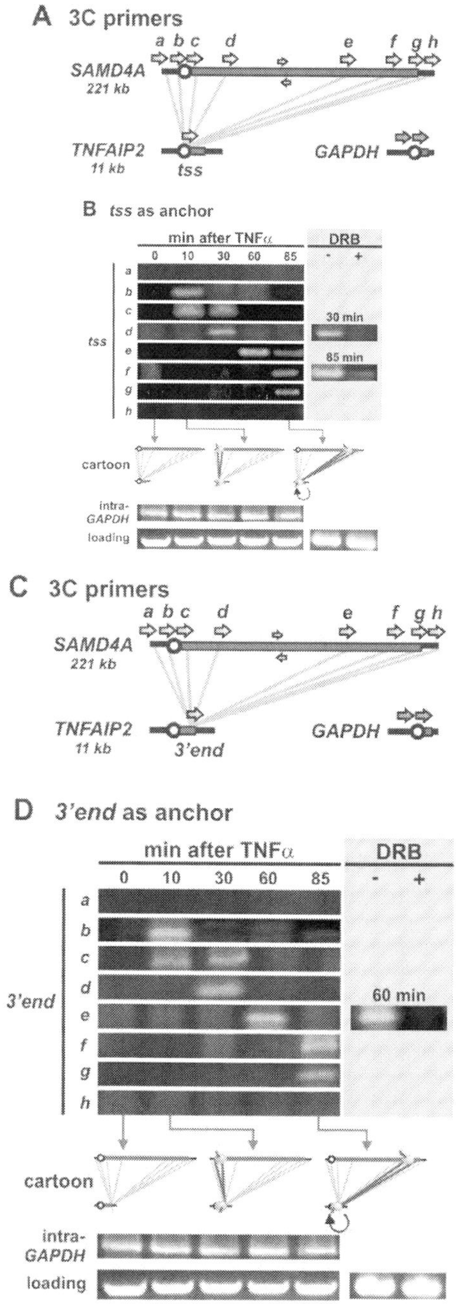

Figure 3. Contacts between two TNFα-responsive genes 50 Mbp apart on the same chromosome follow engaged polymerases. (A) Positions of 3C prim-

ers on SAMD4A and the tss of TNFAIP2 (orange arrows) and GAPDH (grey arrows). Grey lines: 3C interactions monitored. White arrows: primers used for loading controls. (B) 3C. HUVECs were treated with TNFα for 0–85 min, 3C templates prepared using SacI, and PCR conducted using equal weights of DNA and the primer pairs indicated; after gel electrophoresis and SYBR green staining, images of resulting gels are shown. The presence of a band reflects a high contact frequency between respective primer targets. Cartoons illustrate where polymerases are bound at different times and the interactions analyzed (grey lines); red lines indicate interactions yielding bands, and these always correlate with the presence of a polymerase on both contacting partners. In selected cases, DRB was added 20 min prior to harvesting cells (grey box); this reduces band intensity, indicating that contacts depend on transcription. GAPDH primers yield uniform levels of amplimers, as do loading controls. (C) Positions of 3C primers on SAMD4A and the 3′ end of TNFAIP2. (D) Changing contacts between SAMD4A and the 3′ end of TNFAIP2. The pattern is essentially the same as that in panel (B). Panels (B) and (D) share the same pair of loading and intra-GAPDH controls (excluding ± DRB), so the same image is shown in both panels.

Changing Contacts Between TNFα-Responsive Genes on Different Chromosomes

PTRF is a 21 kbp gene that lies on a different chromosome (i.e., 17) from SAMD4A (on 14). The pattern of interactions between the two is much the same as those seen between SAMD4Aand TNFAIP2 , which is again consistent with the model involving fixed polymerases (Figure 1, right).

A more complex pattern of changing contacts is seen between SAMD4A and a 60 kbp gene on chromosome 11, SLC6A5 (Figure 4); this pattern suggests that polymerases must be present on both contacting sequences. Thus, as before, no contacts are seen between the tss of SLC6A5 (the anchor) and regions upstream or downstream of SAMD4A (a, h) at any time, or between the anchor and any region of SAMD4A at 0 min—when no polymerases are engaged on either gene (Figure 4B, cartoon). Again as before, contacts appear between the anchor and SAMD4A region c (which includes the tss and the beginning of intron 1) after 10 min (Figure 4B), when polymerases are first found on both. But after 30 min (when contacts with region dwere seen in Figure 3B), essentially no contacts are found (Figure 4B). This is consistent with pioneering polymerases leaving the tss of the anchor so that they are now transcribing the 3′ end of this ~60 kbp gene, as data in Figure 2 indicate. By 60 min (when a second polymerase is just initiating on the tss of SLC6A5; Figure 2), we see a strong (second) contact with the region on SAMD4A that its pioneering polymerase is now transcribing (i.e., e in Figure 4B). This interaction is DRB-sensitive (Figure 4B, grey box), and so depends on continuing transcription. No prominent interactions are seen at 85 min (Figure 4B) even though we know SAMD4A is still being transcribed. Moreover, the contact seen with region f in Figure 3B is missing, presumably because the second polymerase on SLC6A5 has left the tss used as the anchor and is now transcribing the 3′ end (Figure 2). An almost identical pattern with analogous missing contacts is seen if HindIII replaces SacI during preparation of the 3C template.

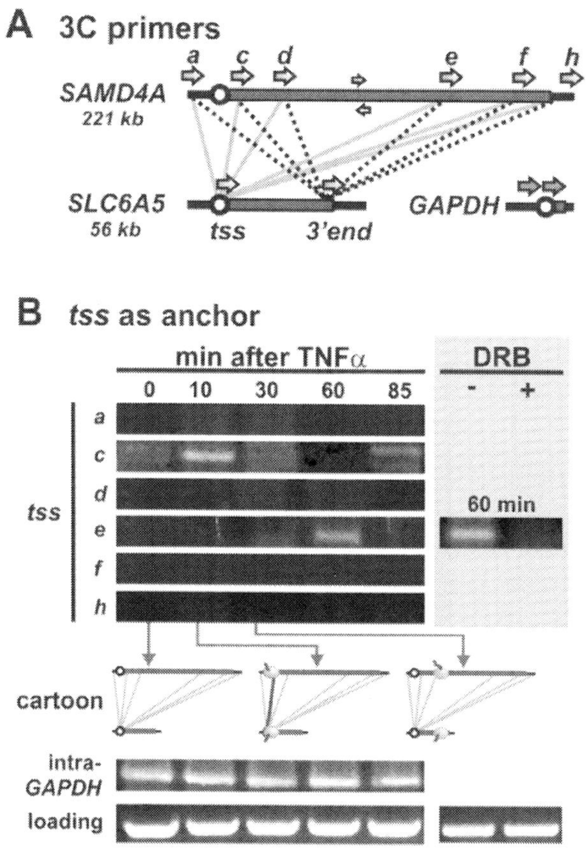

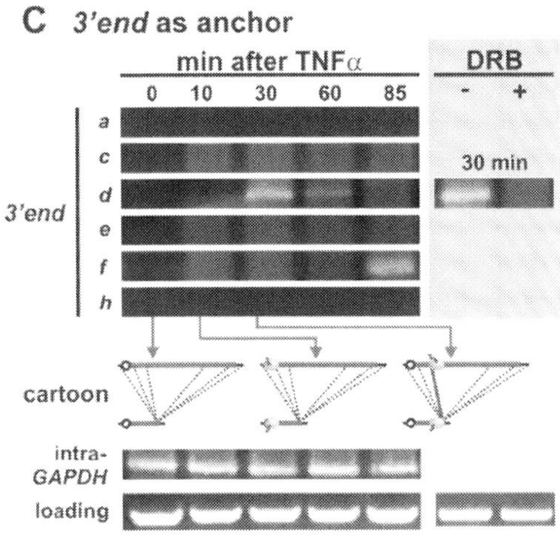

Figure 4. Contacts between two TNFα-responding genes on different chromosomes (14 and 11) follow engaged polymerases. (A) Positions of 3C primers and the interactions screened (grey and dotted black lines). (B) Contacts between the tss of SLC6A5 (the anchor) and different parts of SAMD4A. Contacts/bands are only detected when polymerases are on both contacting partners. (C) Contacts between the 3′ end of SLC6A5 (the anchor) and different parts of SAMD4A. As in (B), two strong bands are seen, but they are in different positions. We suggest this is because it takes a polymerase 20–30 min to reach the 3′ end of SLC6A5 now used as an anchor; then, contacts/bands are again only detected when polymerases are on both contacting partners. Panels (B) and (C) share the same pair of loading and intra-GAPDHcontrols (excluding ± DRB), so the same image is shown in both panels.

If the above explanation is correct, with contacts only being seen if active polymerases are present on both contacting partners, then use of the 3′ end of SLC6A5 as an anchor should change the pattern as follows. The two bands seen in Figure 4B should disappear (as polymerases at the relevant times are on the tss and not the 3′ end now used as the anchor), while the two "missing" bands should reappear (as polymerases have now reached the 3′ end); they do. For example, comparison of Figure 4B and C shows that the first missing band/contact (with d at 30 min in Figure 3B) reappears in Figure 4C, as does the second (with f at 85 min). Bands/contacts are also sensitive to DRB (Figures 4B,C, grey boxes).

This interpretation is reinforced by an analysis involving 5′ and 3′ anchors on another gene (of similar length as SLC6A5) that lie on the same chromosome as SAMD4A. Thus, GCH1 is ~0.8 Mbp away from SAMD4A and responds as rapidly to TNFα. When its 5′ and 3′ ends are used as anchors, a complex set of changing contacts (and missing bands) is again seen.

We also confirmed that the tss of GCH1 lay next to the tss of TNFAIP2 at 10 min but not at 0 min. This is consistent with responding promoters coming together to the same factory when active. As all other contacts analyzed involve SAMD4A, these results also indicate that such reorganization is not peculiar to one long gene.

Nascent RNAs Also Colocalize at the Appropriate Times

If responding regions only lie together when transcribed, their nascent transcripts should also only be together at the appropriate times. To test this we used RNA FISH with pairs of probes each able to detect an intron within a single nascent transcript copied RNA transcript at its transcription site; colocalization of nascent transcripts copied from the two different genes then yields a yellow focus [9],[19]. Yellow foci were given by the TNFAIP2 probe (red) and SAMD4Aprobes c, d, and e/f (green) at 10, 30, and 60 min post-induction (Figure 5A–C). No such colocalization was seen at other times, when relevant regions were not being transcribed. As a control, we analyzed nascent transcripts copied from a non-responsive (constitutively-active) gene—RCOR1—that lies between SAMD4A and TNFAIP2; no yellow foci were detected (Figure 5D). Just as 3C showed the templates lie together (Figure 2), RNA FISH confirms their transcripts also colocalize.

We also investigated inter-chromosomal contacts 30 min post-induction, using probes targeting (green) SAMD4A region d and (red) SLC6A5 intron 1 (close to the tss) or intron 10 (close to the 3′ end). When no 3C contacts between SAMD4A region d and the tss of SLC6A5 were seen (Figure 4B), no yellow foci were detected (Figure 5E). But the “missing” 3C band was seen at 30 min using the 3′ terminus as anchor (Figure 4C), and then yellow foci are seen (Figure 5F). As a control, we analyzed nascent transcripts copied from another non-responsive (constitutively-active) gene—EDN1—that lies on a different chromosome; again, no yellow foci were seen (Figure 5G).

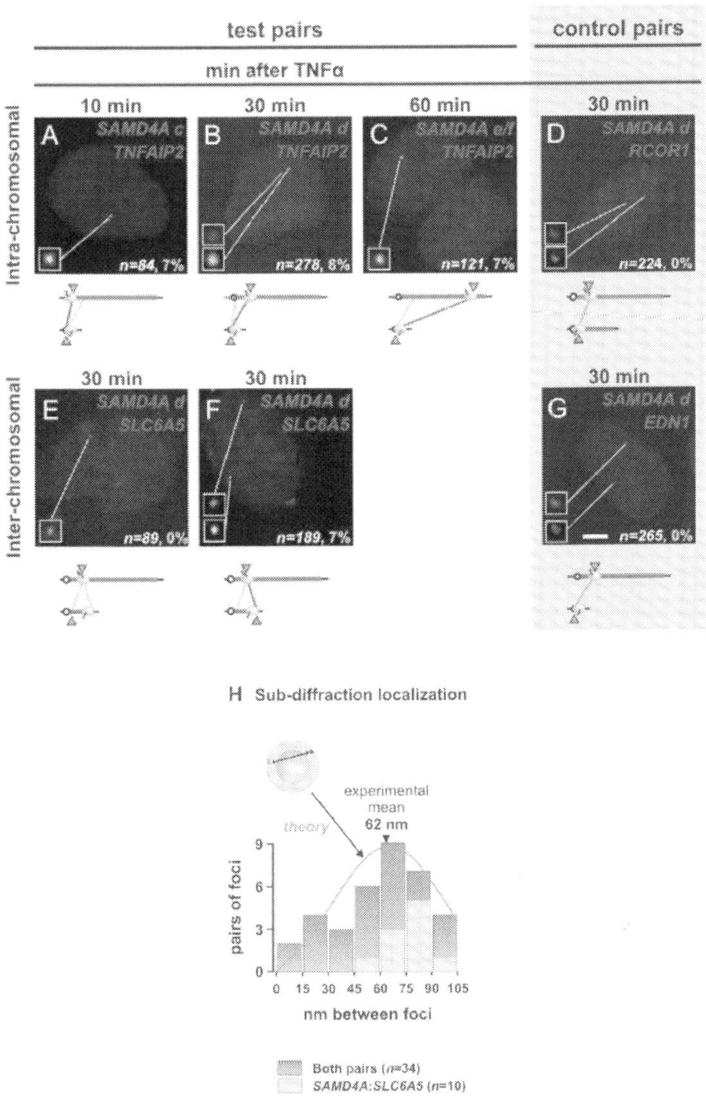

Figure 5. Colocalization of intronic RNA demonstrated by RNA FISH. HUVECs were treated with TNFα for 10, 30, or 60 min, and nascent RNAs copied from test and control pairs of genes detected by RNA FISH. (A–C) Colocalization of nascent RNAs encoded by genes on the same chromosome. The two probes target RNA copied from intron 2 of TNFAIP2 (red) and intronic region c, or d, or e/f of SAMD4A (green); representative images of DAPI-stained nuclei are shown (insets provide magnifications of foci indicated). Red and green foci mark (non-colocalizing) nascent transcripts copied from one (or both) allele, and yellow foci colocalizing ones; n gives the number of alleles

active in all cells analyzed that have ≥1 green focus plus ≥1 red focus. Numbers in yellow give the fraction of colocalizing red and green foci (where >75% pixels in one focus share red and green signal) expressed as a percentage of n; values were significantly different from those seen in (D) with a control gene ($p<10^{-3}$, Fischer's exact test, one-tailed). The cartoon illustrates the targets of red and green probes (triangles), and the positions of polymerases; red lines between targets indicate that yellow foci were detected (grey lines: no yellow foci detected). (D) Yellow foci were never seen with probes targeting intronic RNA copied from SAMD4A and a (non-responsive) control gene (RCOR1) that lies between SAMD4A and TNFAIP2. (E–F) Colocalization of nascent RNAs encoded by genes on different chromosomes. The two probes target RNA copied from intronic region d of SAMD4A (green) and either intron 1 or 10 of SLC6A5 (red); only the latter yields yellow foci (the number of yellow foci was significantly different from that seen in (G) with a control gene; $p<10^{-3}$, Fischer's exact test, one-tailed). (G) Yellow foci were never seen with probes targeting intronic RNA copied from SAMD4A and a (non-responsive) control gene (EDN1) that lies on a different chromosome from SAMD4A. Bar: 5 μm. (H) Sub-diffraction localization of peaks of red and green signal within yellow foci. Gaussian curves were fitted to the intensities of the red and green signals, and distances between peaks determined with a precision of 15 nm (see Methods for details). Dark grey bars illustrate distances obtained from 34 yellow foci seen in images like those in (B) and (F); the mean distance is 62 nm. Light grey bars illustrate similar distances obtained from 10 yellow foci like that in (F). The model shows a red and green point randomly distributed in a 35 nm shell (grey) around an 87 nm diameter factory (orange sphere); simulations using this model yield the distribution indicated (orange line).

Super-Resolution Localization of Nascent Transcripts

Electron microscopy reveals that nascent nucleoplasmic transcripts typically lie on the surface of ~87 nm (protein-rich) factories [20]. To see if colocalizing transcripts encoded by the SAMD4A d:TNFAIP2 and SAMD4A d:SLC6A5 pairs lie this close together, we used a new approach that allows resolution beyond the diffraction limit of the light microscope [21]–[23]. We assume the red and green signals that yield a yellow focus mark two sub-diffraction spots, fit Gaussian curves to their intensities, and measure the distance (with 15 nm precision) between peaks [23]; the distance between the two transcripts ranges from 7 to 102 nm, with a mean separation of 62 nm (Figure 5H). This distribution is much like that seen when a pair of red and green points are repeatedly and randomly distributed in a 35 nm shell surrounding an 87 nm diameter sphere (Figure 5H, orange line). [Subdiffraction-sized red/green fluorescent beads of 110 nm serve as a truly co-localizing control; then, the distance between their red and green peaks is within the uncertainty of our measurements (n=8; not shown).] These results are consistent with nascent transcripts copied from the two different genes lying on the surface of the same transcription factory.

DISCUSSION

We tested the two models illustrated in Figure 1 to address one fundamental assumption of modern molecular biology, namely that a transcribing polymerase tracks along its template as it makes its transcript. SAMD4A has a unique set of properties that make it particularly useful for this analysis; it can be switched on rapidly and synchronously by TNFα (with approximately half the cells in the population responding), its length provides sufficient temporal and spatial resolution (it takes ~70 min to transcribe, and contains many restriction sites that facilitate the use of 3C to discriminate between contacts produced by different parts of the gene), and neither its sense or anti-sense strands encode other transcription units that might complicate analysis. 3C reveals that just the parts of SAMD4A being transcribed at a particular moment lie close to just the parts of three other genes being transcribed at that moment (Figures 3, 4, S3, and S8). These inter-genic contacts occur infrequently, as expected [24]–[26]. RNA FISH confirmed that the relevant nascent RNAs lie together at the appropriate times (Figures 5 and S5), while "super-resolution" microscopy (allowing measurements below the diffraction limit) showed that the distance between the two transcripts is consistent with them lying within 35 nm of the surface of an 87 nm sphere (Figure 5H). Such results are difficult—if not impossible—to explain if polymerases track. Rather, they are consistent with an alternative where two responding genes diffuse to an 87 nm factory to be transcribed by immobilized enzymes. Then, as the two genes are reeled in, only parts being transcribed at a given moment will lie transiently together [5].

These results beg many questions. For example, we were able to detect interacting sequences at a reasonable frequency simply by assuming the existence of factories dedicated to transcribing genes that respond rapidly to TNFα (Figures S2 and S9). If such specialized factories exist [27],[28], how many might there be in a nucleus, and how many are accessible to a gene like SAMD4A? Fortunately, these questions will soon be answered, as techniques for analyzing all contacts made by any gene in a nucleus have been developed [29]. We also note that our results are consistent with others obtained from a recent genome-wide study; after stimulating human cells with estrogen and mapping contacts made by bound estrogen receptor-α (using ChIP, 3C, and "deep" sequencing), contacting partners were often associated with bound RNA polymerase II [30].

MATERIALS AND METHODS

Cell Culture

HUVECs from pooled donors (Lonza) were grown to 80%–90% confluency in Endothelial Basal Medium 2-MV with supplements (EBM; Lonza), starved (18 h) in EBM+0.5% FBS, and treated with TNFα (10 ng/ml; Peprotech) for up to 85 min. In some cases, 50 μM 5,6-dichloro-1-β-D-ribofuranosylbenzimidazole (DRB; Sigma-Aldrich) was added 20 min before harvesting cells.

3C

3C was performed as described [10]. In brief, 10^7 cells were fixed (10 min; room temperature) in 1% paraformaldehyde (Electron Microscopy Sciences), "Dounce"-homogenized, and membranes lyzed (30 min; 4°C) using 0.2% Igepal (Sigma-Aldrich). Nuclei were pelleted and resuspended in the appropriate restriction buffer, incubated (16 h; 37°C) with SacI or HindIII (800 units/10^6 cells; New England Biolabs), diluted to 8 ml in ligation buffer, T4 DNA ligase added (4,000 units/10^6 cells; New England Biolabs), and incubated (48 h at 4°C, then 20 min at room temperature). After reversing cross-links (16 h; 65°C), DNA was purified by phenol extraction and ethanol precipitation, cut with BglII to reduce fragment length, and repurified. 71%–78% restriction sites in the template were cut by SacI or HindIII (determined as in [31]). PCR conditions were adjusted so that reactions were within the linear range of amplification (i.e., ~175 ng template/reaction; 1.75 mM $MgCl_2$, 1% dimethylsulphoxide, 10 pmoles of each primer, and GoTaq polymerase (Promega); 95°C for 2 min, then 34 cycles at 95°C for 55 s, 59°C for 45 s, and 72°C for 20 s, followed by one cycle at 72°C for 2 min); amplimers were resolved on 2.5% agarose gels, stained with SYBR Green (Invitrogen), and scanned using an FLA-5000 scanner (Fuji). Amplification efficiencies were examined using a control template generated by SacI or HindIII digestion of BAC clones covering GAPDH on HSA12 (RP5-940J5; ImaGenes), SAMD4A, GCH1 (RP11-170J16, CTC-775N1, CTD-2586I5, CTD-2378G4; CHORI, Invitrogen), and TNFAIP2 (CTD-2594N9; Invitrogen) on HSA14, SLC6A5 on HSA11 (RP11-120F6; CHORI), and PTRF on HSA17 (RP11-194N12; CHORI) followed by ligation. This synthetic template was spiked (to reach 175 ng/μl) with HUVEC DNA cut with the relevant restriction enzyme and ligated. Other control templates included non-digested/ligated DNA and digested/non-ligated DNA (both from 10^6 cells). Results shown were reproduced using at least two independently obtained templates.

ACKNOWLEDGMENTS

We thank Francisco Ramirez, Mika Kobayashi, and Akashi Izumi for their help, and Dirk Eick and Hiroshi Kimura for kindly providing antibodies.

AUTHOR CONTRIBUTIONS

The author(s) have made the following declarations about their contributions: Conceived and designed the experiments: AP PRC. Performed the experiments: AP JDL. Analyzed the data: AP JDL PRC. Contributed reagents/materials/analysis tools: JDL YW YO SI TK. Wrote the paper: AP PRC.

REFERENCES

1. Alberts B, Johnson A, Lewis J, Raff M, Roberts K, Walter P (2002) Molecular biology of the cell 4th edition. New York: Garland Science. 1463 p.B. AlbertsA. JohnsonJ. LewisM. RaffK. RobertsP. Walter2002Molecular biology of the cell 4th editionNew YorkGarland Science1463
2. Jackson D. A, McCready S. J, Cook P. R (1981) RNA is synthesised at the nuclear cage. Nature 292: 552–555.D. A. JacksonS. J. McCreadyP. R. Cook1981RNA is synthesised at the nuclear cage. Nature292552555
3. Cook P. R (1999) The organization of replication and transcription.

Science 284: 1790–1795.P. R. Cook1999The organization of replication and transcription.Science28417901795

4. Sexton T, Umlauf D, Kurukuti S, Fraser P (2007) The role of transcription factories in large-scale structure and dynamics of interphase chromatin. Semin Cell Dev Biol 18: 691–697.T. SextonD. UmlaufS. KurukutiP. Fraser2007The role of transcription factories in large-scale structure and dynamics of interphase chromatin. Semin Cell Dev Biol18691697
5. Cook P. R (2009) A model for all genomes; the role of transcription factories. J Mol Biol 395: 1–10.P. R. Cook2009A model for all genomes; the role of transcription factories.J Mol Biol395110
6. Sutherland H, Bickmore W. A (2009) Transcription factories: gene expression in unions? Nat Rev Genet 10: 457–466.H. SutherlandW. A. Bickmore2009Transcription factories: gene expression in unions?Nat Rev Genet10457466
7. Hoffmann A, Baltimore D (2006) Circuitry of nuclear factor kappaB signaling. Immunological Reviews 210: 171–186.A. HoffmannD. Baltimore2006Circuitry of nuclear factor kappaB signaling. Immunological Reviews210171186
8. Bradley J. R (2008) TNF-mediated inflammatory disease. J Pathol 214: 149–160.J. R. Bradley2008TNF-mediated inflammatory disease.J Pathol214149160
9. Wada Y, Ohta Y, Xu M, Tsutsumi S, Minami T, Inoue K, Komura D, Kitakami J, Oshida N, Papantonis A, et al. (2009) Visualizing a wave of transcription as it sweeps along activated human genes. Proc Natl Acad Sci U S A 106: 18357–18361.Y. WadaY. OhtaM. XuS. TsutsumiT. MinamiK. InoueD. KomuraJ. KitakamiN. OshidaA. Papantonis2009Visualizing a wave of transcription as it sweeps along activated human genes.Proc Natl Acad Sci U S A10618357l8361
10. Miele A, Gheldof N, Tabuchi T. M, Dostie J, Dekker J (2006) Mapping chromatin interactions by chromosome conformation capture. Curr Protoc Mol Biol 21, Unit 21.11.A. MieleN. GheldofT. M. TabuchiJ. DostieJ. Dekker2006Mapping chromatin interactions by chromosome conformation capture.Curr Protoc Mol Biol21, Unit 21.11

11. Simonis M, de Laat W (2008) FISH-eyed and genome-wide views on the spatial organisation of gene expression. Biochim Biophys Acta 1783: 2052–2060.M. SimonisW. de Laat2008FISH-eyed and genome-wide views on the spatial organisation of gene expression. Biochim Biophys Acta178320522060
12. Göndör A, Ohlsson R (2009) Chromosome crosstalk in three dimensions. Nature 461: 212–217.A. GöndörR. Ohlsson2009Chromosome crosstalk in three dimensions. Nature461212217
13. Ling J. Q, Li T, Hu J. F, Vu T. H, Chen H. L, Qiu X. W, Cherry A. M, Hoffman A. R (2006) CTCF mediates interchromosomal colocalization between Igf2/H19 and Wsb1/Nf1. Science 312: 269–272.J. Q. LingT. LiJ. F. HuT. H. VuH. L. ChenX. W. QiuA. M. CherryA. R. Hoffman2006CTCF mediates interchromosomal colocalization between Igf2/H19 and Wsb1/Nf1.Science312269272
14. Würtele H, Chartrand P (2006) Genome-wide scanning of HoxB1-associated loci in mouse ES cells using an open-ended chromosome conformation capture methodology. Genome Res 14: 477–495.H. WürteleP. Chartrand2006Genome-wide scanning of HoxB1-associated loci in mouse ES cells using an open-ended chromosome conformation capture methodology.Genome Res14477495
15. Degenhardt T, Rybakova K. N, Tomaszewska A, Moné M. J, Westerhoff H. V, Bruggeman F. J, Carlberg C (2009) Population-level transcription cycles derive from stochastic timing of single-cell transcription. Cell 138: 489–501.T. DegenhardtK. N. RybakovaA. TomaszewskaM. J. MonéH. V. WesterhoffF. J. BruggemanC. Carlberg2009Population-level transcription cycles derive from stochastic timing of single-cell transcription.Cell138489501
16. Ashall L, Horton C. A, Nelson D. E, Paszek P, Harper C. V, Sillitoe K, Ryan S, Spiller D. G, Unitt J. F, Broomhead D. S, et al. (2009) Pulsatile stimulation determines timing and specificity of NF-kappaB-dependent transcription. Science 324: 242–246.L. AshallC. A. HortonD. E. NelsonP. PaszekC. V. HarperK. SillitoeS. RyanD. G. SpillerJ. F. UnittD. S. Broomhead2009Pulsatile stimulation determines timing and specificity of NF-kappaB-dependent transcription.Science324242246

17. Kimura H, Sugaya K, Cook P. R (2002) The transcription cycle of RNA polymerase II in living cells. J Cell Biol 159: 777–782.H. KimuraK. SugayaP. R. Cook2002The transcription cycle of RNA polymerase II in living cells.J Cell Biol159777782
18. Chapman R. D, Heidemann M, Albert T. K, Mailhammer R, Flatley A, Meisterernst M, Kremmer E, Eick D (2007) Transcribing RNA polymerase II is phosphorylated at CTD residue serine-7. Science 318: 1780–1782.R. D. ChapmanM. HeidemannT. K. AlbertR. MailhammerA. FlatleyM. MeisterernstE. KremmerD. Eick2007Transcribing RNA polymerase II is phosphorylated at CTD residue serine-7.Science31817801782
19. Femino A. M, Fay F. S, Fogarty K, Singer R. H (1998) Visualization of single RNA transcripts in situ. Science 280: 585–590.A. M. FeminoF. S. FayK. FogartyR. H. Singer1998Visualization of single RNA transcripts in situ.Science280585590
20. Eskiw C. H, Rapp A, Carter D. R, Cook P. R (2007) RNA polymerase II activity is located on the surface of protein-rich transcription factories. J Cell Sci 121: 1999–2007.C. H. EskiwA. RappD. R. CarterP. R. Cook2007RNA polymerase II activity is located on the surface of protein-rich transcription factories.J Cell Sci12119992007
21. Thompson R. E, Larson D. R, Webb W. W (2002) Precise nanometer localization analysis for individual fluorescent probes. Biophysical J 82: 2775–2783.R. E. ThompsonD. R. LarsonW. W. Webb2002Precise nanometer localization analysis for individual fluorescent probes.Biophysical J8227752783
22. Yildiz A, Forkey J. N, McKinney S. A, Ha T, Goldman Y. E, Selvin P. R (2003) Myosin V walks hand-over-hand: single fluorophore imaging with 1.5-nm localization. Science 300: 2061–2065.A. YildizJ. N. ForkeyS. A. McKinneyT. HaY. E. GoldmanP. R. Selvin2003Myosin V walks hand-over-hand: single fluorophore imaging with 1.5-nm localization.Science30020612065
23. Larkin J. D, Publicover N. G, Sutko J. L (2010) Photon event distribution sampling: an image formation technique for scanning microscopes permits tracking of sub-diffraction particles with high spatial and temporal resolution. J Microscopy. J. D. LarkinN. G. PublicoverJ. L. Sutko2010Photon event distribution sampling:

an image formation technique for scanning microscopes permits tracking of sub-diffraction particles with high spatial and temporal resolution.J Microscopyadvanced online publication DOI: 10.1111/j.1365-2818.2010.03406.x. advanced online publication DOI: 10.1111/j.1365-2818.2010.03406.x.

24. Osborne C. S, Chakalova L, Brown K. E, Carter D, Horton A, Debrand E, Goyenechea B, Mitchell J. A, Lopes S, Reik W, Fraser P (2004) Active genes dynamically colocalize to shared sites of ongoing transcription. Nat Genet 36: 1065–1071.C. S. OsborneL. ChakalovaK. E. BrownD. CarterA. HortonE. DebrandB. GoyenecheaJ. A. MitchellS. LopesW. ReikP. Fraser2004Active genes dynamically colocalize to shared sites of ongoing transcription. Nat Genet3610651071

25. Zhao Z, Tavoosidana G, Sjölinder M, Göndör A, Mariano P, Wang S, Kanduri C, Lezcano M, Sandhu K. S, Singh U, et al. (2006) Circular chromosome conformation capture (4C) uncovers extensive networks of epigenetically regulated intra- and interchromosomal interactions. Nat Genet 38: 1341–1347.Z. ZhaoG. TavoosidanaM. SjölinderA. GöndörP. MarianoS. WangC. KanduriM. LezcanoK. S. SandhuU. Singh2006Circular chromosome conformation capture (4C) uncovers extensive networks of epigenetically regulated intra- and interchromosomal interactions.Nat Genet3813411347

26. Dhar S. S, Ongwijitwat S, Wong-Riley M. T (2009) Chromosome conformation capture of all 13 genomic loci in the transcriptional regulation of the multisubunit bigenomic cytochrome C oxidase in neurons. J Biol Chem 284: 18644–18650.S. S. DharS. OngwijitwatM. T. Wong-Riley2009Chromosome conformation capture of all 13 genomic loci in the transcriptional regulation of the multisubunit bigenomic cytochrome C oxidase in neurons.J Biol Chem2841864418650

27. Pombo A, Jackson D. A, Hollinshead M, Wang Z, Roeder R. G, Cook P. R (1999) Regional specialization in human nuclei: visualization of discrete sites of transcription by RNA polymerase III. EMBO J 18: 2241–2253.A. PomboD. A. JacksonM. HollinsheadZ. WangR. G. RoederP. R. Cook1999Regional specialization in human nuclei: visualization of discrete sites of transcription by RNA polymerase III.EMBO J1822412253

28. Xu M, Cook P. R (2008) Similar active genes cluster in specialized transcription factories. J Cell Biol 181: 615–623.M. XuP. R. Cook2008Similar active genes cluster in specialized transcription factories.J Cell Biol181615623
29. Lieberman-Aiden E, van Berkum N. L, Williams L, Imakaev M, Ragoczy T, Telling A, Amit I, Lajoie B. R, Sabo P. J, Dorschner M. O, et al. (2009) Comprehensive mapping of long-range interactions reveals folding principles of the human genome. Science 326: 289–293.E. Lieberman-AidenN. L. van BerkumL. WilliamsM. ImakaevT. RagoczyA. TellingI. AmitB. R. LajoieP. J. SaboM. O. Dorschner2009Comprehensive mapping of long-range interactions reveals folding principles of the human genome.Science326289293
30. Fullwood M. J, Liu M. H, Pan Y. F, Liu J, Xu H, Mohamed Y. B, Orlov Y. L, Velkov S, Ho A, Mei P. H, Chew E. G, et al. (2009) An oestrogen-receptor-alpha-bound human chromatin interactome. Nature 462: 58–64.M. J. FullwoodM. H. LiuY. F. PanJ. LiuH. XuY. B. MohamedY. L. OrlovS. VelkovA. HoP. H. MeiE. G. Chew2009An oestrogen-receptor-alpha-bound human chromatin interactome. Nature4625864
31. Hagège H, Klous P, Braem C, Splinter E, Dekker J, Cathala G, de Laat W, Forné T (2007) Quantitative analysis of chromosome conformation capture assays (3C-qPCR). Nat Protocol 2: 1722–1733.H. HagègeP. KlousC. BraemE. SplinterJ. DekkerG. CathalaW. de LaatT. Forné2007Quantitative analysis of chromosome conformation capture assays (3C-qPCR).Nat Protocol217221733

CITATIONS

CHAPTER 1

R. D. Astumian, How molecular motors work – insights from the molecular machinist's toolbox: the Nobel prize in Chemistry 2016, DOI: 10.1039/C6SC04806D.

CHAPTER 2

Kristina Konstas, Steven J. Langford, and Melissa J. Latter, Advances Towards Synthetic Machines at the Molecular and Nanoscale Level, doi:10.3390/ijms11062453.

CHAPTER 3

Kai Y Soo, Manal Farg and Julie D. Atkin, Molecular Motor Proteins and Amyotrophic Lateral Sclerosis, doi:10.3390/ijms12129057.

CHAPTER 4

Robert P. Erickson, Zhiyuan Jia, Steven P. Gross, Clare C. Yu; How Molecular Motors Are Arranged on a Cargo Is Important for Vesicular Transport, https://doi.org/10.1371/journal.pcbi.1002032.

CHAPTER 5

Mattia Gazzola, Christoph J. Burckhardt, Basil Bayati, Martin Engelke, Urs F. Greber , Petros Koumoutsakos, A Stochastic Model for Microtubule Motors Describes the In Vivo Cytoplasmic Transport of Human Adenovirus, https://doi.org/10.1371/journal.pcbi.1000623.

CHAPTER 6

Kristen M. Bartoli, Darryl L. Bishop, and William S. Saunders, "The Role of Molecular Microtubule Motors and the Microtubule Cytoskeleton in Stress Granule Dynamics," International Journal of Cell Biology, vol. 2011, Article ID 939848, 9 pages, 2011. doi:10.1155/2011/939848.

CHAPTER 7

Yusuke V. Morimoto and Tohru Minamino; Structure and Function of the Bi-Directional Bacterial Flagellar Motor; doi:10.3390/biom4010217.

CHAPTER 8

Johanna Haiko and Benita Westerlund-Wikström; The Role of the Bacterial Flagellum in Adhesion and Virulence; doi:10.3390/biology2041242.

CHAPTER 9

Matthew A. Watson and Scott L. Cockroft, Man-made molecular machines: membrane bound, DOI: 10.1039/C5CS00874C.

CHAPTER 10

L. Niels Cornelisse, Evgeni Tsivtsivadze, Marieke Meijer, Tjeerd M. H. Dijkstra, Tom Heskes, Matthijs Verhage; Molecular Machines in the Synapse: Overlapping Protein Sets Control Distinct Steps in Neurosecretion; https://doi.org/10.1371/journal.pcbi.1002450.

CHAPTER 11

Taizo Mori, Ken Okamoto, Hiroshi Endo, Keita Sakakibara, Jonathan P Hill, Satoshi Shinoda, Miki Matsukura, Hiroshi Tsukube, Yasumasa Suzuki, Yasumasa Kanekiyo and Katsuhiko Ariga, Mechanical tuning of molecular machines for nucleotide recognition at the air-water interface; https://doi.org/10.1186/1556-276X-6-304.

CHAPTER 12

Argyris Papantonis, Joshua D. Larkin, Youichiro Wada, Yoshihiro Ohta, Sigeo Ihara, Tatsuhiko Kodama, Peter R. Cook, Active RNA Polymerases: Mobile or Immobile Molecular Machines?, https://doi.org/10.1371/journal.pbio.1000419.

INDEX